GENERATION AND DIFFUSION OF AGRICULTURAL INNOVATIONS: THE ROLE OF INSTITUTIONAL FACTORS

The World Employment Programme (WEP) was launched by the International Labour Organisation in 1969, as the ILO's main contribution to the International Development Strategy for the Second United Nations Development Decade.

The means of action adopted by the WEP have included the following:

— short-term high-level advisory missions;

— longer-term national or regional employment teams; and

— a wide-ranging research programme.

Through these activities the ILO has been able to help national decision-makers to reshape their policies and plans with the aim of eradicating mass poverty and unemployment.

A landmark in the development of the WEP was the World Employment Conference of 1976, which proclaimed inter alia that 'strategies and national development plans should include as a priority objective the promotion of employment and the satisfaction of the basic needs of each country's population'. The Declaration of Principles and Programme of Action adopted by the Conference will remain the cornerstone of WEP technical assistance and research activities during the 1980s.

This publication is the outcome of a WEP project.

Generation and Diffusion of Agricultural Innovations: The role of institutional factors

Edited by
IFTIKHAR AHMED
and
VERNON W. RUTTAN

A study prepared for the International Labour Office within the framework of the World Employment Programme

1988

Gower

Aldershot · Brookfield USA · Hong Kong · Singapore · Sydney

The designations employed in ILO publications, which are in conformity with United Nations practice, and the presentation of material therein do not imply the expression of any opinion whatsoever on the part of the International Labour Office concerning the legal status of any country, area or territory or of its authorities, or concerning the delimitation of its frontiers.

The responsibility for opinions expressed in studies and other contributions rests solely with their authors, and publication does not constitute an endorsement by the International Labour Office of the opinions expressed in them.

Reference to names of firms and commercial products and processes does not imply their endorsement by the International Labour Office, and any failure to mention a particular firm, commercial product or process is not a sign of disapproval.

Published by
Gower Publishing Company Limited
Gower House
Croft Road
Aldershot
Hants GU11 3HR
England

Gower Publishing Company
Old Post Road
Brookfield
Vermont 05036
USA

ISBN 0 566 05679 8

Printed and bound in Great Britain by
Athanaeum Press Limited, Newcastle upon Tyne

Contents

Tables

Figures

Preface

Despite three decades of sustained international and national programmes of agricultural research and extension aimed at the generation and diffusion of agricultural innovation in the Third World countries, mass poverty and unemployment, particularly in rural areas, still remain acute. This volume attempts to develop a more coherent analytical framework for the generation and diffusion of agricultural innovations. The theories of induced technological and institutional innovations and their refinements provided the basis for this analytical work which takes into account the institutional biases in the current agricultural research and extension programmes.

In view of the fragmentary evidence and data limitations, the volume, edited by Dr. Iftikhar Ahmed of the ILO World Employment Programme and Professor Vernon W. Ruttan of the Department of Agricultural and Applied Economics of the University of Minnesota, makes a contribution to improving the data and knowledge base and attempts to include in the analysis neglected (by research and extension) agricultural commodities produced and consumed by the

Third World's poor in both free-market and non-market contexts. The empirical chapters of the volume would facilitate policy-making and help refine the theoretical framework.

The research was financed by a grant from the Swedish Government (Swedish Agency for Research Co-operation with Developing Countries) to the ILO World Employment Programme. The volume confirms the existence of a general process of induced technological innovations based on variations in national factor endowments for a range of countries at different levels of development, research and extension capacities and economic systems. The empirical case studies provide evidence on the technology-induced institutional innovations simultaneously at the national and grass-roots levels. The existence of causal interrelationships between induced technological and institutional innovations has been noted for both a centrally-planned economy and the free-market system. These insights would enhance the policy-making capacity in a wide and diverse range of countries.

For the formulation of effective policies for rural poverty alleviation in Third World countries, the empirical analysis of the three major forms of biases relating to agricultural extension, commodity choice in agricultural research and dominant role of the public sector in agricultural research and extension would be valuable. However, further empirical research is needed before statistically significant conclusions can be drawn about the hypotheses concerning (a) farm-size distribution and demand for innovations and (b) decentralisation of research and technology generation.

A.S. Bhalla,
Chief,
Technology and Employment Branch,
International Labour Office,
Geneva

Acknowledgements

This volume reflects the efforts of a number of individuals who have contributed in various ways at different stages of the work. Martha Loutfi, Peter Peek and Armand Pereira provided many useful comments. We are greatly indebted to Ajit Bhalla without whose encouragement and guidance this research could not have been completed.

We are very thankful to Susan Swannell and Michèle Bhunnoo who helped with the typing of the initial drafts. The tedious work of placing the entire final typescript on word processor was undertaken by Michèle Bhunnoo. We are also very grateful to her for the long hours of strenuous proof-reading. We also benefited immensely from the editorial comments provided by P.M.C. Denby.

Finally, we are grateful to Bruce Johnston, Barbara Harris, Paul Richards and Lawrence Busch for their comments on Chapter 2.

Contributors

IFTIKHAR AHMED is a Development Economist with the ILO World Employment Programme and has worked on agricultural and rural development. Previously, he was a Post-Doctoral Associate at the Iowa State University of Science and Technology, a, Visiting Fellow at the Institute of Development Studies, University of Sussex, and Associate Professor of Economics, Dhaka University, Bangladesh. He is author of <u>Technological change and agrarian structure: A study of Bangladesh</u> (Geneva, ILO, 1981), co-editor of <u>Farm equipment innovations in eastern and central southern Africa</u> (Aldershot, Gower, 1984) and editor of <u>Technology and rural women: Conceptual and empirical issues</u> (London, George Allen and Unwin, 1985).

STEPHEN D. BIGGS is an Agricultural Economist who has worked with agricultural research and extension systems in Bangladesh, India and Nepal. He is particularly interested in irrigation and agricultural research policy, linkages between researchers and resource poor rural clients, and processes of informal R & D. He is currently a lecturer in the School of Development Studies, University of East Anglia, Norwich.

EDWARD J. CLAY is a Fellow of the Institute of Development Studies at the University of Sussex and Director of the Relief and Development Institute in London. Whilst an Associate of the Agricultural Development Council he worked as an economist at the Bangladesh Rice Research Institute and has been an adviser to the Bangladesh Agricultural Research Council.

MICHAEL J. DORLING is an Agricultural Economist with a particular interest in the horticultural sector. He has taught and undertaken research at the University of British Columbia, Wye College (London University), and the University of Nairobi. More recently he was in Jordan doing research on indigenous technology in agriculture. At present he teaches Economics at the University of Kent, Canterbury, and acts as a consultant in agrarian development.

WAHIDUDDIN MAHMUD is Professor of Economics at the University of Dhaka, Bangladesh. He was head of the macro-modelling group at the Bangladesh Planning Commission (1978-80), Director of the Bureau of Economic Research, Dhaka University (1980-84), and Visiting Fellow at the Institute of Development Studies, University of Sussex (1984-85). He has worked as a consultant to various international organisations including the World Bank, the UN Economic and Social Commission for Asia and the Pacific, the Asian Development Bank, and the Asian Employment Programme of the ILO. He has written on a wide range of topics in economic development.

SUDHIN K. MUKHOPADHYAY is a Development Economist with major interest in the area of agricultural and human resource development. He has worked in India, USA and Japan, and has participated in leadership roles in national and international workshops, seminars and conferences. He has developed and conducted multi-institutional and international collaborative research projects. His teaching and research experiences include multidisciplinary studies of problems of growth, technology, population and poverty. Currently he is Professor-Director, Centre for Human Resource Development, Department of Economics, University of Kalyani, West Bengal, and Visiting Professor, Indian Statistical Institute, Calcutta.

MUHAMMED MUQTADA is a Development Economist whose recent contributions include <u>Hired labour and rural labour markets in Asia</u> (1986) and <u>Bangladesh: Selected issues in employment and development</u> (1987). He was a faculty member in Economics at the Dhaka University, Bangladesh, and he is currently serving as a Senior Development Economist with the ILO's Asian Regional Team for Employment Promotion, based in New Delhi.

NGOC-LUU NGUYEN is a Lecturer in Development Studies who has specialised in rural planning and on technologies for rural development. He has undertaken research and published works on technological innovation in food production and processing. He is currently with the Institute of Social Studies, The Hague, The Netherlands.

VERNON W. RUTTAN is a Regent's Professor of Economics and Agricultural Economics and an Adjunct Professor in the Hubert H. Humphrey Institute of Public Affairs at the University of Minnesota. His research has been in the fields of agricultural development, resource economics, and research policy. He is the author of <u>Agricultural Research Policy</u> (Minneapolis, University of Minnesota, 1982) and (with Yurijo Hayami) <u>Agricultural Development: An International Perspective</u> (Baltimore, Johns Hopkins University Press, rev. ed. 1985).

Abbreviations

AA	Agricultural Assistant
ADC	Agricultural Development Council
AER	Agro-Economic Research (Ministry of Agriculture)
AICRP	All India Co-ordinated Research Projects
ARTEP	Asian Regional Team for Employment Promotion
BADC	Bangladesh Agriculture Development Corporation
BBS	Bangladesh Bureau of Statistics
BIDS	Bangladesh Institute for Development Studies
BKB	Bangladesh Krishi Bank
BRRI	Bangladesh Rice Research Institute
BWDB	Bangladesh Water Development Board
c.i.f.	cost, insurance, and freight
CARS	Coast Agricultural Research Station
CGIAR	Consultative Group on International Agricultural Research
CIAT	Centro de Investigación Agrícola Tropical
CIMMYT	International Maize and Wheat Improvement Centre
CIP	International Potato Centre
CSIRO	Commonwealth Scientific and Industrial Research Organisation

DAEM	Directorate of Agricultural Extension and Management
DANIDA	Danish International Development Agency
DAO	District Agricultural Officer
EAAFRO	East African Agriculture and Forestry Research Organisation
FAO	Food and Agriculture Organisation of the United Nations
FTC	Farm Training Centre
GNP	Gross National Product
HCDA	Horticultural Crops Development Authority
HCU	Horticultural Cooperative Union Limited
HILAC	High Level Advisory Committee on Horticultural Research
HPC	Horticultural Production Centre
HYV	High-Yielding Varieties
IADP	Integrated Agricultural Development Programme
IADS	International Agricultural Development Service
IAEA	International Atomic Energy Agency
ICAR	The Indian Council of Agricultural Research
IDRC	International Development Research Centre
IDS	Institute for Development Studies
IFPRI	International Food Policy Research Institute
ILO	International Labour Office
IRDP	Integrated Rural Development Programme
IRRI	International Rice Research Institute
IRS	Integrated Rural Survey
ISNAR	International Service for National Agricultural Research
JAA	Junior Agricultural Assistant
KARI	Kenya Agricultural Research Institute
KSS	Krishi Shamabaya Samity
MNC	Multinational Company
MSU	Michigan State University
NHRS	National Horticultural Research Station
ODA	Overseas Development Administration
OECD	Organisation for Economic Cooperation and Development
OLS	Ordinary Least Squares
PARAC	Provincial Agricultural Research Advisory Committee
PDA	Provincial Director of Agriculture

R & D	Research and Development
RWP	Rural Works Programme
SAREC	Swedish Agency for Research and Economic Co-operation
SFTI	Soil Fertility Testing Unit
SIDA	Swedish International Development Agency
STC	Specialist Technical Committee
T & V	Training and Visit
TA	Technical Assistant
TCCA	Thana Central Co-operative Association
TIP	Thana Irrigation Programme
TTDC	Thana Training and Development Centre
TVA	Tennessee Valley Authority
UN	United Nations
UNESCO	United Nations Educational, Scientific and Cultural Organisation
UNICEF	United Nations International Children's Emergency Fund
UNRISD	United Nations Research Institute for Social Development
USA	United States
USAID	United States Agency for International Development
USDA	United States Department of Agriculture

1 Introduction

IFTIKHAR AHMED and VERNON W. RUTTAN

BACKGROUND

Despite three decades of sustained international and national programmes of agricultural research and extension aimed at the generation and diffusion of agricultural innovations in the Third World countries, mass poverty and unemployment, particularly in rural areas, still remain acute. These programmes have not succeeded in reducing widespread malnutrition and the recurrence of famines have become chronic in some regions of the world. Based on a review of these programmes, the literature points to two developments. The first development relates to the formulation of the theories of induced technological and institutional innovations. The second development concerns certain institutional biases in the generation and diffusion of agricultural innovations having important policy implications for poverty alleviation.

INDUCED TECHNOLOGICAL AND INSTITUTIONAL INNOVATIONS

Hayami and Ruttan (1971) articulated and empirically tested the model of induced technological innovation which critically hinged on the operation of the market forces in the macro-economic context of aggregate national factor endowments. The Hayami-Ruttan model essentially postulates that changes in relative factor scarcities are reflected in changes in relative factor prices which in turn guide technological progress towards saving on the factors that become relatively more expensive. Since technology generation in the Third World is almost entirely public sector activity (elaborated in a later section), it is the State which is expected to respond to market signals on the assumption that markets do work. Indeed, the analysis of the empirical evidence on Third World agricultural research and extension programmes over the past decades clearly suggests that a process of induced innovation was actually at work with a significant public sector response to the changing needs of greater emphasis on adaptive research of imitative extension with respect to agricultural technology.

The Hayami-Ruttan induced innovation theory was not without its critics. Major criticisms were[1]: (a) inability to explain adequately technology generation where markets tend to fail or do not exist (certain classes of innovations are price inelastic); (b) rural factor market imperfections arising from inequalities in agrarian structure often do not reflect shadow values of factors (net prices are not the same across social groups); (c) the theory explains more in the long run when relative factor scarcities have become more fully reflected in relative factor prices; (d) it explains more technological change in the aggregate for the entire agricultural sector than for specific crops, farms and regions; (e) it explains more the _adoption_ than the generation of agricultural innovations and private sector research than public since both of those are more directly guided by price signals and profitability considerations than public sector research (which, as noted above, is virtually the only source of technology generation and dissemination in the Third World); (f) it works better for advanced capitalist democracies where markets are more developed and the State relatively less interventionist; since the public sector is the dominant source of technology generation in the Third World, the unequal distribution of assets, wealth and political power results in unequal influences on the State (we have already seen this being reflected in the commodity bias of technology generation); and (g) imported technology or the activities of multi-national companies with multi-country interests may not correspond to relative factor endowments of individual countries.

As a sequel to the theory of induced technological innovation (Hayami-Ruttan 1971), Ruttan and Hayami (1984) developed the related theory of induced institutional innovation. Under the former, technological change was induced by relative resource endowments and growth of demand. Now, institutional change is postulated to be an economic response to changes in resource endowments and technical change. The latter clearly suggests the existence of an interrelationship between induced technological and institutional innovations.

The public sector bias (analysed later) in agricultural technology generation and extension in the Third World has been interpreted by Ruttan and Hayami (1984) as a major institutional innovation, as it amounted to socialisation of research. Following the critics of their earlier induced technological innovation theory, it is argued that markets fail to allocate resources efficiently and equitably for the supply of a public good (agricultural technology) for a large, unidentifiable clientele group. The public good attributes of agricultural technology and the stochastic nature (risk and uncertainty) of outcome of the research required for its generation, make the reliance on the public sector for its generation socially desirable. But, as we have noted from the evidence provided above, the level and composition of public sector research and extension activity is still largely influenced by a minority of the population representing politically powerful social groups. Moreover, the commodity bias discussed above also confirms agricultural technology assuming the character of an imperfect public good (as explained on page 6).

More striking is the institutional innovation induced by changes in relative factor endowments and technical change at the grass roots level in the Philippines (Ruttan-Hayami 1984). Historical 20 years micro-economic time series village-level data from the Philippines revealed an increase in labour land ratio (following population growth) accompanied by the use of high yielding varieties of rice, chemical fertiliser and irrigation (which increased output per unit of land and permitted double cropping). Both of these developments induced institutional innovations in the land rental market and the labour market. In the former, land reforms replaced share tenancy by lease tenancy which, in turn, was informally replaced by subtenancy. Subtenancy was considered an institutional innovation because the rent paid to landlords reflected higher yields with the new technology and lower wage rates following increases in the man-land ratio.

In the labour market, the traditional practice of hiring labour "humisan" (labour hired for harvesting and threshing and paid one-sixth of the harvest) was replaced by another contractual arrangement, the "gamma" system (under which labourers engaged for wages in harvesting were only those who participated in weeding as unpaid workers). The gamma system was interpreted as an institutional innovation because it brought the wage rate for harvesting to a level equal to the marginal productivity of labour.

The gamma system and subtenancy turned out to be institutional innovations which not only improved the efficiency of resource allocation, but also promoted equity. Without subtenancy, the landless workers could not have become farm operators, and without the "gamma" system, which lowered the wage rate closer to the equilibrium level, technological change induced would have a labour-saving bias (e.g. mechanisation of threshing).

Similar influence of the institutional biases of labour use on technological change is observed for

Indonesia (Reddy et al 1985). Under the traditional
"bawon" system of hiring labour in Java, the
individual workers are engaged for the combined work
of harvesting, threshing and transportation of rice
for which each worker received a package payment
amounting to 9 to 14 per cent of the harvested
output. The adoption of a pedal operated thresher
which was affordable by these resource-poor hired
labour, permitted a continuation of this "bawon"
system under which the individual worker could
continue to be engaged for the above series of
harvest and post-harvest operations and receive a
single payment. In contrast, the non-existence of
the "bawon" system in West Sumatra permitted the
hiring of wage-labour separately for each segment of
the above harvesting and post-harvest operations.
This facilitated the adoption of an imported (but
locally adapted) mechanical thresher for which labour
was specifically hired for threshing and paid
corresponding to his work of threshing only. Under
the "bawon" system, such an institutional separation
of the labour input for harvesting and post-harvest
operations was not possible as it created accounting
problems of disaggregating the package payment
corresponding to individual operations. Therefore,
the "bawon" institution of hiring labour induced a
technological change having a labour-using bias in
Java and conversely, the absence of this institution
in West Sumatra induced a labour-saving bias in the
technological change.

INSTITUTIONAL BIASES

From an assessment of the research and extension
programmes for Third World agriculture, one is struck
by three distinct biases: (1) an extension bias, (2)
a commodity bias, and (3) a public sector bias.
These are examined and analysed below.

Extension bias

Empirical evidence shows that lower-income countries
spend a larger share of the value of agricultural

product on extension than on research and higher income countries spend a smaller share (table 1.1). Several reasons are offered for this bias:

(a) given current patterns of distribution of operational holdings in developing countries, Third World agriculture is primarily composed of numerous small farms operated by resource-poor illiterate farmers, and it is argued that lower income countries and regions are compelled to spend a higher share of the value of agricultural product on extension, simply because there is a preponderance of small farmers (with higher educational need) to be reached in isolated, scattered and remote rural areas;

(b) the extension bias was particularly more intense during the earlier decades because it was mistakenly assumed that agricultural technology was transferable among countries so that national extension programmes could simply screen internationally available technologies for effectiveness and extend these technologies to farmers[2];

(c) another incentive for this extension bias was the fact that most developing countries (particularly in Asia) were able to train extension workers at low cost and to staff extension programmes at very low costs per extension worker (M.A. Judd et al. 1986).

However, the percentage of the value of agricultural product spent by low-income developing countries on research was higher in 1980 compared to that of 1959 (table 1.1). A major reason for a shift in priorities was the recognition that agricultural technology was not transferable across countries and within sub-regions of individual countries due to differences in soil and climatic conditions. It is little surprising that the national strategy consisted of an innovative mix of imitative extension and minimal adaptive research. One observes a further bias in favour of applied research as opposed to basic research which is supported by research policy-makers generally, but not universally (Dieter Elz 1984). While applied research originates from basic research, the latter generates a new field of

Table 1.1
Research and extension expenditures as percentage of the value of agricultural product by sub-region and level of development at the end of decades[a, b]

Sub-region/Level of development	Agricultural Research Expenditures			Agricultural Extension Expenditures		
	1959	1970	1980	1959	1970	1980
Northern Europe	.55	1.05	1.60	.65	.85	.84
Central Europe	.39	1.20	1.54	.29	.42	.45
Southern Europe	.24	.61	.74	.11	.35	.28
Eastern Europe	.50	.81	.78	.32	.36	.40
Soviet Union	.43	.73	.70	.28	.32	.35
Oceania	.99	2.24	2.83	.42	.76	.98
North America	.84	1.27	1.09	.42	.53	.56
Temperate South America	.39	.64	.70	.07	.50	.43
Tropical South America	.25	.67	.98	.34	.71	1.19
Caribbean and Central America	.15	.22	.63	.09	.18	.33
North Africa	.31	.62	.59	1.27	2.21	1.71
West Africa	.37	.61	1.19	.58	1.24	1.28
East Africa	.19	.53	.81	.67	.88	1.16
Southern Africa	1.13	1.10	1.23	1.64	.67	.46
West Asia	.18	.37	.47	.25	.57	.51
South Asia	.12	.19	.43	.20	.23	.20
South-east Asia	.10	.28	.52	.24	.37	.36
East Asia	.69	2.01	2.44	.19	.67	.85
China	.09	.68	.56	N.A.	N.A.	N.A.
Level of development:[c]						
Low-income developing	.15	.27	.50	.30	.43	.44
Middle-income developing	.29	.57	.81	.60	1.01	.92
Semi-industrialised	.29	.54	.73	.29	.51	.59
Industrialised	.68	1.37	1.50	.38	.57	.62
Planned	.33	.73	.66	...	...	...
Planned, excluding China	.45	.75	.73	.29	.33	.36

cont'd ...

7

a   Source: M. Ann Judd, James K. Boyce and Robert E. Evenson
     (1986).
b   Use is made only of the public sector expenditure since these
     cover more than 95 per cent of total research and extension
     expenditures of developing countries (elaborated in a later
     section). Data has to be interpreted bearing in mind the
     difficulty of currency conversions, the perception of what
     constitutes a research scientist varies from country to
     country and forestry and fisheries research is funded through
     the same agency as agricultural research in some of the
     countries.
c   Industrial countries are members of OECD, except for Greece,
     Portugal, Spain and Turkey; planned economies consist of
     Eastern European countries, the Soviet Union and China;
     semi-industrialised countries are other countries with annual
     per capita income above US$1,050; middle-income developing
     countries are other countries with annual per capita income
     between US$360 and US$1,050; low income developing countries
     are the remaining countries with annual per capita income
     below US$360.

basic research. The two are assumed to be
interrelated and nourish each other.

Despite this shift in priorities from extension to
research, the growth in extension intensity in the
Third World national programmes remains unchecked.
The extension intensity (number of extension workers
per unit of agricultural product) has increased
steadily over the three decades (table 1.2).

Commodity bias

A pronounced bias (table 1.3) in research and
extension programmes[3] is observed with respect to
two categories of commodities: (a) bias in favour of
traded (specially export crops) commodities as
opposed to non-traded items, and (b) bias against

Table 1.2
Research and extension intensities by sub-region and level of development at the end of decades[a]

Sub-region/Level of development	Research intensity (scientist man-years per US$10 million agricultural product)[b]			Extension intensity (extension worker per US$10 million agricultural product)[b]		
	1959	1970	1980	1959	1970	1980
Northern Europe	1.05	2.01	3.14	2.76	2.56	2.61
Central Europe	.80	1.21	1.56	2.19	2.77	2.73
Southern Europe	.93	1.17	.96	2.00	2.76	2.69
Eastern Europe	1.44	2.97	2.84	2.36	2.88	3.13
Soviet Union	1.38	2.37	2.34	2.26	2.33	2.50
Oceania	1.91	2.64	2.43	2.26	2.17	2.11
North America	.84	.89	.84	1.44	1.31	1.08
Temperate South America	.46	1.15	1.32	.26	1.19	1.26
Tropical South America	.41	1.41	1.77	1.71	3.95	6.46
Caribbean and Central America	.53	.86	1.20	.82	1.53	3.12
North Africa	.91	1.44	4.24	18.83	28.45	22.23
West Africa	.33	.61	1.42	7.61	14.01	18.08
East Africa	.32	.77	1.76	16.28	22.41	26.64
Southern Africa	1.90	1.96	2.47	8.73	5.94	5.62
West Asia	.33	.84	.88	4.39	7.25	6.54
South Asia	.50	.65	1.29	20.83	19.51	19.53
South-east Asia	.47	1.28	2.07	9.81	13.07	19.72
East Asia	3.80	5.29	5.72	6.57	7.05	6.13
China	.22	1.66	1.49	N.A.	N.A.	N.A.
Level of development:[c]						
Low-income developing	.43	.67	1.40	18.14	18.61	20.43
Middle-income developing	.69	1.31	2.40	8.89	14.68	15.98
Semi-industrialised	.70	1.21	1.36	2.80	4.95	5.21
Industrialised	1.24	1.71	1.85	2.37	2.31	2.12
Planned	1.02	2.27	2.13	...	...	...
Planned, excluding China	1.40	2.54	2.50	2.29	2.49	2.63

[a] Source: M.A. Judd, et al (1986), table 4, p.88.
[b] At constant 1980 prices.
[c] Same classification as in table 1.1.

9

Table 1.3
Research expenditure as percentage of the value of product by commodity, by region and by source: 1972–79 average*

Commodity	Region			Source	
	Africa	Asia	Latin America	All countries	International centres
Wheat	1.30	.32	1.04	.51	.02
Rice	1.05	.21	.41	.25	.02
Maize	.44	.21	.18	.23	.03
Cotton	.23	.17	.23	.21	...
Sugar	1.06	.13	.48	.27	...
Soya beans	23.59	2.33	.68	1.06	...
Cassava	.09	.06	.19	.11	.02
Field beans	1.65	.08	.60	.32	.04
Citrus	.88	.51	.57	.52	...
Cocoa	2.75	14.17	1.57	1.69	...
Potatoes	.21	.19	.43	.29	.08
Sweet potatoes	.06	.08	.19	.07	...
Vegetables	1.56	.41	1.13	.73	...
Bananas	.27	.20	.64	.27	...
Coffee	3.12	1.25	.92	1.18	...
Groundnuts	.57	.12	.60	.25	.005
Coconuts	.07	.03	.10	.04	...
Beef	1.82	.65	.67	1.36	.02
Pork	2.56	.39	.60	1.25	.02
Poulty	1.99	.32	1.12	1.64	...
Other livestock	1.81	.89	.42	.71	...

*Source: M.A. Judd et al. (1986), table 6, p.92.

food (specially subsistence) crops. Clearly, export crops like citrus, coffee and cocoa, are generally given heavy emphasis. Balance of payments considerations often contribute to this bias.

Important subsistence food crops like cassava and sweet potatoes are neglected in every region of the world and field beans in Asia. These represent critical subsistence crops for peasants but are not consumed in urban areas and have little industrial use. These crops produced by the resource-poor peasantry are neglected because the growers are unable to voice their demand for new technology at the level of the state (virtually the only source of agricultural technology generation in Third World countries).[4] In view of short-term considerations, the longer gestation period in technology generation in such neglected commodities, as opposed to established crops (like rice and wheat), also acts as a disincentive to allocating scarce research resources to these crops.

In addition to commodity-by-commodity type of research there is now a trend towards farming systems research which more broadly takes into account overall farming conditions. The basic core of the approach involves using research-resource-minimising procedures to identify those factors that affect farmers' decisions with respect to the use of cropping technologies.

Public sector bias

Technology generation and diffusion in the agricultural sector rests almost exclusively in the hands of Third World governments. In the resource-poor countries of Asia and Africa, over 97 per cent of the entire agricultural research is under the domain of the public sector (table 1.4). Furthermore, a positive correlation is observed between the proportion of agricultural research by the private sector and the level of national income of countries.

Given that nearly the entire responsibility for agricultural technology generation and diffusion in the Third World countries rests in their public

Table 1.4
Share of total agricultural research by source,
sectors, region (1974) and level of development
(1971)*
(percentage of total research)

Source Region, level of development	Sector	
	Public	Private
North America and Oceania	74.6	25.4
Western Europe	89.2	10.8
Eastern Europe and USSR	91.7	8.3
Latin America	94.9	5.1
Africa	97.1	2.9
Asia	97.8	2.2
Level of development (GNP per capita)		
Less than US$150	94.8	5.2
US$150 – US$400	97.2	2.8
US$400 – US$1,000	92.6	7.4
US$1,000 – US$1,750	93.0	7.0
US$1,750 and more	76.0	24.0

* Source: J.K. Boyce and R.E. Evenson (1975).

sector, it is interesting to examine the determinants
of decision-making in national agricultural research
programmes. The first significant feature to note is
that priorities in national programmes are guided by
the strengths of different interest groups.
Available evidence reveals that poor countries with
large agricultural labour forces invest less in
agricultural research and extension reflecting the
group's weak political power. Similarly, urban bias
adversely affects agricultural research and extension
spending, as is confirmed by a negative correlation
between urbanisation (proportion of population living
in large urban areas) and the level of agricultural

research and extension expenditures (M.A. Judd et al 1986). More than one-half of the resources of the research and extension system have been poured into plant breeding and related activities followed by development of better systems of land and water management (Alain de Janvry and J.-J. Detheier 1985). Such a thrust of public sector research is a direct consequence of the patent laws which can protect mechanical and chemical inventions more easily than biological inventions. Although agricultural technology is valuable to all producers, rich or poor, it assumes the character of an imperfect public good since it is generated under state auspices. That is to say, while, in principle, no individual can be excluded from deriving the benefits of new technology, the benefits vary sharply across commodities, regions, agrarian structures, resource-bases of producers. Different social groups having different technological demands can through their political power and lobby, induce a public-sector response to their demands.

OBJECTIVES OF THE VOLUME

Although this volume has been inspired by the Hayami-Ruttan theories of induced technological (Hayami-Ruttan 1971) and institutional (Ruttan-Hayami 1984) innovations, its purpose is to develop a more coherent, theoretical and methodological framework of analysis of the generation and diffusion of agricultural innovations on the basis of criticisms levelled against these two basic principles and in the context of the biases in the current agricultural research and extension programmes. More specifically, the volume attempts an empirical verification of the theories of induced technological and institutional innovations, both at the macro and micro levels and in market and non-market economies. This task is made difficult by a paucity of data and empirical evidence provided even by critics is fragmentary.[5] The volume examines whether institutional biases observed for international cross-section data are reinforced by country data.

This volume widens the geographical basis of the empirical analysis, includes a centrally planned economy in order to provide valuable insights from a non-market context, and focuses on a specific but important group of commodities to trace, in greater depth, implications for neglected commodities produced and consumed primarily by the poor of the Third World. Following Ruttan and Hayami (1984), the interrelationships, if any, between technological and institutional innovations are explored by the volume through case studies, particularly using micro-data at the grass roots level, both in free-market (including mixed-economy) and non-market contexts. The volume also attempts a verification of two important hypotheses about the influences of farm size distribution and decentralised research on the supply and sustained demand for agricultural innovation (conceptual bases elaborated in Chapter 3). Finally, the literature is vague and the evidence fragmentary on the relative priorities assigned to two functional components of basic and applied research. Therefore, this volume examines whether the mix of priorities noted from international cross-sectional evidence is supported by country data and attempts its rationalisation.

METHODOLOGY AND DESIGN

The volume may be more appropriately described as a contribution to the "second generation" impact studies (Horton 1986) of the post-Green Revolution era, focusing on agricultural research including the impact of international programmes on domestic research and extension capacity building in developing countries.[6] Most of the earlier "first generation" impact studies are concerned with the diffusion of new varieties of cereals at the farm level and with their impact on output, income, employment and income distribution.[7] Since national programmes generate technology under different policy and socio-economic settings, divergent cropping patterns and agro-climatic conditions, inequalities in agrarian structures,

differing ideological and external contexts, regional
disparities, diversity of strategies and types of
institutional impact, there is no practical
alternative to the case study method, particularly in
the absence of an analytical framework for assessing
institutional impacts analogous to the production
function framework for assessing production
technologies at the farm level under the "first
generation" impact studies and because it is
impossible to find adequate long time-series data to
trace the process of institutional innovations which
is often evolutionary.

In the light of the background and objectives
enumerated above, the volume brings together
conceptual and empirical analyses emphasising
multi-disciplinary approaches, often with the authors
having to tailor their qualitative analyses to the
nature and availability of historical data and
background material.

STRUCTURE OF THE VOLUME

The book is divided into two distinct parts. Part I
deals with conceptual, analytical and theoretical
approaches and consists of two chapters. It begins
with a comprehensive survey (Chapter 2) of the
alternative methodological and conceptual approaches
to the understanding of the institutional factors
contributing to the generation and diffusion of
agricultural innovations. As a result, it includes
in the review, a critical appraisal of the
Hayami-Ruttan theory (1971) of induced technological
innovation which is subsequently elaborated and
broadened (in Chapter 3) to include the presentation
and discussion of the process of induced
institutional innovation, and the interrelationship
between these two types of innovations is analysed.
This chapter also provided the analytical framework
for the empirical case study authors (often proposing
specific hypotheses testing). The survey chapter
(Chapter 2) was completed towards the concluding
stages of the preparation of the volume.

Part II consists of an empirical piece (Chapter 4)

from India which draws on inter-state cross-sectional data somewhat analogous to the international cross-country data (reported in Chapter 1) to verify the major hypotheses of the Hayami-Ruttan theory of induced technological and institutional innovation in the light of the biases (enumerated in Chapter 1). Given the commodity biases (observed in Chapter 1) of technology generation and diffusion programmes, two empirical case studies (Chapters 5 and 6) focus on a cereal (rice, which is an important component of the diet of a vast majority of the world population) and horticultural crops (whose cultivation is labour-intensive and whose consumption could significantly boost the dietary and nutritional standards of the poor). The final empirical piece on Viet Nam (Chapter 7) makes a significant contribution in two respects: (a) tests the induced innovation theory in the absence of market mechanisms of resource allocation, and (b) uses micro level socio-economic data to analyse the interrelationships, if any, between induced technological innovation and induced institutional innovation. The latter is to a minor extent also examined in a free market context (in Chapter 5), as has been done more thoroughly in Chapter 1.

The concluding chapter (Chapter 8) attempts an overview and synthesis of the findings of the empirical chapters to facilitate policy-making in relation to both national and international agricultural research and extension programmes. It also discusses analytically the highlights of the major themes of the volume at both conceptual and empirical levels so that the contribution made by this volume to refining the theoretical framework and formulation of development plans is clearly understood.

<u>NOTES</u>

[1] These are summed up from Alain de Janvry and J.-J. Dethier (1985), Nordhaus (1973), Mooney (1979) and Dorling (1985).

16

[2] Due to strong interactions of improved technology with soil and climatic factors, it impeded the diffusion of technology across broad regions (M. Ann Judd et al 1986).

[3] The international programmes are not immune from this bias and as a consequence have not contributed to rectifying the imbalance (table 1.3), and the limited size of their programmes may not even permit this.

[4] This is amply demonstrated by pressures successfully applied on national research programmes by powerful producers' lobbies for sugar in Colombia and maize in Argentina (Alain de Janvry and J.-J. Dethier 1985, pp.36, 43, 74-76).

[5] A recent empirical effort has a limited regional focus on Latin America (Alain de Janvry and J.-J. Dethier 1985).

[6] The significance of both institutional capacity and institutional impact have been described as important features or hallmarks of the "second generation" studies (Horton 1986, and Alain de Janvry and J.-J. Dethier 1986, pp.16 and 35). Conceptually, Horton makes a distinction between production technology, whose effect is observed at the direct farm level, and R&D technology, which is generated by international programmes and transferred to national programmes aimed at strengthening their capacity to generate new or adapted production technology (Horton 1986, p.454).

[7] For a review of the innumerable "first generation" impact studies, see the survey by Feder, Just and Zibberman (1985).

2 Generation and diffusion of agricultural technology: Theories and experiences

STEPHEN D. BIGGS and EDWARD J. CLAY

INTRODUCTION

Since the 1950s there has been a growing awareness about the importance for developing countries of formal (institutionalised) agricultural research systems. Major efforts to promote agricultural research have been made by strengthening research institutions in developing countries and the establishment of a network of International Agricultural Research Institutes (Oram and Bindlish 1981; World Bank 1981).

The policy practice of these 30 years has been characterised by an emphasis on dramatic modernisation. Science policy makers sought to by-pass the slow incremental process by which Western and some countries' agricultural research systems evolved, by the large-scale transfer of technology and institutional models for technology generation (Ruttan 1975). The creation of the International Agricultural Research Institutes represents the single most important effort rapidly to expand research capacity, filling the gap on the neglected subject of food crops. During the colonial period agricultural research in developing countries had been focused (but with important exceptions) on export and non-food crops (Busch and Sachs 1981).

These transfer and modernisation actions have had a wide range of consequences and generated considerable controversy. That there now exists a very large and growing literature about technical change and science policy in Third World agriculture is a reflection of this continuing controversy about the role of formal science, technical and institutional transfers in the generation and diffusion of technologies for Third World agriculture.[1] This literature is inevitably very diverse. However, in reviewing this literature we have identified a number of themes or preoccupations. These preoccupations all represent responses to the theory and practice of modernisation (or transformation) of agriculture in developing countries through the transfer of technology and institutions.

For example, some analysts mainly focus on national factor endowments (of land and labour, and sometimes capital), and factor prices (Hayami and Ruttan 1971; Griffin 1974). Others are concerned with the world economy and focus attention on the political and international capital and dependency dimensions of technical change in agriculture (Cleaver 1972; Dalrymple 1979).

These different themes do not represent a school of thought where there is a consensus as regards theory and prescription. In fact the opposite is frequently the case. However, they do represent the preoccupations by different groups of writers. One finds, for example, that economists tend to write more in the theme area of factor endowments at a macro (national or sectoral) or micro level. Anthropologists and sociologists are more preoccupied with social relationships and transactions between socio-economic groups in rural communities. Geographers and natural resource scientists are more preoccupied with technological change in the context of environmental balances and ecological crises in a spatial context. Not only are disciplinary thematic preoccupations different but these preoccupations typically imply a different level of focus in terms of the actual organisation and the observation and measurement of human productive and reproductive activity.

What is interesting and important for this review is that writers preoccupied with one theme appear to either dismiss, by-pass or misrepresent the preoccupations of the analysis of writers on other themes, or to be unaware of the potential benefits which would be gained from a broadening of their interests.

OBJECTIVES

In our attempt to map and interpret the very wide range of theories and experiences concerning the generation and diffusion of agricultural technology we have identified six major themes which appear to dominate the literature. These are: (i) modernisation; (ii) factor endowments; (iii) distribution and equity; (iv) world economy; (v) environmental themes and ecological crisis; and (vi) research and extension institutions. In each case we have tried to avoid giving our opinions about which theories appear to be better or worse, or more or less useful than others. It is not the purpose of this review to enter the debate on these topics. Nor is it our claim that we have reviewed all the literature. However, we hope that we have illustrated as fairly as possible the preoccupations and concerns of writers in these six major areas.[2] We recognise that our categorisation is a simplification of the breadth and content of some analysis. However, as with all representation through model building, simplification and judgement is needed. We hope that our attempts to stand back from the internal debates between different theorists will help focus attention on a subject which, as a result of the actions of the last 30 years, most writers would agree is of central concern to agricultural research. These concern the structure, organisation and control in agricultural research systems. For 30 years the modernisation and transfer concepts prevailed. Today it is recognised that more analysis and research is needed on the structure and organisation of formal research institutions.

MODERNISATION

Thinking about the role of agricultural science in development was dominated until around 1975 by the concept of modernisation.[3] Policy practices based unequivocally on this concept continue, notably at the extensive frontier of capitalist agriculture development, for example in Amazonia (Maxwell 1980), and in application of socialist principles to tropical agriculture, for example in Mozambique. Elsewhere in the Third World and amongst practitioners and analysts in developed countries there has been an important period of questioning and redefining of the role of agricultural science again, approximately since the emergence of the Green Revolution in Asia in the mid-1960s. The themes which have preoccupied the critics of modernisation, provide the main part of this review. However, to place the debates in context, those have to be related back to aspects of modernisation theory and practices which the critics came to regard as problematic or unhelpful.

In retrospect, there was no single model of modernisation, but a cluster of concepts which provided a vocabulary for modernisation.[4] This vocabulary provided the analytical underpinning of the practice of scientists, governments of Third World countries emerging from European, United States and Japanese colonialism, and the multilateral and bilateral aid community. Amongst these key concepts which subsequently became the focus of critical analysis were: (a) the dichotomy of the modern and the traditional; hence (b) the <u>transformation</u> of the traditional into the modern; (c) <u>agricultural development</u> as the transformation which is <u>measurable</u> in <u>agricultural product growth</u>; (d) <u>stages</u>: for many there were clearly demarcated phases on the road from the traditional to the modern. The notion of stages applied not only to the overall development process, but it was also an organising concept for the transformation of activities in an institutional context. There is an underlying notion of linearity in the transformation process; (e) the _transfer_ of technology, know-how and the institutional models, play a key role in modernisation thinking and practice because they offer the opportunity to accelerate transformation. Transfer is potentially, an accidental or a purposive, planned activity. It can be achieved by borrowing and in doing so even possibly by-passing stages in what might otherwise be an evolutionary process; and (f) the _diffusion_ of improved ways of doing things through the _adoption_ of new practices including use of improved or new techniques both in directly productive _and_ in research processes would bring about the upward movement from the traditional with higher productivity (of labour, land, capital, all factors).

Central to the process of transformation as it relates to agriculture (and probably other sectors) are two preoccupations. The first is new and improved technology.

"The sharp growth in diffusion studies in developing countries was due to the fact that technology was assumed to be the heart of development, and innovativeness was thought to be one of the best single indicators of the multi-faceted dimension called modernisation." (Rogers 1980, p.3)

Inevitably, this preoccupation must bring into a more prominent place the formal research process which is the source of technology generation in a modern economy and society.

The second preoccupation is with institution building. A traditional, underdeveloped economy lacked the appropriate institutional forms which would facilitate, even be a pre-condition for, rapid transformation. The linearity of the development process implied that relevant institutional forms could be identified within the successful modernisers and that these forms could also be _transferred_ to facilitate the development process.[5]

Focusing on agricultural science, transfer was especially important, perhaps the key concept. The history of agricultural practice underscored the enormous importance of transfers, particularly of biological material, plants and animals. Many of the more significant changes in production practices had been associated with the haphazard patterns of transfer. Transfers of biological technology which extend the genetic pool of a species available in a specific location had long been recognised as potentially the most important source of improvement in animal and plant types. The identification of gaps between traditional and modern practice associated with using specific genetic technologies therefore offers opportunities to accelerate the process of transformation by direct transfer. The limited past exploration of the gene pool for many plants and animals provides opportunities for innovation to produce improved plant types.[6] This points not merely to the transfer of usable hardware but of material for further scientific research. Such perceptions of the possibilities in agricultural technology transfer are well founded. However, modernisers constructed much more ambitious notions of transfer.

The _gap_ between the traditional and the modern agriculture had been seen as an opportunity to accelerate the process of transformation by direct transfers. There was no uniform model of the transfer process or classification of transfers. However, the typology of transfers suggested by Hayami and Ruttan, themselves amongst critics of aspects of earlier thinking, is suggestive of the different emphasis in thinking about transfers.[7]

Many were, and still are, preoccupied with material transfer, the direct transfer of hardware for use in developing country agriculture. The technologies to be transferred include genetic material, chemical inputs, especially fertilisers, pesticides, etc. Ruttan includes amongst these direct transfers of technology specific practices in cultivation. An example would be East Asian transplanting techniques for rice which were in vogue in extension work in India and possibly elsewhere in the 1950s (Nair 1961).

Second, there is design transfer of know-how and blue-prints which can be copied, sometimes material which can be tested and multiplied for local use.

There is no hard and fast distinction for <u>most</u> genetic material, because local use typically involves testing and selection (Brammer 1980; Rogers 1980).[8]

Ruttan's third and more advanced form of transfer is that of technology generation <u>capacity</u> through institution building and the migration of highly trained scientific personnel. The emphasis in this notion of transfer is strongly on technical co-operation and the copying of appropriate institutional models for organising agricultural scientific research.

The notion of transfer is powerful, as we have noted in an agricultural context, because historically, apart from the usually slow informal process of adaptation, the haphazard transfers – especially of genetic material but also machine technologies, e.g. the mould board plough, the water mill, the Persian wheel – have all had significant non-marginal impact on agricultural production systems. However, the attempt to provide a typology of transfers and the implicit emphasis given to transfer has proved to be problematic, as much of the more recent critical literature has demonstrated.

First, discussion of transfer needs to be located precisely in the context of the recipient systems. The notion of transfer of technology embodied in genetic material or machines is unambiguous. However, the use, the reproduction, replication and modification of these material transfers raise questions about the innovative capacity of the systems into which transfers are made. We have suggested elsewhere (Biggs and Clay 1981) that Third World agricultural systems are not and have never been in a state of equilibrium. There is always research within the (often serious) limitations imposed by available pool of genetic material, the equipment and testing procedures of the agriculturalists as informal scientists. The emphasis on transfer linked to diffusion of innovation models of change de-emphasised or excluded, unconsciously or by assumption, the informal research within agricultural production systems.[9]

A second problematic aspect of the preoccupation with transfer as the engine of transformation is that it is inherently centralist in its linear practices. This centralism is, as Rogers (1980) suggests, implicit in the diffusion model. It is also explicit in practice based on the transfer model, particularly institution building to establish capacity for technology generation _for_ and within developing countries. The most striking example of this practice was the establishment and early work of the international agricultural research centres.

The focus initially in the crop investment work of the first international centres was on wheat, maize and rice. This focus reflected perceptions that there had been considerable advances through agricultural research in developed countries and East Asia. There was therefore a gap which offered possibilities for accelerated land productivity increases in other developing countries. _Centres_ were established to undertake not merely transfer and testing of material but the development of new technology that would be widely adapted and therefore appropriate for widespread transfer and adoption. The centres established outreach programmes to assist local testing and scientific work organised around the improved material being developed at the centre.

There is now a widespread recognition of certain problematic aspects of the original model underpinning this practice based on hard learning. First, ecological diversity restricts the possibilities for diffusion of widely adapted plant types.[10] Institutions organised on a centralist, transfer model are not designed to learn systematically through built-in feedback procedures or to be highly flexible and responsive to lack of fit or different priorities. For example, the eventual establishment of research programmes for deep water or upland rices were therefore part of the second generation response to partial fit of the widely adapted semi-dwarf rice technology (Barker 1981).

Second, a centre-periphery model does not automatically foster technology generation capacity at the periphery. The models were rather intended to accelerate the process of technical change in production through material transfer and design testing. This has been recognised. There has been a shift of emphasis towards strengthening national research capacities within the <u>CGIAR</u> system, symbolised by the establishment of ISNAR (International Services for National Agricultural Research). This shift of emphasis makes it particularly important to look closely therefore at the modernisation models of institution building.

There is nothing modern about the notion of building institutions based on imported or transferred models. For example, the construction of educational systems in many colonial situations was based consciously on the use of imported models, the German Hochschule, the English public school, college and university. However, in part because of the relatively late development of publicly financed agricultural research institutions, many colonial research systems, e.g. those of the British Empire, evolved <u>in situ</u>, reflecting different contexts.[11]

Theorising and practice after the Second World War have been in sharp contrast to this earlier unsystematised <u>evolution</u> of institutions in different contexts. Undoubtedly modernising preconceptions dominated thinking and practice in the earlier period. However, the practice was unsystematic and pragmatic.

The power of linear modernising models in the post-Second World War period is reflected in the conscious exportation of the American land-grant college institutional model as the appropriate way to organise training, research and extension.[12] There are detectable shifts of emphasis in the institution building literature. Many decolonised countries lacked any significant capacity in higher level agricultural training and research (e.g. Indonesia). Hence there was an early emphasis on the building-up of training and research institutions. The American based agencies drew upon American experience, models and personnel, just as British and other national colonial and subsequently development assistance agencies drew upon their metropolitan country experience.[13]

Parallelling the typology of transfer which emphasises an upward progression, thinking about institutional models has emphasised the movement from the establishment of component institutions to the creation of a national research capability.[14] Again, however, thinking and practice has been strongly transferrist. Thus Moseman (1970) proceeds from the identification of the need to build up a national capability to draw consciously on American experience.[15]

What is striking about modernisation theory and practice is the presumption that components or whole systems will transfer from a unique economic, social political context to another context and that the positive functioning attributes will likewise transfer, at least to some degree. The possibilities and problems are more obvious in the case of material transfers, especially plants and animals. There are well established procedures for testing which will clarify _to some degree_ what the consequences of transfer will be in terms of plant or animal growth and reproduction in definable circumstances. These may, of course, not necessarily be production conditions. Hence the significance of informal producer research and the need for feedback or more ambitiously interaction between the research system and the producers. These problems of testing and interaction have been interpreted as either implying finer tuning of centralised research or a reconstruction on more decentralised lines, i.e. even the problematic aspect of material transfer has institutional implications.

The problems which arise are potentially infinitely more complex, when what is entailed is institutional transfer or reorganisation to allow transfer. These issues are taken up in more detail in a later section. However, it should be noted that an important part of critiques of transfer of institutional models is that the consequences are invariably something other than what was intended (Hart 1962). The land-grant colleges are part of a complex and _decentralised_, highly successful system integrating training, research and extension. The institutions established on this model have invariably proved to be centralist and elitist, whatever the developing country context, i.e. whatever else quite unlike the original model in a fundamental sense.[16] Perhaps there is a contradiction in the notion of transferring – through the use of high profile, high cost development assistance resources, (a highly centrist activity) – decentralised institutional models.

What is strikingly absent from modernisation theory or practice is the attempt to build on the alternative postulate of evolving institutions appropriate or relevant, and therefore potentially different from the metropolitan models. The Chinese socialist models provide another more recent example of alternative non-centrist conceptualisation (Rogers 1980; Stavis 1978, 1979). However, the practices were primarily concerned with production and social organisation of production, rather than agricultural science.[17]

A review of modernisation thinking and practice in relation to agricultural science becomes an exploration of institutional rather than technical questions. The technologising thrust of modernist practice is, in terms of its conceptualisation of the institutional, wholly problematic.

While we describe different critical reactions to the practice of modernisation, it should be pointed out that this is a way to organise our presentation and, to some extent, to follow a historical perspective.

In dwelling on critical responses, we do not say that some modernisation strategies have not achieved various objectives. For example, Dalrymple (1979) justifies and supports the major thrust of the Green Revolution strategy by pointing to the fact that countries such as India have increased food production so significantly since the 1950s and 1960s that it is no longer dependent, politically and economically, on the United States and other food producers. Similarly China has used improved cereal grains transferred from international agricultural institutes to increase food production. Recent food production data shows that the countries in South and South-East Asia, are increasingly becoming less dependent on food imports and are even exporting (Clay 1983).

In addition, the themes of demonstrating and transferring new institutions are not only a characteristic of modernisation strategies of the 1960s; some of the new Farming Systems Research (FSR) methodologies[18] have similar preoccupations (CIMMYT 1977). However, even in this subject area we have a responsive literature coming out of these experiences and pointing to the need for a more sensitive and evolutionary approach to the development of overall agricultural research institutions (Collinson 1982; Biggs 1983; Gilbert et al. 1980).

FACTOR ENDOWMENTS

Perhaps the most extensive literature concerning technical change in agriculture is that of writers concerned with issues of factor endowments, factor prices and choice of techniques. We are portraying their themes as a reaction to modernisation theory because much of this literature came in response to the indiscriminate actions of transferring methods of farming from one country to another, irrespective of local factor endowments. Most often this is characterised by the transfer of capital-intensive technologies to labour-abundant economies.

Their responses came in various ways. Schumacher (1973) promoted the philosophy of transferring and developing intermediate technology which was appropriate to local factor endowments and the choice of technique literature has been a central theme in the economic planning literature for many years.[19]

In an important contribution to the theory on agricultural technological change Hayami and Ruttan (1971)[20] have developed an induced innovation model. Drawing originally on comparative historical data for the United States and Japan they illustrated that where factor prices appeared to reflect national factor scarcities, technology appropriate to those resource endowments was generated and used. In the United States, where labour was short, and land and capital was relatively abundant, capital-intensive

technology had been induced. In Japan where capital
was scarce, but labour was abundant, labour intensive
technology had been induced. These appropriate
technologies had been induced because the market
price of capital, labour and other factors of
production had reflected relative national scarcities
for these factors. Figure 2.1 illustrates this
situation. Given a hypothetical iso-product line II'
Japan, with cheap labour and expensive capital, would
be at point J, using K_J and L_J of capital and
labour respectively. The United States on the other
hand would be at point U, using more capital (K_U)
and less labour (L_U).

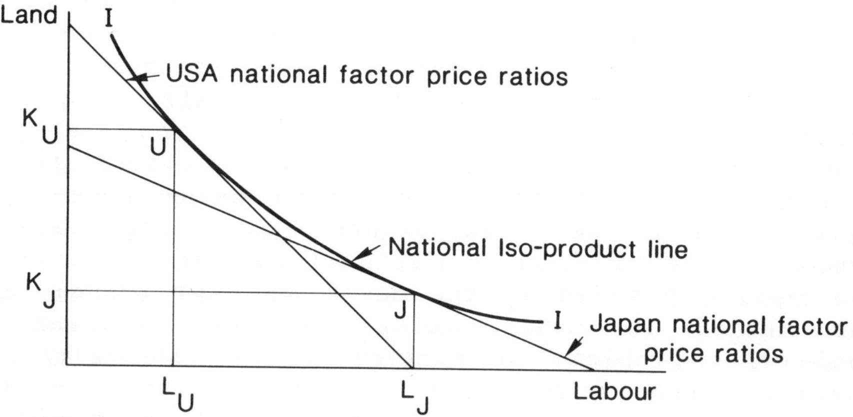

Figure 2.1: Factor prices and positions of appropriate
technology

This theory is not only an important response to
modernisation theory, but also a critical alternative
to the Malthusian model. In a Malthusian model
technology is given a parameter. As population
increases, relative to land and other factors, real
wage rates decline and the population stabilises
after famines, etc. at a subsistence wage. In the
induced innovation model technology is a variable.
This enables population densities to be sustained at
wage rates higher than a subsistence wage.
The induced technical innovation theory has been
further developed to encompass induced institutional
change (Binswanger and Ruttan 1978). In Japan and

the United States, the behaviour of research institutions is seen as a normal and <u>expected</u> institutional response by society to meet the needs of producers for appropriate technology.

The emphasis as regards technology generation and diffusion in the induced innovation theory is on the importance of factor prices. Policy actions which follow from the induced innovation model are focused on getting factor market prices (or surrogates for market prices) right so as to reflect relative factor scarcities. If this is done, then appropriate technologies (and/or institutions to produce them) will be induced into existence and spread.

The other major theme of writers preoccupied with factor endowments has been the problem of appropriate technology creation, development and extension. For example, in a review of wheat and maize adoption studies Perrin and Winkelmann (1976) found that it was not so much a matter of factor prices being 'wrong' than that the technologies being generated and promoted were inappropriate to the factor endowments (and other conditions) of farmers. This perception has led to the development of principles and methods for understanding the factor endowments and other problems of farmers before technology is developed and extended (Byerlee and Collinson 1980). Farming Systems Research is the broad heading given to this type of work.

Another type of research which has followed from the lack of adoption of high yielding rice varieties (Evenson 1974; Farmer 1979) has been the use of ex-ante benefit-cost analysis for allocating research resources within commodity programmes. Barker (1981) has shown that the expected social returns to investments in research on rice varieties for rainfed conditions would be far higher than investments in either extending irrigation facilities or investments in more research on rice varieties for irrigated conditions.

Again we see an emphasis on looking at the resource endowments in order to understand, or direct technological change.

An emphasis on appropriate technology has also been the major concern of the Intermediate and Appropriate

Technology (AT) movement. In its work focus has been placed on making available information on technologies which describe the range of known techniques which might be useful to farmers.

Manuals describing different types of tools, and directories of manufacturers of small-scale farm equipment are examples of this behaviour.[21] Most farmers in developing countries are portrayed by the AT movement as being small, with plenty of labour and little capital. Consequently most emphasis has been given to the dissemination of information about small, labour-intensive production techniques. While AT institutions have disseminated technical knowledge on production methods they have not, until recently, been on the whole concerned with analysing price policy and institutional issues, _which affect the development and diffusion of technology_.[22]

In summary, we can say that the major preoccupation of writers in this area has been directly related to issues of factor endowments and the development of technology which is appropriate to those endowments. Policy analysis and action revolve around getting the factor prices right and in developing and extending knowledge about appropriate techniques. The great strength of this analysis is that these are crucial aspects in understanding processes of technological change in mixed and socialist economies. However, amongst some of the issues which are not of central concern to this set of preoccupations are problems associated with (a) the reconciliation of short- and long-term factor prices, (b) the analysis of commodity (as opposed to factor) prices and the impact of income distribution on effective demand, and (c) the effects of a skewed distribution of land and other resources on technology generation and diffusion. These issues make up some of the themes of distributionalism, the environment, etc.

DISTRIBUTION AND EQUITY

Practice and thinking about science and agricultural development were, we have suggested, dominated by modernisation models of change or transformation.

The considerable productivity gains in terms of land or labour resulting from the shift to a science based agriculture in developed countries implied an opportunity for the rapid transformation of Third World agriculture through a similar process. The initial emphasis on direct transfer was unsuccessful because the developed country technology was not directly applicable to the conditions of tropical agriculture, either through extension to farmers (Moseman 1970) or attempts at radical transformation of the production process (the Groundnuts Scheme type mechanisation approach). One set of responses to this unsuccessful experience was to emphasise the importance of local factor endowments and factor prices. A second set of themes concerns equity and distribution issues.

The subsequent shift of emphasis towards providing such technology initially for cereals through internationally organised research capacity involved planned public sector generation and subsequent _transfer and diffusion_. The planned purposive nature of the activity, the initially limited availability of genetic material resulting from such programmes of research, entailed decisions about the distribution of the benefits of new technology. These decisions were therefore about the opportunity to use, to have _access_ to, to _acquire_ this technology. From this perspective planned change through the generation and diffusion of improved technological capability could dramatically increase land productivity but had unavoidable _income_ distributional and equity implications. These consequences might initially be to increase spatial or interpersonal inequality within agriculture or the rural economy. This was recognised from the outset, at least in some countries, in which the new science-based technology was given a major role in the agricultural development strategy, e.g. India (Hopper 1978). This awareness and concern for the potential tension between agricultural growth and equity is epitomised in the debate on consequences of the Green Revolution that has dominated all subsequent thinking about science and Third World agricultural development. However, this protracted and inconclusive debate suggests that the initial conceptualisation of distributional issues, even the concept of a Green Revolution, are themselves problematic.

The Green Revolution as a phenomenon attracted analysts from the whole range of social science disciplines and from across the political spectrum. There was, nevertheless, a broad similarity in the conceptualisation of technical change and the equity problem by most analysts within the first generation of Green Revolution studies. The Green Revolution was about the rapid movement from a resource-based to a science-based agriculture (Hayami and Ruttan 1971). Pearse (1980) characterises the Green Revolution as "an international campaign aimed at increasing the productivity of land by means of the introduction of a science-based technology in the production of foodgrains", adding with the benefit of hindsight "referred to throughout for convenience rather than accuracy as the new technology" (op. cit., 1).

This focus on the introduction of <u>new technology</u> implies a conceptualisation of exogenously-generated technical change, transferred from outside, impacting on agricultural practice and on the rural economy and society. This focus had a number of restricting consequences. The dichotomised notion of traditional or local as against new, modern, or HYV encourages a research agenda in terms of adopters and non-adopters of these specific technology components and further elaboration of the diffusion models characteristic of modernisation approaches. The appropriate unit of observation is the production unit – the farm, the cultivating household or holding. This approach similarly suggests a specific agenda of issues. This is illustrated by perhaps the two largest research projects in the early 1970s looking at the implications of the new cereal technology, UNRISD Global Two and the IRRI coordinated project on Changes in rice farming in selected areas of Asia (henceforward called the UNRISD GII and IRRI Changes project).

The IRRI project was concerned with the following issues: adoption, constraints on adoption, physical or socio-economic, and the consequences in terms of productivity change.[23] Some constituent case studies were extended to explore more fully questions on employment and the partitioning of income streams from innovation and socio-economic consequences in the farming communities affected by technical change. The project relied largely on cross-sectional survey data for cultivating households or holdings which could be characterised in terms of size and tenurial status. Issues such as employment effects could be picked up through cross-sectional comparison of labour requirements of modern and local varieties, but the fuller ramifications in terms of socio-economic change are difficult to quantify where, for example, the landless are not directly considered.[24]

The approach has the advantage of throwing up issues for further research – on constraints to adoption, on consequences which could not be fully explored, and the feedback raised questions too for agricultural science on the when and where of adoption, the possible limitations in terms of extent of adaptability of the new technology. The cross-sectional methods of data collection and, in effect, comparative static analysis of with and without adoption is not appropriate for calibrating in any sensitive way the longer-term changes which provide the context of the snapshot of innovation. In the context of the IRRI research programme this exploratory research led on to further elaboration of concepts of constraints or adoption of modern varieties and higher productivity in agronomic and related social science research (IRRI 1979), and articulation of the complex anatomy of communities (Hayami 1978). The thrust of the research effort is towards fine tuning – knowing more about the environment, the socio-economic system, delivery institutions which can provide feedback to agricultural research policy or policy on agricultural production support systems. The wider consequence of this and similar programmes has been the development of on-farm agro-economic research concepts and programmes as an integral part of the work of scientific institutions. Particularly in Africa and Latin America this has generated a whole new sub-discipline of farming systems research. The articulation of this more complex model of feedbacks into formal research is indicated by terms such as farmer-back-to-farmer research (Rhoades and Booth 1982) and interactive research (Barker 1981).

The UNRISD Project had an overlapping set of concerns (Pearse 1980, pp.1-2).[25] The model is again one of the impact of exogenously generated technology. Most of the empirical evidence on which the country case studies were based were again cross-sectional snapshots of recently initiated processes of innovation. Acquisition of technology implies a more purposive role for the eventual controller and user in contrast to diffusion, thus placing emphasis on the market and bureaucratic mechanisms for distribution of inputs and sale of output and a complex of associated access issues (Griffin 1974). The broadening of perspectives, at least retrospectively, into a concept of livelihoods rather than a focus on a partitioning of income streams from technical change makes the community and all households (and members of those households differentiated by sex and age) the object of study. However, this very broadening of the approach in relation to a narrow, simplified notion of technical change makes it more difficult for the analyst to provide data of a specificity that can feed back into agricultural science. The alternative to increased specificity has been to provide a more generalised sense of the socio-economic consequences of observed technical change, the broader technical options that might imply different outcomes (Dasgupta 1977) and especially a focus on the social institutions - the structures that are the context for marginalisation of some communities and classes and, growing spatial and interpersonal inequalities (Pearse 1980). At the same time the focus is still micro - on the rural community (Pearse 1980).

The major contribution of the distributionalist literature has been to draw attention to the highly inegalitarian and potentially economically irrational (in the factor endowments sense) consequences of actual observed patterns of technical change. The actual structure of demand for technology from within agriculture is seen to reflect the distribution of productive assets. In terms of comparative dynamics, the long run consequences in terms of factor proportions changing asset distribution and capitalisation of agricultural production processes may therefore be radically different depending on earlier differences in asset structures between economies despite earlier similarities in factor endowments. Factor endowments contribute significantly to the explanation of broad cross sectional differences in paths of technical change as between economies with wide differences in factor endowments such as East Asia and North America. However, there are significant observable cross sectional differences in patterns of asset distribution, capitalisation and factor proportions in regions which earlier had similar factor endowments but differences in asset distribution and structures of transactions in commodities (e.g. Denmark, the Netherlands and eastern England).[26] There is evidence of a consensus growing on this issue.

The thrust of the distributionalist concern is now widely accepted. For example, induced innovation theorists who initially focused on the historical significance of factor proportion in long run patterns of technical change now explicitly acknowledge the significance of asset distribution as potentially establishing actual price structures which are not in accord with factor endowments (e.g. Hayami 1982). This allows _ex ante_ in current circumstances for the possibility of perverse patterns of technical change which can only be remedied by the incorporation of market surrogates reflecting factor shadow prices into bureaucratic decision processes or asset redistribution. These are the policy prescriptions of the distributionalist literature, (Griffin 1974, Pearse 1980). What is different about current circumstances is of course the possibility and actuality of transfer between economies with very different agricultures and factor price ratios.

An important aspect of the distributionalist concern has been to draw attention to the already

severe and potentially rapidly growing pressure on
the livelihoods of the landless and marginal
cultivating households; so severe as to place in
doubt the scope of longer run processes of technical
change to prevent politically unacceptable
neo-Malthusian adjustments and/or the build-up of new
population structures economically unsustainable
within the resources of nation states unprecedented
in historical experience. Such concerns are
reflected in Barraclough's introduction to Pearse's
1980 retrospective on UNRISD II:

"The food systems that have maintained humankind
throughout most of its history are disintegrating
before other forms of economic activity are able to
offer alternative means of livelihood to the
displaced peasantry." (op. cit., p.vii)

In conclusion it is important to note that a
consensus appears to be emerging now as regards the
planned development of agricultural technology for
small and resource-poor farmers.[27] Many of the
equity criticisms of the Green Revolution have been
taken up by the farming systems researchers. This
orientation can be seen as a food production approach
where emphasis is on providing technology for
sustained incomes (or livelihoods) to
self-provisioning small farmers (or peasants).

In other words, helping to provide basic needs to
resource-poor farmers is a growing focus for
agricultural research. This, in turn, affects the
commodity composition of agricultural research
budgets because greater emphasis is given to food
crops for consumption in rural areas (Ruttan 1980).

The effective demand for food is now more
adequately taken into account because emphasis is on
growing crops for direct consumption by poor
farmers. Employment problems are taken account of
because the client group for research are small
farmers who have plenty of labour, and scarce capital
and land. How to use their resources efficiently is
the major goal. However, this approach is only a
partial analysis which represents agriculture as a
closed system where most farmers have small holdings
and they primarily eat what they grow.

However, we are still left with the problem of many

rural areas where landless labourers make up a significant proportion of the rural poor and certain types of small farmer technology could, in fact, increase rather than reduce employment opportunities for labourers. The other unresolved related problem concerns the demand for agricultural commodities. Pinstrup-Andersen et al. (1976) bring this point out very forcefully in the analysis to derive commodity priorities in agricultural research in Colombia. They take a planner's perspective and instead of putting all the emphasis on the problems of resource-poor farmers, take a broader policy basic needs view and estimate agricultural commodity research priorities, if the government's primary objective is to use national agricultural resources to improve the nutritional status of all poor people. They take into account the distribution of incomes between different socio-economic groups (in and out of agriculture) and the effects of different income elasticities of demand.

This study by Pinstrup-Andersen is important because it illustrates well the dilemma for agricultural science policy where preoccupations are only with factor endowments, factor prices and with distributional and equity issues within the agricultural sector. For national policy the perspective must be broader. In addition, composition of demand for agricultural commodities is a central theme of writers concerned with issues of the world economy. In colonial situations, export crops received priority because of the demand from importing countries. More recently multinational corporations are playing an increasing role in affecting and controlling the demand for agricultural commodities.

WORLD ECONOMY

Another set of responses to the theory and practice of modernising Third World agriculture has been those of the writers concerned with dependency relationships in the world economy. In particular, they have been concerned with modernisation strategies which promote agri-business and the integration of agrarian economies within the world capitalist system (Agri-business Council 1975). This group of writers[28] has been preoccupied with analysing such issues as: the role of multinational companies or corporations (MNC) in the Third World (Burbach and Flynn 1980; Dinham and Hines 1982), aid as a vehicle for promoting the interests of international capital (Burch, 1980; Feder 1977), food aid as a political weapon (Cleaver 1972), and the promotion of Third World food production as a means to counter or forestall rural social unrest in developing countries.

One of the earliest authors to take up some of these issues was Cleaver. Referring to Borlaug's receipt of the Nobel prize for peace and not for plant breeding and genetics Cleaver concludes:

"It is woven into the fabric of American foreign policy and is an integral part of the postwar effort to contain social revolution and make the world safe for profits ... This association between food production and anti-Communism was quite conscious." (Cleaver 1972, p.81)

At the centre of the analysis of this group of writers is the overriding preoccupation with the consequences of the spread of international capitalism. Control over resources and dependency issues are recurring themes in this literature. An example of the work in this area is that of Franke and Chasin (1980). In reviewing the socio-economic causes of the 1968-74 Sahel drought and famine, they argue that the causes lie in the colonial promotion of cash-cropping (such as peanuts and cotton) and the further development of the international capitalist system after independence in many African countries. For them:

"Dependency theory, then, is not merely a statement of the internationally _interdependent_ character of the modern world economy. Rather, what distinguishes countries with dependent economies is a set of political and economic relationships that render the dependent country relatively powerless to control or affect the course of development." (Franke and Chasin 1980)

For them technological change – whether it be the expansion of peanut cultivation for export, or the spread of soil erosion – can only be understood when the role of international capital and its search for short-term profits is placed at the centre of the analysis. They argue this position after comparing alternative explanations of the Sahel famine,[29] including over-population, land mismanagement, weather change, migration and overgrazing.

In the context of this review we are not attempting to assess the validity of these arguments. For us the critical point is that writers such as Franke and Chasin are preoccupied with a different set of issues from, say, those exploring the implications of national factor endowments.

Other writers in this theme area look at the growing role of the MNCs in vegetable and flower production in Third World countries (Burbach and Flynn 1980 and Dinham and Hines 1982, and Feder 1979). Different types of out-grower schemes, producer contracts etc. are diagnosed in a centre/periphery economic framework, with the multinational company being seen as the centre of a world capitalist system. Thus the extraction of an economic surplus by the multinational company is viewed in the context of changes in real wage rate (by sex) in the Third World countries. For example, in "Strawberry imperialism" Feder (1979) analyses the way in which United States capital dominates and controls the vertically integrated Mexican strawberry industry which exports to the United States. Young plants from the United States are sent to Mexico for propagation and direct distribution to producers with little opportunities for adaptation to Mexican conditions. So although strawberries are an export crop, Mexico does not have its own R & D capability which might give it some <u>control</u> over production practices and markets. Who controls, and who is able to use and is responsible for, centres of research and development (R & D) capability is a principal concern of the literature in this theme area.

The primary rationale for the establishment of the International Agricultural Research Institutes was a response to fill the gap left by the emphasis in public sector, colonial and MNC private sector research on the development of export and cash crops. Frequently these crops (tobacco, bananas, tea, cocoa, cotton, rubber) had been managed under colonial regimes and the surpluses had been accumulated by colonialists. Under these circumstances, there was often a strong research capability working on export crops where international capital had strong interests. However, research on food for local consumption was often minimal (Busch and Sachs 1981; Ruttan 1982).

Recently, Plant Breeders Rights have become an increasingly important theme in discussions about the generation and spread of agricultural technology. Mooney (1979), for example, emphasises the implications of the actions of the MNCs, who have complementary interests in the commercial use of seeds, pesticides and fertilisers in Third World countries. These companies have large investments in agricultural technology R & D and are primarily interested in commercial strategies which are profit motivated. Amongst other things this behaviour is identified as contributing to a worldwide reduction in the pool of genetic material (genetic erosion). While Mooney's broad sweeping analysis is technically incorrect in places, it still represents a distinctive contribution. He focuses on issues of who controls and directs R & D — the implications of which have international ramifications. Most writers on international genetic resources have not sought to analyse in depth the motivation of different interest groups behind the establishment of international germplasm banks, international agricultural research centres, etc. The point which Mooney emphasises is that these are all political problems, and not purely technical issues susceptible to purely technical solutions.

Another example of the work and types of analysis of writers in this area is research on the reasons for the transfer of tractors to Sri Lanka (Burch 1980). In this case study emphasis is placed on the role of the British Aid Programme and the promotion of the interests of British tractor manufacturers. In the name of modernising agriculture this type of

programme could appear as a sensible aid strategy.
However, if a more thorough analysis of local farming
conditions had been conducted then it would have been
seen that tractors were inappropriate for the
declared development objective of Sri Lanka and the
farming systems for which they were destined. Burch
in his analysis studies how and why the interests of
international capital influence the behaviour of aid
agencies. The implications go further than just the
short run transfer of tractors. This type of
agricultural technology transfer has long-term
implications for energy use and the future
composition of agricultural inputs. Some of these
may need importing and provide markets for external
investments.

Finding explanations for the opening up and
development of markets for agricultural inputs by
international capital is a recurrent and growing
theme in the world economy literature. It not only
covers seeds and machinery, but increasingly looks at
pesticides which are often developed and promoted by
multinational firms. There is a growing coincidence
of interests between writers who analyse the role of
multinational companies and agricultural technology
in both rich and poor countries. The world economy
literature on agro-chemicals is paralleled by writers
who concentrate on the role of pharmaceutical
companies in the generation, promotion and spread of
drugs (Melrose 1982). Again we find there is a
preoccupation with issues which focus on the short
and long run behaviour and interests of the MNCs as a
critical determinant in understanding processes which
have shaped the specific characteristic patterns of
technological generation, transfer and diffusion.

In summary, the theme of the world economic system
includes those preoccupied with the role of
international capitalism. Writers in this theme
area, while they take technological illustrations
from different parts of the world, do not place their
major emphasis on the specific analysis of
technological change in both socialist and mixed
economies. It is a serious limitation of this
perspective. In some situations problems caused by
international capital in mixed economies also occur

in socialist countries. An example is genetic erosion. This is also occurring in China for cereal crops as in the rest of East and South-East Asia (National Academy of Sciences 1975; Stavis 1979). The promotion of a few dominant varieties - which led to genetic vulnerability - caused a major reduction both in tobacco and sugar production in Cuba in the late 1970s. In addition modernisation strategies involving the use of tractors, improved cereals, agro-chemicals, fertilisers, etc. have been caused by a variety of reasons and had very mixed outcomes in socialist economies. Such examples suggest that some important aspects of the process of technological change are missing from world economy analysis and they halt their exploration of cause and effect with the MNC instrument of transfer. Yet the policy preoccupations of writers in this area are generally broad sweeping in nature and emphasise the need for radical change, socialist transformation and commitment as the preconditions for better policy practice.

This is the paradox of the work of many of those writing in the world economy theme area. Their critical research from outside the institutions engaged in, and notionally responsible for, technology generation and transfer has unearthed important and powerful data on specific and consequential policy issues. This data, as with the work on distribution themes and the environmental problems, is evidence that there are no grounds for complacency in science and technology policy practice. These are areas in which controls will lead to different, probably more desirable outcomes. There is above all a world-wide problem of responsible practice in agricultural, scientific and technological policy: a theme developed in the volume edited by Anderson et al, 1982. Yet many of those who have contributed under the world economy theme offer sweeping policy prescriptions rather than engaging in an exploration of how to enforce responsible policy practice. This disengagement allows the dismissal of both the data and its authors as respectively unprofessional and ideological. It is this de-emphasis of the specific technical and

motivational issues of agricultural science and
technology policy, and a limited concern for better
policy practice, that distinguishes this group of
writers from those concerned with environmental and
ecological issues.

ENVIRONMENTAL THEMES AND ECOLOGICAL CRISIS

A response to the modernisation theories has been
that writers have taken up environmental and
ecological issues. This analysis is a critical
response to the simple modernisation emphasis of
perceiving agricultural development as a race between
increasing food production in developing countries
and population growth. It is a response which
analyses situations where the expansion and
sustaining of food production and other economic
activities can produce environmental stress and
threaten crisis. Writers[30] in this area focus
attention on data demonstrating the decline of soil
fertility, soil erosion and abandonment,
desertification, deforestation, overgrazing,
siltation of irrigation systems and reservoirs,
flooding and the depletion of fishery resources.
  Distinguishing features of the work in this area
are the exploration of (i) the actual or potential
long-term destructive consequences of specific
short-term agricultural production activities, (ii)
the inter-relationships of economic activities in one
geographical area and other areas, (iii) problems of
ecological balance, and (iv) the use of renewable and
non-renewable resources, and the interaction of
geographically specific physical, demographic and
social relationships.
  Much of the writing in this area appears to fall
into two categories. On the one hand there is more
practical and technical literature which records the
incidence of such things as desertification and sets
out technocratic causes. Analysts primarily look at
the implications of climatic or production phenomena
such as overgrazing, which lead to a loss of soil
cover and erosion (UN Conference on Desertification
1977). In the literature on this area there is a

strong practical and policy bias. For example policy analysis has resulted in an emphasis on the need for, and practicalities of wood-lots, social forestry and community action. Analysts also consider the significance and opportunities for bilateral and international agreements to address environmental issues on a regional (inter country) and on a global scale (Johnson and Blake 1980).

On the other hand there is literature which attempts to integrate environmental issues into a wider political economy analysis (e.g. Franke and Chasin 1980; Blaikie 1983a; and Bernstein 1979).

In developing a theory for the political economy of soil erosion Blaikie (1983a) highlights some of the key elements which are central to theorising about this subject: (a) the interaction between social and physical systems in a specific geographical region (a place-based orientation); and (b) the importance of the specific historical and other contextual settings of the region.

In a Marxist analysis of some of the causes of local soil erosion Bernstein describes the process as a simple reproduction squeeze — terms of trade deteriorate against peasants, which leads to lower consumption and/or intensification of commodity production. This initiates the cultivation of poorer or more distant lands which results in lower returns to labour and a possible deterioration of the resource base.

Some of the problems of developing a theory relating to environmental issues are also well expressed by Blaikie. While many writers on environmental problems see capitalism as a major cause of the environmental crisis, writers of a Marxist orientation are either strongly divided or have not addressed the issue in a policy or contemporary setting. The following quotation illustrates well part of the dilemma in the Marxist literature on the twin problems of rapid population growth and environmental deterioration:

"Perhaps one reason is that the population problem has often been seen to be associated with neo-Malthusianism, and state-run family planning programmes as no more than class warfare (perpetrated through international capitalist ideology by various class alliances upon the most deprived and vulnerable who usually have the highest birth rates) and as a diversion from the fundamental reasons behind scarcity and deprivation. The UN Population Conference at Bucharest in 1973 was a turning point in that it made explicit and public the ideology of population programmes. However, even in some socialist states, high population growth-rates have been sufficiently recognised as a problem and prompted state action (e.g. in China and Cuba)." (Blaikie 1983a).

There are clear parallels in theoretical work with the paradoxical character of the world economy literature. The general analysts amass data which could imply a specific and powerful policy agenda, but do not engage in the attempt to construct specific and feasible agendas for today's policy makers.

Although the emphasis in the work of the two groups of writers in this area may differ, their overriding preoccupation concerns the conflict (at local, national and international level) between short-run production activities and long-run environmental and ecological results. Short term prices, market surrogate prices, or planners'directives may arise from the activities of socialist or mixed economies. Their production activities may be for a variety of reasons - e.g. for providing good livelihoods for resource-poor farmers, for producing an export crop in an MNC system or commercial farming in California. Whatever the reasons, and whatever the political environment, the environmentalist is concerned with the long term environmental balance and the sustaining of productive systems.

Three of the major contributions of people concerned with these themes are: (i) their often equal handedness as regards whether countries of different political persuasions are better or worse on environmental issues, (ii) their pointing to major areas where new research is needed, for example, Third World energy conservation (wood-lots and efficient stoves) and alternative and sustainable farming systems, and (iii) their concern with the irreversible consequences of many of today's production and consumption activities. For example, once certain types of plants and environments are lost there is no way of replacing them. Do we know what the long run cost of these activities will be? The environmentalists would say that we need to do more analysis on these issues before we can even come to an argued position.

RESEARCH AND EXTENSION INSTITUTIONS

During the last few years there has been a growing literature on agricultural research and extension facilities.[31] This includes descriptive and analytical material on the historical development of systems in developed and developing countries and in the field of international agricultural research. There is also a growing literature on principles, techniques and methods for organising and managing agricultural research.[32] Because it is beyond the scope of this review to try and cover such an enormous literature we shall concentrate on material which describes past situations and attempts to draw up analytical frameworks for understanding those situations.[33] It would appear logical that such analysis is prerequisite for the application of management techniques.

In reviewing the literature we have found a concentration on different themes in explaining why the institutions of agricultural research and extension systems have different characteristics and exhibit different behaviour in different countries at different points in time. One approach to reviewing this literature is to illustrate some of the alternative approaches by examples, and then ask what in these approaches might be useful to policy makers seeking to improve the performances of an agricultural research policy in a given situation.

Decentralised innovation and diffusion: an alternative model

Recently Rogers (1980) has proposed an alternative model of the processes of technology generation and diffusion. He juxtaposes this decentralised model with the ideal centralised normative model of diffusion. In the centralist model, technology is generated at the centre and disseminated through an extension system. The Training and Visit (T & V) system (Benor and Harrison 1977) is a recent example of the application of this approach in developing countries.

The decentralised diffusion model seeks to take account of behaviour and documented diffusion processes which do not fit the centralist-periphery model. Rogers goes on to show how this approach helps in understanding the source of the innovation and how government policy promoted the spread of 1) the barefoot doctors in China; 2) the "three don't wants" self-reliance philosophy of the Tachai Brigade in China (the "three don't wants" were: don't want state funds for recovery, don't want state grain, don't want relief materials) and 3) the energy conservation innovations in Davis, California, USA. This California town, which developed viable energy conservation methods, became a major source of national advice for other communities.

The importance of this decentralised model is two-fold. A role is accorded to the centre and to the periphery as sources of ideas and innovations. The actual generation and diffusion of innovations is not a linear process (research — extension — farmer). Nevertheless it is such a linear model which characterises the <u>practice</u> in many agricultural research and extension systems. We emphasise the word <u>practice</u> because feed-back and the collection of local knowledge is often accorded recognition, but commands few resources, if any. Second, the contrast of two opposition constructions of generation and diffusion processes each reflected in actual practice suggests that there is a need for relativism and sensitivity to the specific context in characterising the processes of particular systems. There is no universal model which offers prescriptions for all circumstances.

It would be useful to apply this model to an analysis of the actual behaviour of some major national and international research and diffusion systems.

For policy purposes, the model is helpful as it highlights different and complementary sources of innovations and a reversal of roles which different institutions in a system might play. This has important implications for the allocation of scarce research resources.

Stages of growth models

Another approach for understanding why a system has certain characteristics is to see it as a specific stage in the historical evolution of systems in general. For example Evenson (1978, p. 240) suggests four stages in the evolution of agricultural research: 1) the pioneering stage of institution development; 2) the simple adaptive system stage; 3) the advanced adaptive system stage; and 4) the technology-potential capacity stage.

In another case Ruttan (1980, p. 42) gives three stages which are characterised by different activities: 1) innovative activities of individual farmers and inventors; 2) organisation of agricultural experiment stations staffed by research workers; and 3) evolution of integrated national agricultural research systems.

Using these models we could say that a certain specific situation exists because it is at a certain stage of a linear development path. In a description of the stages which have characterised the last 30 years of international development in agricultural research in developing countries, Esman (1978) gives four stages: 1) extension model (emphasis on transferring known knowledge to farmers); 2) Land Grant College (LGC) model (transfer of the LGC institutional structure to developing countries); 3) Big Science model (setting up of the international agricultural research system to generate technology for transfer to farmers and national systems); and 4)

Interactive model (methods for providing appropriate technology to farmers bypassed by the Green Revolution and strengthening the linkages between farmers and researchers and extension workers).[34]

As regards the future, it would be interesting to ask where Esman sees us going next, because his stages encompass the international agricultural system and the appropriate role of international institutes has been open to debate for many years (Wade 1975).

Such stage models are, at least as with Esman (1978), an attempt at a stylised history, to draw out significant developments. The other stage models carry a normative loading in implying that this is the inevitable or desirable evolutionary profile through which systems pass.[35] Such linear models also encourage the possible attempt to bypass stages, thus becoming a recommendation of the modernisation through an institution building approach as a prescription for agricultural scientific capacity creation. For policy makers concerned with problems of specific countries or regions (big or small?) specific technologies (complex or simple, rapidly or slowly changing?), and in a specific political setting these normative stage models give relatively little help.

Induced institutional change model

The induced technological change model (Hayami and Ruttan 1971) has been extended to cover issues of induced institutional change. As with the induced technical change analysis, national factor proportions, and relative factor prices (or scarcity) are at the heart of the model. Ruttan (in Chapter 3 of the volume) elaborates the approach. According to this approach, the relationship between technical and institutional change has represented a continuous source of concern to economists and other social scientists interested in the historical and institutional dimensions of development. Technical change and institutional change are highly interdependent and must be analysed within the context of interdependence.

The significance of the proposed theory of institutional change is that it suggests an economic theory of induced institutional change that is capable of generating testable hypotheses regarding (a) alternative paths of institutional change over time for a particular society, and (b) divergent patterns of institutional change among countries at a particular time.

Several writers have sought to test and elaborate this model in case studies of research mobilisation in mixed (Chapter 6 of this volume) and socialist economies (see Chapter 7) and for international agricultural research (Ruttan and Binswanger 1978). We now look a little closer at these.[36]

One of the problems, as the author recognises, is that the system is indeterminant (Chapter 3 of the volume). The direction of causation is specific to any given situation. This means that the model is difficult to test, especially in the short run because it does not specify under what conditions one would expect the causation to be in one direction or in the other. For example, what does the model explain or lead us to expect in a mixed economy like Kenya where horticultural growers appear to exert little direct pressure on the horticultural research system? Rather it is left mostly to government officials and organisations to decide what needs doing. Nevertheless horticultural research has responded well in the last two decades to supplying recommendations for technical innovation (Chapter 6).

Do we understand that the right research institutional structures have been induced or should fast or slow change be expected? Does the fact that a) horticulture is important to the livelihoods of the majority of farmers (who have small farms) and b) public expenditure on research and development in horticulture is less than proportional to the industry's contribution to the total crop value in agriculture (op.cit.) mean that in fact the system has not responded well in the past and institutional change will be induced shortly to increase horticulture's share? If this is so, what specific types of political activities would we expect to occur in Kenya?

In the case of Viet Nam, Luu (Chapter 7) explains that one stage of innovation was dominated by a transformation of production relations so the government could buy time for agricultural sciences to develop and produce some useful and practical discoveries. This was in response to conditions of the problems of previous modernisation campaigns, a farming system which would not allow for multiple cropping without new technology, a weak and atomised peasantry, and the inability of the socialist government to provide investment to support take-off in agricultural production.

In this context we have to ask what are the specific cultural endowments which differentiate say, Kenya from Viet Nam and which lead to mistakes being made, or for new directions of technical and institutional change to be established. Ruttan and Binswanger (op. cit.) express some apprehension about the capacity of technical change to induce institutional innovation. This is particularly true of countries which have experienced the Green Revolution.

It would seem then that in certain applications of the model it could be that cultural endowments may be used to represent the degree of equity in the society.[37] In this reformulation we would then have a general (as opposed to a partial) economic model which took explicit account of the distribution of income and the effects of this on the composition of aggregate demand and commodity prices.[38] These types of innovation, though, would bring the model closer to the preoccupations and models of social scientists who look at broader, social, economic, bureaucratic and political determinants for understanding the existence and changing character of research and extension systems. (Brass 1982; Clark 1980).

There would be near universal acceptance of a proposition that institutional behaviour is sensitive to and reflects its socio-economic, political and cultural context. In this sense, there is always induced institutional innovation. If cultural endowment becomes a black box to explain everything other than the influence of factor prices then the power of the theory is much diminished.

While the work of de Janvry (1977) is used by Ruttan and is helpful in making more explicit the fact that such things as socio-economic structure and politico-bureaucratic structure are important, these components are diagrammatically placed in boxes in their work. They do not analyse what goes on inside the boxes.

In summary, we should conclude that our criticism of the induced innovation theory is two-fold. Firstly, it is a very important contribution to the theory because of the issues it raises. However, part of our response is a sense of frustration in trying to understand the model! However, the second point is more one of where the model appears to place emphasis.

The induced institutional innovation theory, in focusing on relative factor prices, <u>de-emphasises</u> another equally important dimension of the research policies process – the commodity composition of research and the allocations and establishments which this represents. Here too there is induced innovation in the same way that allocations definitionally reflect the influences on allocation decisions. For example, we have recently observed a response to the recognition of an energy crisis in research. However, it would be churlish to ask in terms of sustainable rural livelihoods what the costs have been of the failure of short-run prices and market surrogates for these prices to reflect long-run scarcities until after a discontinuous shift in 1972-74. This line of questioning would be highly relevant in looking at the long-run environmental implications of allocations in research and production reflecting short-run prices (and priorities) in both mixed and centrally planned economies. Because today's technology is potentially so powerful, it means that induced legislation or other institutional change may be ineffective or too late to prevent irreversible changes which have very high social costs.

Political science and sociology models

Under this very broad category we place the work of
social scientists who attempt to diagnose and
understand the internal workings of research
institutions and the different external political,
economic, foreign donor and other determinants which
influence the behaviour of research and extension
institutions. For the sake of this analysis we can
split writers into two groups. Studies which have
looked at the causes and results of modernisation and
transfer of institutional experiments, and those
situations where external modernisation elements are
less important.

An example of the former type of work is the
article by Brass (1982), which analyses the transfer
of the Land Grant College model to establish the G.B.
Pant University of Agriculture and Technology,
Pantnagar, Uttar Pradesh, India. In a political
science analysis Brass investigates the motivating
forces behind the building and funding of the
University and seeks to understand why only parts of
the institutional model were transferred in
practice. He then explains why an eventual crisis at
the University was inevitable. It is significant to
note that as early as 1961 other social scientists
were predicting the problematic nature of this type
of attempt to create research capability through
transfer of a specific institutional model (Hart
1961). Similar concerns were echoed by Hunter
(1969).[39]

An example of the second type of analysis is the
work of Stavis (1978) which reviews the growth and
development of the agricultural research and
extension system in China. In this case he
explicitly analyses political social structures which
result in the close control over research
capabilities by potential clients of research.
Recently literature by Hightower (1976) and McCalla
(1978) on the American agricultural research system
is another example of where analysts have tried to
review the actual processes by which different
interest groups influence agricultural research
policy and practice. Hightower looks at the control

of agri-business over public sector research capabilities while McCalla, in a refreshing analysis, compares the way we are told the system works (or should work) and the way it actually works. Not only do a few people have very great power in the system but the research establishment has a tremendous tendency to maintain the status quo and to continue its lumbering disjointed movement along well-trodden paths (McCalla 1978). In an assessment he feels though that this is a suitable system for the very specific political, economic and historical situation of the United States.[40]

Another type of analysis by sociologists and political scientists looks at the actual professional behaviour of scientists within institutions.[41] For example they ask questions about rice plant breeders in Asia who rank important production problems in one order, but carry out research on items ranked in a different order (Hargrove 1977). In the United States the same type of behaviour was found amongst all agriculturalists (Busch et al. 1980). A particularly important factor is the process by which researchers are socialised in the particular scientific discipline in which they specialise. For example the relevance of key-fellow scientists, the content and structure of training and education programmes, the development of new theoretical orientations and methods, the historical development of the various agricultural disciplines, and the process of faculty selection (Busch and Lacy 1981, p. 119). These types of factors are an internal influence and differentiate from external influences such as financial and political support.

Another area of growing importance for analysis and prescriptive actions is concerned with the role of private and public sector research.[42] There are different views on the role of private sector research. In Latin America Trigo and Pineiro (1981) show how private sector research which specialises in the problems of commercial farmers of the modern sector is attracting, by higher wages, national scientists - often trained at public expense, away from the public sector into private activities. The result is a decline in the capability of public sector research and more research effort being allocated to the technological problems of resource rich farmers. Pray (1983) on the other hand, in reviewing the Asian experience, where private sector agricultural research is small, sees a far greater role for private sector research. He advocates that Asian governments should probably do far more to encourage greater private sector involvement. Incentives to local and multinational companies may be important ingredients in such a strategy.

The significance of the political science, and public administration approaches to understanding research institutions is that it is a perspective where not merely specific interest groups, specific classes, funders, etc. are identified in each specific situation but the internal dynamics of bureaucratic institutions are <u>explicitly</u> part of the analysis. Actual policy actions directed to improving institutional performances have to grapple with the <u>specifics</u> of a system, allocations, and establishments, more of this and less of that in the short run.

It would appear to us therefore that these types of preoccupations and models have a major potential for policy makers.

Organisational and management models

The last area of literature on research institutions concerns the normative literature on how research and extension systems should be organised efficiently on a functional basis. It generally covers principles, methods and techniques for the collection and analysis of data, as well as procedures on how different components of a research and extension system should be organised and managed.[43] The strength and usefulness of this literature lies in its emphasis on sensible and logical approaches to cost-effective problem solving ways of allocating research resources, given financial constraints and defined objectives. Its limitations lie in the problem that to say what should be done assumes that one has a model which adequately explains why things happened in the past, and what is viable for policy action in the future. Preoccupation with this type of management literature can divert attention away from i) an analysis of the socio-economic situations of specific scientific systems where the techniques are supposed to be used, and ii) an awareness that political and institutional factors always have, and always will, determine when techniques are used in a ritualistic or real sense in a given situation.

Now that we have identified four types of research and extension analysis we can examine why each type may be important to contemporary research policy makers; policy makers who are concerned with feasible actions which might improve the sustained generation and spread of socially desirable technology.

First, we can examine what issues each type of approach raises. For example, the decentralised innovation and diffusion approach would always be asking questions about the way the centre learns from, uses the technology, and complements the work of decentralised units. As yet these types of questions have not been the major focus of the actions of people analysing the relationship between national and international agricultural research systems. The classic central-peripheral model and stages of growth perception still dominate the actions and documentation in this area (ADC and IRRI 1980).

Second, the sociologist, with a dominant interest in what determines the professional behaviour of scientists would first look at career structures, etc. of an institution before advocating that a new set of multidisciplinary research methods be tried. Some of the problems now being faced by those trying to transfer to developing country farming systems research methods, show that an analysis of these issues is a prerequisite for the local development of such approaches (Collinson 1982; Biggs 1983).

Finally, for broad general policy analysis the induced institutional innovation model might help policy makers to encourage research on labour-intensive techniques in a labour abundant economy and discourage private or public sector research and extension activities on capital-intensive technologies. The issue here would appear to be how to provide more effective market surrogates which impinge on administrative allocations and distributions.

The point we are making is that there is no one model or set of institutional concerns — as represented by the four types of analysis above — which can, or should, _a priori_ play a dominant role in policy analysis.

CONCLUDING REMARKS

In this review of theories of agricultural technology generation and diffusion we have attempted to identify major sets of themes and preoccupations which have resulted in major different types of theory and policy action. We have concentrated on events of the last 30 years because the political, communications and many other factors of these years make them relatively more important to future policy action, than earlier events. In order to structure the review we identified five different themes — world economy, distributionism, national factor endowments, environmental, research and extension institutions — which came in response to the preoccupations of and the policy action in the 1950s and 1960s of modernisation and transfer theory. This structure of the report is not meant to imply that many current policy actions are not justified and rationalised on modernisation theories. In fact they are; however we now have the benefit of learning from the experiences and lessons of some of these early actions.

While recognising that we ourselves are involved in some of the debates which centre around alternative theories for technology generation and diffusion, we have tried in this review to stand back from entering these debates and identify the preoccupations of different groups of analysts.

NOTES

[1] Reviews of the literature and large
    bibliographies can be found in such sources as
    Ruttan 1982a; Anderson et al. 1982; Busch and
    Lacy 1981; Scobie 1979; Schuh and Tollini 1979;
    Binswanger and Ruttan 1978; Biggs 1981; Gilbert
    et al. 1980; Arndt et al. 1977; Brokensha, et al.
    1980.
[2] At the outset a recognition that different
    preoccupations relate to different ways of
    conceptualising the organisation of activity –
    the sector, the community, the agro-ecological
    space – suggests one reason for non-interaction.
    The data of analysis are organised in different
    overlapping sets.
[3] See for example the developments in the Soviet
    Union as discussed in Lapidus and Ostrovityanov
    1929.
[4] Such a vocabulary was not unique to the work of
    agricultural science or problems of agricultural
    development, but this area represented a sector
    in which ideas on modernisation were given
    specific application in theory and practice.
    Examples of texts of the wider modernisation
    literature are: Apter 1965; Eisenstadt 1966;
    Swerdlow 1963; Nettle 1967; Almond and Coleman
    1960; Esman and Blaise 1966; Woods 1972.
[5] On institutional building see Esman and Blaise
    1966. For a critique of these perceptions and
    the consequent practice in public administration,
    see Schaffer 1969.
[6] See for example Auckland, A.K. 1980, who reports
    on how exploratory collecting expeditions in
    probable source areas for Corchorus caprularis,
    quickly established that jute breeding had for a
    century worked within a small gene pool based
    only on early unsystematic collections and highly
    restricted informal transfer of material for
    Corchorus olitorius and Corchorus caprularis.
[7] This typology is first elaborated in Hayami and
    Ruttan (1971) and there is a succinct
    presentation in Ruttan (1975).

[8] Rare exceptions, of course, are material which is only vegetatively propagated by producers, e.g. sugarcane, Ruttan's (1975) example, or bananas.

[9] Cf. Rogers (1976, 1976a, 1980) who argues that the model behind earlier generation of diffusion studies was unable to accommodate. On the other hand Bunting (1979) has put emphasis on the point that formal research systems must evolve out of local knowledge systems.

[10] For example see Evenson 1975; Farmer 1979.

[11] Cf. the Indian public bureaucratised agricultural research system of central research institutions for food crops and its network of substations with rubber research in Malaysia funded and dominated by estate producer interests (Ruttan 1982).

[12] It must be emphasised that such modernising ways of thinking were far more widespread, and in no sense confined to Western development assistance agencies or academics, and those whom they influenced. Edward Goldstucker, then a senior Czech government official and Communist party member described how Czech agriculture was the beneficiary, in the late 1940s and early 1950s, of technical co-operation from the USSR. Personnel from collective farms and Soviet agricultural institutions were sent to advise on how to emulate the more advanced Soviet model. All this despite, as Goldstucker points out, the subsequently available evidence that Czech agriculture had significantly higher overall factor productivity than the Soviet model. Personal communication.

[13] For example, the survey of agriculturally related research and training capacity in selected African countries by Rinlinger 1982 provides interesting examples of the contrasting consequences of different British and French colonial models of <u>administration</u> and patterns of recruitment of cadres into public service roles.

[14] "Many international technical assistance programs have been concerned with some aspect of

institution building. Few however have been directed towards the formation of effective systems of agricultural research." (Moseman 1970, p. 19).

[15] "The continuous increase in productivity of US agriculture, permitting the doubling and redoubling of yields of many crops during the past three decades has come from a highly integrated system of institutionalised agricultural science. But technical assistance programmes to date have emphasised primarily attempts to transfer the _fruits_ of this system, or to transfer only _selected components_ of the institutions within the system" (op. cit., pp. 19-20).

[16] See for example Brass 1982, for an Indian example, or Fitzgerald 1981, on one of the earliest attempts at an institutional transfer of the land-grant model to Mexico in the 1940s.

[17] See the extremely interesting reflective historical essay by A.T. Mosher 1976 on fashions in rural development.

[18] For a fuller set of references on Farming Systems Research and on-farm research methods see the section on research and extension institutions.

[19] For example, see Schumacher 1973, Sen 1968, Eckaus 1955, Stewart 1978 and Jéquier 1976.

[20] Besides the members of the induced innovation school who are well documented in Ruttan 1982a and Binswanger and Ruttan (eds.) 1978, other writers who have a similar theme in their work are Boserup 1965, 1981, and Geertz 1963. In the economics literature the induced innovation theme is derived from Hicks 1932, pp. 124-125.

[21] The Appropriate Technology Journal, published by the Intermediate Technology Development Group, London, well illustrates the actions of people involved in intermediate technology.

[22] This does not mean that some appropriate technology writers have not analysed policy and institutional issues. They have, for example Holtermann 1979 and Whitcombe and Carr 1982. However, the recent work by Jéquier 1983

represents the most signficant and comprehensive treatment of AT and institutional questions.

[23] "- To what extent have the modern or so-called high-yielding varieties been accepted in the study areas? Were changes in other farm practices associated with the introduction of modern varieties?
- What are the major obstacles to further growth in rice production? Are these a function of physical environment, socio-economic factors, or a combination of both?
- How has the new technology affected the level and structure of employment? Has the labour requirement increased? Is labour-saving equipment being adopted?
- Who has benefited from the new technology, and how have these benefits been spent? To what degree are profits capitalised into rising land values? How has the relationship among various tenure groups (landowner, tenant, landless labourer) changed?" (IRRI 1975, p. 5).

[24] See for example the excellent case study by Parthasarathy West Godavari: Andra Pradesh, in IRRI 1975. Barker and Herdt, in IRRI 1978, p. 98 in the follow-up to the 1975 volume acknowledge that a more thorough investigation of the impact of new technology on employment requires more information than that within the scope of this study.

[25] i. identification of the factors facilitating or obstructing the acquisition and use of a 'genetic-chemical technology';
ii. identification of the economic and social changes that follow the large-scale introduction of the technology to be observed in the agrarian structure, in the level and quality of livelihood of the participants and in the social structure of the rural society;
iii. assessment of measures and programmes proposed and carried out by governments in order to manage or modify the processes set off by technological change.

[26] The significance of such concerns could be characterised in terms of an analysis which sees

the outcome of technical change, even where this reflects prices in less reflecting factor proportions, as a dynamic 'corner solution' with unemployment or absence of <u>ethically</u> acceptable livelihoods for a large proportion of the rural population.

[27] For example, see Lipton 1978, Dasgupta 1977, Collinson 1981, Harwood 1979, Norman 1980.

[28] Writers who illustrate the preoccupations of this group are Franke and Chasin 1980; Feder 1976, 1977 and 1979; Mooney 1979; Frank 1981; Cleaver 1972; Bernstein 1979; Perelman 1976; George 1976; Busch and Sachs 1981; Burch 1980; Burbach and Flynn 1980; Dinham and Hines 1982; Kaufman and Cook 1982.

[29] For a critical review of the Franke and Chasin book see Ryan 1981.

[30] Examples of literature in this area are Eckholm 1976; Blaikie 1981, 1983, 1983a; Franke and Chasin 1980; Redclift 1982; UN Conference on Desertification 1977; Johnson and Blake 1980; Stocking 1981; Gibbon 1981; Dahlberg 1979; Bernstein 1979; Sandbach 1980; Wisner 1977; Norris 1982.

[31] For collections of articles, reviews of the literature etc. see Arndt et. al. 1977; Biggs 1982; Binswanger and Ruttan 1978; Ruttan 1980 and 1982; Hayami and Ruttan 1971; Fishel 1971; Brokensha 1980; Busch 1981; Scobie 1979; Whyte 1981; Schuh and Tollini 1979; Gilbert et al. 1980; Pinstrup-Andersen 1982; Shaner et al. 1982; Biggs and Grosvenor-Alsop 1983; Stavis 1979; Nortan and Davis 1981; Anderson et al. 1982; Rogers 1976 and 1980.

[32] For example Daniels and Nestel 1981; ISNAR and IADS 1982; Byerlee and Collinson 1980; Arnon 1975.

[33] We are not reviewing the literature on rates of return to investments in agricultural research (e.g. Ruttan 1980 and 1982; Evenson 1975; Schuh and Tollini 1979; Scobie 1979). One of the reasons for this is because the themes of this literature do not seek to explain why good or bad investment decisions were made in the past.

While some of the original work in this area was done by Grilliches 1958 on maize in the USA, there have been many studies since then which have looked at different commodities in developing countries. Generally, high rates of return have been estimated (e.g. 30 per cent and above) although a few studies have taken examples where there have been low and negative rates of return (Wennergren and Whitaker 1977). In the literature on this subject, there are fierce debates about whether, and how, costs and benefits from technological change can be meaningfully estimated and allocated between different categories. For example see Wise 1977 who questions the logic and internal consistency of the Grilliches exercises and the work of Lindner and Jarret 1978 and Norton et al. 1982.

[34] Taken from Barker 1981.

[35] In an important study by Boyce and Everson 1975 they found that where per capita incomes were high, then the proportion of research funds allocated to different commodities tended to be close to the proportion of each crop in total agricultural output. In a linear approach to resource allocation, one might use this model as a primary reason for explaining the past or for directing future policy actions. We have to ask if this is sensible.

[36] We ask these questions of the induced innovation model because several of the other ILO studies in this series have used the induced innovation model as their primary focus for analysis.

[37] However this may not be what Ruttan means by cultural endowment because he gives as an example, (a nation's) ability to organise and sustain the institutional changes conducive to research in agricultural science and the development of agricultural technology (Ruttan 1980, p. 40).

[38] It is a standard proposition in welfare theory that there is a Pareto optimum - i.e. a point at which resources are efficiently allocated for every different initial distribution of assets (Meade 1964). When this concept of 'efficiency'

is extended to a dynamic context, with different asset distributions, it will produce different growth paths. Ex-post it may be difficult to rank such paths in terms of efficiency in relation to factor endowments since these will in turn change.

[39] More widely see Schaffer 1969, on the crisis in development administration resulting from such institution building practices.

[40] It is interesting to note Dorling's similar comments on horticulture in Kenya.

[41] Some writers who discuss these complex subjects are Gomory 1983, Beveridge 1959, Ladd 1979, Shumway 1981, Binswanger 1978, Blaug 1980, Busch 1981, Busch et al. 1980.

[42] For example Ruttan 1982, Pray 1983, Trigo and Pineiro 1981. This type of analysis is concerned with formal (institutionalised) private and public sector research and does not relate to the informal research by farmers, rural artisans as described and discussed by Biggs and Clay 1981, Brammer 1980, Bell 1979 and some of the articles in Brochensha 1980.

[43] For example, the farming systems research and resource allocation literature coming from the International Agricultural Research Institutes.

3 Generation and diffusion of agricultural technology: Issues, concepts and analysis

VERNON W. RUTTAN

INTRODUCTION

The capacity to develop and manage technology in a manner consistent with a nation's physical, human and cultural endowments is the single most important variable accounting for differences in agricultural productivity among nations. The development of such capacity is dependent on many factors. These include the capacity to organise and sustain the institutions that generate and transmit scientific and technical knowledge, the ability to embody new technology in equipment and materials, the level of husbandry skill and the educational accomplishments of rural people, the efficiency of input and product markets, and the effectiveness of social and political institutions.

The objective of this chapter is to spell out the issues and to formulate the analytical framework for empirical case studies aimed at achieving a better understanding of the economic and political factors which influence the generation and diffusion of more productive agricultural technologies or of technologies that are more appropriate to the resource endowments of developing countries.

Analysis will specially focus on the factors that have made some countries more successful than others in generating productive labour-intensive technologies suited to the very high labour-land ratios that characterise some of the poorest developing countries.

This knowledge is of critical importance to both the national governments of the developing countries and to the national and international aid agencies. During the 1950s and 1960s substantial resources were devoted by the aid agencies to the development of agricultural research and extension institutions. In some countries, India for example, these efforts were highly successful. In a large number of countries, initial success was followed by failure of the emerging research institutions to obtain sustained support or to become productive or efficient sources of new technology or for the diffusion of new technology.

By the late 1960s, many national and international aid agencies, discouraged by their attempts to assist in the development of national institutions, shifted their support to the development of a new complex of international agricultural research institutes modeled on the International Rice Research Institute (IRRI) and the International Centre for Corn and Wheat Improvement (CIMMYT). By the late 1970s however, after the new international system had been put in place, it again became apparent that the limitation in the capacity of national agricultural research and extension institutions was a serious constraint on the effectiveness of the new international system. The ability to take advantage of the prototype technology being generated by the international centres is dependent on the development of national research and extension capacity.

It is hypothesised that the factors that determine the political and economic viability and scientific and technological productivity of national agricultural research and extension systems can be grouped under two broad headings: (a) the structure and organisation of the research and extension system itself and (b) the economic and political organisation of the society in which the research and extension service is being developed. The induced innovation perspective, which is outlined in the next section, provides a useful conceptual framework for linking differences and changes in the external environment to behaviour and performance.

AGRICULTURAL RESEARCH SYSTEMS: AN INDUCED INNOVATION FRAMEWORK[1]

The last two decades have been highly productive in advancing both our analytical capacity and our empirical knowledge of the role of technical change in agricultural development and of the sources of productivity growth in agriculture. When growth is based on a more intensive use of traditional inputs little surplus becomes available to improve the well-being of rural people or to be transferred to the rest of the economy. Little surplus has been generated by simple resource reallocation within farms, communities, or regions in the absence of technical change embodied in less expensive and more productive inputs. Only as the constraints on growth imposed by the primary reliance on indigenous inputs – those produced primarily within the agricultural sector – are released by new factors whose productivity is augmented by the use of new technology is it possible for agriculture to become an efficient and equitable source of growth and an important source of employment in a modernising economy.

Initially, growth in productivity – output per unit of total input – has typically been dominated by growth in a single partial productivity ratio. In the United States, and the other developed countries of recent settlement, growth in labour productivity – output per worker – has typically "carried" the initial burden of growth in total productivity. In countries characterised by relatively high man/land ratios at the beginning of the development process, the Federal Republic of Germany and Japan for example, growth in land productivity – output per hectare – was largely responsible for growth in total productivity during the early years of modernisation. As modernisation has continued, there emerged a tendency for total productivity growth to be fed by a more balanced growth in the partial productivity ratios, i.e., growth in output per worker and per hectare (see figure 3.1).

For a number of countries, however, the model outlined above has little meaning. The 20th century has been characterised by a massive, and continuously widening, disequilibrium in the efficiency of resource use and the welfare of rural people between rich and poor countries. Since the Second World War, output per hectare has been growing at approximately the same rate of about 2 per cent per year in the developing countries as in the developed countries. But output per worker in the developing countries has been growing at only one-third the rate of that in the developed countries – about 1.5 per cent per year compared to about 4.5 per cent per year.

Any attempt to evolve a meaningful perspective on the process of agricultural development must abandon the view of agriculture in pre-modern or traditional societies as essentially static. Historically, the problem of agricultural development is not that of transforming a static agricultural sector into a modern dynamic sector, but of accelerating the rate of growth of agricultural output, productivity and employment generation consistent with the growth of other sectors of a modernising economy. Similarly, a theory of agricultural development should provide insight into the dynamics of agricultural growth, i.e., into the changing sources of growth, in economies ranging from those in which output is growing at a rate of 1.0 per cent or less to those in which agricultural output is growing at an annual rate of 4.0 per cent or more.

The literature on agricultural development can be organised along the lines of several general models of which two models provide the conceptual framework relevant to this volume. The first is the high-payoff input model which offers significant insight into the role of research in opening up new opportunities for productivity and employment growth. The second is the induced innovation model which helps to clarify the relationship among

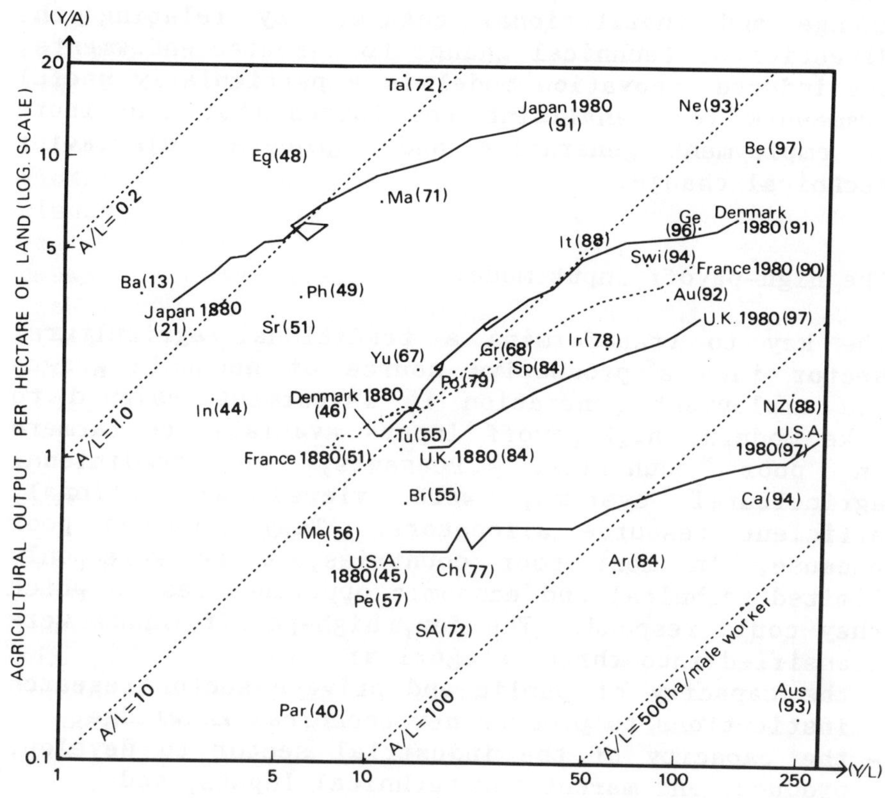

Key: See Figure 3.4.

Source: Data from Yujiro Hayami and Vernon W. Ruttan: _Agricultural development: An international perspective_, 2nd ed. Baltimore, Md. Johns Hopkins University Press: 1985, Appendices.

Figure 3.1: Historical growth paths of agricultural productivity of Denmark, France, Japan, the United Kingdom, and the United States for 1880–1980, compared with inter-country cross sectional observations of selected countries in 1980. (Values in parentheses are percentages of male workers employed in nonagriculture.)

resource endowments, cultural endowments, technical
change and institutional change. By relating the
direction of technical change to resource endowments,
the induced innovation model is a particularly useful
framework for canalising the forces that contribute
to employment generating and employment displacing
technical change.

The high-payoff input model

The key to transforming a traditional agricultural
sector into a productive source of economic growth
and employment generation is investment-designed to
make modern, high-payoff inputs available to farmers
in poor countries. Peasants, in traditional
agricultural systems, were viewed as rational,
efficient resource allocators. They remained poor
because, in most poor countries, there were only
limited technical and economic opportunities to which
they could respond. The new, high-payoff inputs were
classified into three categories:
- the capacity of public and private sector research
  institutions to produce new technical knowledge;
- the capacity of the industrial sector to develop,
  produce, and market new technical inputs; and
- the capacity of farmers to acquire new knowledge
  and use new inputs effectively.
The enthusiasm with which the high-payoff input
model has been accepted and translated into economic
doctrine has been due in part to the proliferation of
studies reporting high rates of return to public
investment in agricultural research. It was also due
to the success of efforts to develop new,
high-productivity grain varieties suitable for the
tropics. New, high-yielding wheat varieties were
developed in Mexico, beginning in the 1950s, and new,
high-yielding rice varieties were developed in the
Philippines in the 1960s. These varieties were
highly responsive to industrial inputs, such as
fertiliser and other chemicals, and to more effective
soil and water management. The high returns
associated with the adoption of the new varieties and
the associated technical inputs and management

practices have led to rapid diffusion of the new
varieties among farmers in several countries in Asia,
Africa and Latin America.

The high-payoff input model remains incomplete as a
theory of agricultural development. Typically,
education and research are public goods not traded
through the market place. The mechanism by which
resources are allocated among education, research,
and other alternative public and private sector
economic activities was not fully incorporated into
the model. It does not explain how economic
conditions induce the development and adaptation of
an efficient set of technologies for a particular
society. Nor does it attempt to specify the
processes by which input and product relationships
induce investment in research in a direction
consistent with a nation's particular resource
endowments.

An induced innovation model of technical change

These limitations in the high-payoff input model led
to efforts to develop a model of agricultural
development in which technical change is treated as
endogenous to the development process, rather than as
an exogenous factor that operates independently of
other development processes. The induced innovation
perspective was stimulated by historical evidence
that different countries had followed alternative
paths of technical change in the process of
agricultural development.

The productivity levels achieved by farmers in the
most advanced countries in each productivity grouping
(figure 3.1) can be seen as arranged along a
productivity frontier. This frontier reflects the
level of technical progress and factor inputs
achieved by the most advanced countries in each
resource endowment classification. These
productivity levels are not immediately available to
farmers in most low productivity countries. They can
only be made available if there is investment in
agricultural research capacity needed to develop
technologies that are appropriate to the natural and
institutional environments of the low productivity

countries and in physical and institutional infrastructure needed to realise the new production potential opened up by advances in technology.

<u>Alternative paths of technological development</u> There is clear evidence that technology can be developed to facilitate the substitution of relatively abundant and hence cheap factors for relatively scarce and hence expensive factors of production. The constraints imposed on agricultural development by an inelastic supply of land have, in economies such as Japan and Taiwan, China, been offset by the development of high-yielding crop varieties designed to facilitate the substitution of fertiliser for land. The constraints imposed by an inelastic supply of labour, in countries such as the United States, Canada, and Australia, have been offset by technical advances leading to the substitution of animal and mechanical power for labour. In both cases the new technology – embodied in new crop varieties, new equipment, or new production practices – may not always be substituted for land or labour by themselves. Rather, the new technologies may serve as catalysts to facilitate the substitution of the relatively abundant factors, such as fertiliser or mineral fuels, for the relatively scarce factors.

In agriculture, two kinds of technology generally correspond to this taxonomy: mechanical technology to "labour-saving" and biological (or biological and chemical) technology to "land-saving". The primary effect of the adoption of mechanical technology is to facilitate the substitution of power and machinery for labour. Typically this results in a decline in labour use per unit of land area. The substitution of animal or mechanical power for human labour enables each worker to extend his efforts over a larger land area. The primary effect of the adoption of biological technology is to facilitate the substitution of labour and/or industrial inputs for land. This may occur through increased recycling of soil fertility by more labour-intensive conservation systems; through use of chemical fertilisers; and through husbandry practices, management systems, and inputs such as insecticides which permit an optimum yield response.

Historically, there has been a close association between advances in output per unit of land area and advances in biological technology; and between advances in output per worker and advances in mechanical technology. These historical differences have given rise to the cross-sectional differences in productivity and factor use illustrated in figure 3.1. Advances in biological technology may also result in increases in output per worker if the rate of growth of output per hectare exceeds the rate of growth of the agricultural labour force.

In the Philippines, for example, growth in output per worker prior to the mid-1950s was due primarily to expansion in the area cultivated per worker. Since the early 1960s, growth in output per worker has been due to increase in output per unit of land area.

<u>Induced technical innovation</u> An examination of the historical experience of the United States and Japan illustrates the theory of induced technical innovation. In the United States it was primarily the progress of mechanisation, first using animal and later tractor motive power, which facilitated the expansion of agricultural production and productivity by increasing the area operated per worker. In Japan it was primarily the progress of biological technology such as varietal improvement leading, for example, to increased yield response to higher levels of fertiliser application which permitted rapid growth in agricultural output in spite of severe constraints on the supply of land. These contrasting patterns of productivity growth and factor use can best be understood in terms of a process of dynamic adjustment to changing relative factor prices.

<u>In the United States</u> the long-term rise in wage rates relative to the prices of land and machinery encouraged the substitution of land and power for labour. This substitution generally involved progress in the application of mechanical technology to agricultural production. The more intensive application of mechanical technology depended on the

invention of technology which was more intensive in its use of equipment and extensive in its use of land relative to labour. For example, the Hussy or McCormick reapers in use in the 1860s and 1870s required the use, over a harvest period of about two weeks, of five workers and four horses to harvest 140 acres of wheat. When the binder was introduced it was possible for a farmer to harvest the same acreage of wheat with two workers and four horses. The process illustrated by the substitution of the binder for the reaper has been continuous. As the limits to horse mechanisation were reached in the early part of this century, the process was continued by the introduction of the tractor as the primary source of motive power. The process has continued with the substitution of larger and more highly-powered tractors and the development of self-propelled harvesting equipment.

<u>In Japan</u> the supply of land was inelastic and its price rose relative to wages. Furthermore, opportunities arising from a continuous decline in the price of fertiliser relative to the price of land were exploited through advances in biological technology. Varietal improvement was directed, for example, towards the selection and breeding of more fertiliser-responsive varieties of rice. The enormous changes in fertiliser input per hectare that have occurred in Japan since 1880 reflect not only the effect of the response by farmers to lower fertiliser prices but the development by the Japanese agricultural research system of "fertiliser-consuming" rice varieties in order to take advantage of the decline in the real price of fertiliser. Relative prices of land and labour and land and fertiliser contributed to higher labour-intensity through greater use of biotechnologies.

The effect of relative prices in the development and choice of technology is illustrated with remarkable clarity for fertiliser in figure 3.2 in which the American and Japanese data on the relationship between fertiliser input per hectare of arable land and the fertiliser/land price ratio is plotted for the period 1880-1960. In both 1880 and

1960 American farmers were using less fertiliser than Japanese farmers. However, despite enormous differences in both physical and institutional resources the relationship between these variables has been almost identical in the two countries. As the price of fertiliser declined relative to other factors both Japanese and American scientists responded by inventing crop varieties which were more responsive to the lower prices of fertiliser – although American scientists always lagged behind by a few decades in the process because the lower price of land relative to fertiliser resulted in a lower priority being placed on yield-increasing technology in the United States than in Japan. It is possible to illustrate the same process for mechanical technology. Variations in the level of tractor horsepower per worker are very largely a reflection of the price of labour relative to the price of power (see figure 3.3).

The effect of a rise in the price of fertiliser relative to the price of land or of the price of labour relative to the price of machinery has been to induce advances in biological and mechanical technology. The effect of the introduction of lower cost and more productive biological and mechanical technology has been to induce farmers to substitute fertiliser for land and mechanical power for labour. These responses to differences in resource endowments among countries and to changes in resource endowments over time by agricultural research institutions, by the farm supply industries, and by farmers, have been remarkably similar in spite of differences in culture and tradition.

During the last two decades, as wage rates have risen rapidly in Japan and as land prices have risen in the United States there has been a tendency for the pattern of technological change in the two countries to converge (see figure 3.4). Between 1960 and 1980 fertiliser consumption per hectare rose more rapidly in the United States than in Japan and tractor horsepower per worker rose more rapidly in Japan than in the United States. Both countries appear to be converging towards the European pattern of technical change in which increases in output per

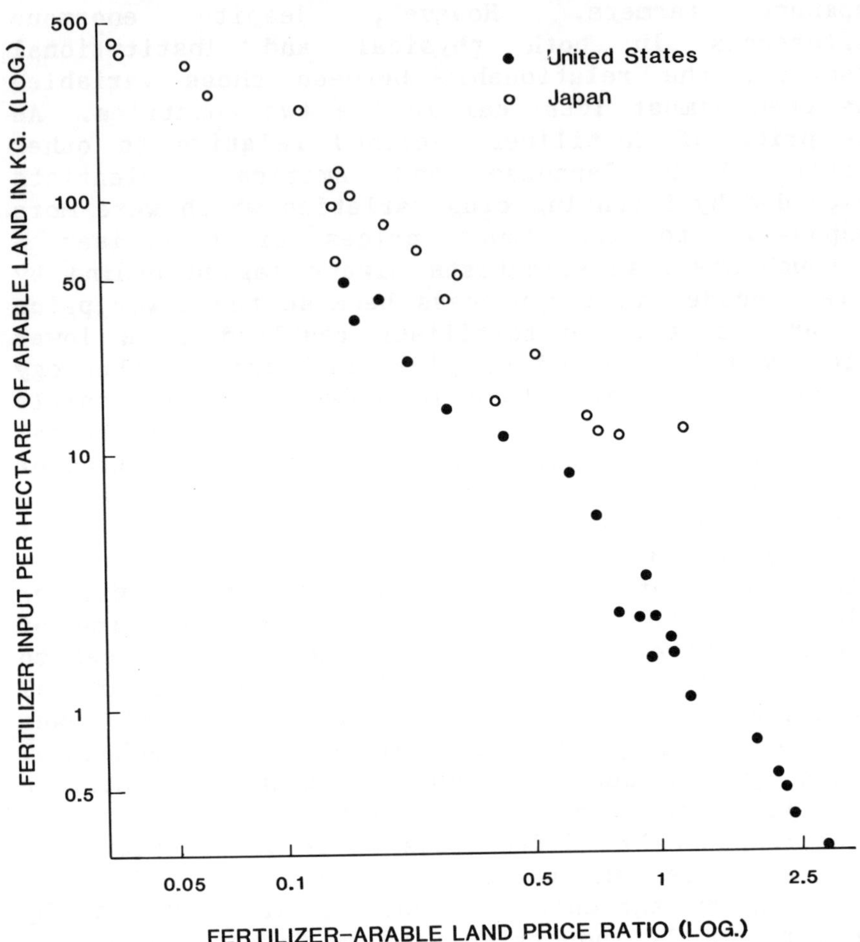

Source: As for figure 3.1.

Figure 3.2: Relation between fertiliser input per hectare of arable land and the fertiliser: arable land price ratio (hectares of arable land which can be purchased by one tonne of commercial fertiliser)

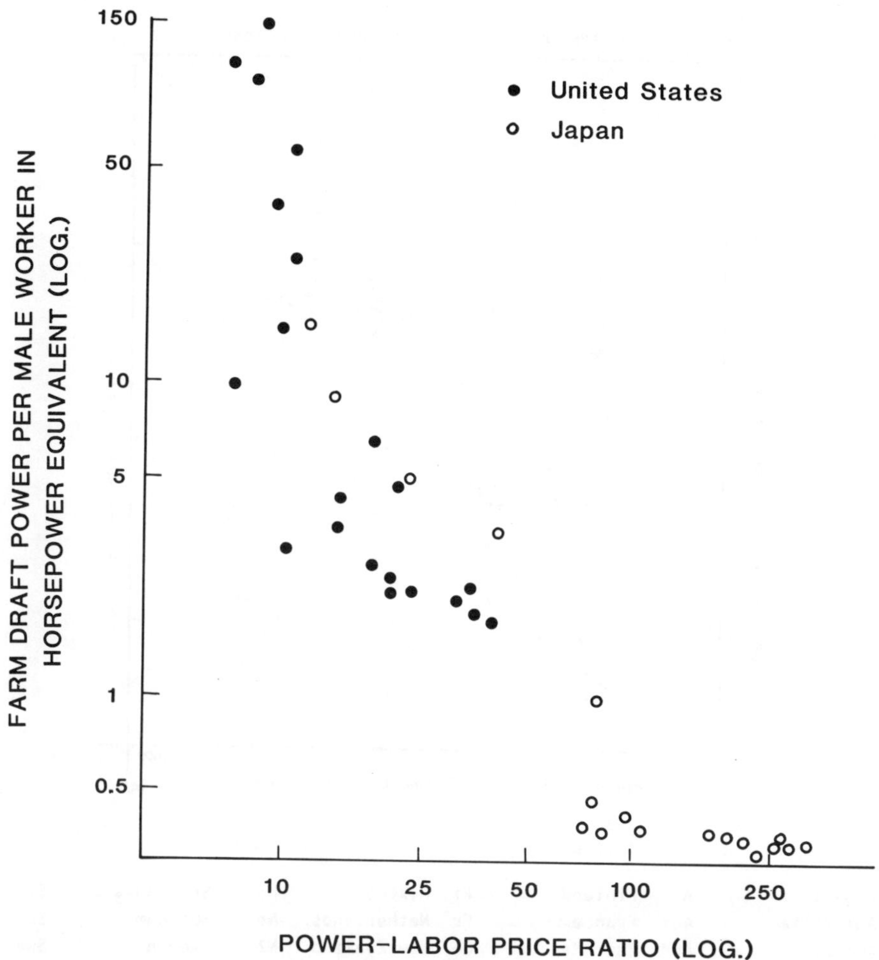

Source: Yamada and Ruttan (1980), p. 540.

Figure 3.3: Relation between farm draft power per male worker and power/labour price ratio (United States and Japan, quinquennial observations for 1880–1980)

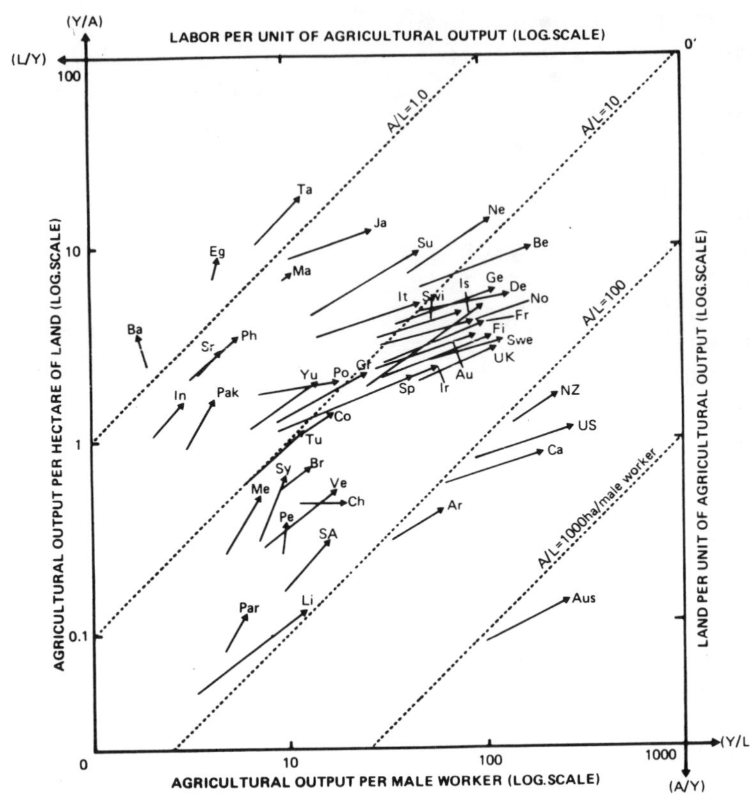

Key:

Argentina	Ar	Finland	Fi	Mexico	Me	Sri Lanka	Sr
Australia	Aus	France	Fr	Netherlands	Ne	Surinam	Su
Austria	Au	Fed. Rep.	Ge	New Zealand	NZ	Sweden	Swe
Bangladesh	Ba	Germany		Norway	No	Switzerland	Swi
Belgium	Be	Greece	Gr	Pakistan	Pak	Syria	Sy
(& Luxembourg)		India	In	Paraguay	Par	Taiwan, China	Ta
Brazil	Br	Ireland	Ir	Peru	Pe	Turkey	Tu
Canada	Ca	Israel	Is	Philippines	Ph	United Kingdom	UK
Chile	Ch	Italy	It	Portugal	Po	United States	USA
Colombia	Co	Japan	Ja	South Africa	SA	Venezuela	Ve
Denmark	De	Libya	Li	Spain	Sp	Yugoslavia	Yu
Egypt	Eg	Mauritius	Ma				

Source: Yamada and Ruttan (1980), p. 546.

Figure 3.4: International comparison of labour and land productivities in agriculture

worker and increases in output per hectare occur at approximately equal rates.

There will be further changes in the future. In the early and mid-1970s the price of energy has risen. This has affected both the price of fuel and the price of fertiliser. It is unlikely that declining fertiliser price will in the future be as important a factor in determining the direction of biological technology as during the past century. Higher fertiliser prices have already induced a substantial increase in the research resources devoted to the investigation of potential biological and organic sources of plant nutrition. It is possible that the momentum of advance in biological technology will, during the next several decades, be faced with the necessity of a transition comparable to the shift from the horse to the tractor as a source of motive power in the area of mechanical technology.

In most developing countries, however, fertiliser prices at the farm level should continue to decline relative to the price of land for some time with the expansion of domestic capacity and the improvement of the marketing and transport systems. Growth of fertiliser use per hectare will continue to make an important contribution to greater intensity in crop production, to employment generation, and to productivity growth in most developing countries during the next several decades.

<u>Induced institutional innovation</u> In discussing the theory of induced technical change in agriculture we need to consider the personal behaviour of individual research scientists or the institutional behaviour of the agricultural experiment stations or research institutes which support their research. In most countries which have been successful in achieving rapid rates of technical progress in agriculture, "socialisation" of agricultural research has been deliberately employed as an instrument of modernisation in agriculture. The induced innovation model of technical change in agriculture implies that both research scientists and research administrators are responsive to differences in resource endowments

and to changes in the economic environment in which they work although in reality such a response may not be forthcoming.

The response of research scientists and administrators represents the critical link in the inducement mechanism. The model does not imply that it is necessary for individual scientists or research administrators in public institutions consciously to respond to market prices, or directly to farmers' demands for research results, in the selection of research objectives. They may, in fact, be motivated primarily by a drive for professional achievement and recognition. Or, they may view themselves as responding to an "obvious and compelling need" to remove the constraints on growth of production or on factor supplies. It is only necessary that there exists an effective incentive mechanism to reward the scientists or administrators, materially or by prestige, for their contributions to the solution of problems that are of social or economic significance.

The response by the scientific community to the recent rise in the price of fossil-fuel-based inputs represents a dramatic example of the induced innovation process. Increases in the price of nitrogen fertiliser have induced a shift in scientific resources towards more intensive and development activity on the biological and organic sources of plant nutrition. The low productivity of agricutural scientists in many developing countries is due to the fact that many societies have not yet succeeded in developing incentives that lead to the focusing of scientific effort on the significant problems of domestic agriculture. Under such conditions scientific skills atrophy or are directed to the reward systems of the international scientific community.

It is not argued, however, that technical change in agriculture is wholly of an induced character. There is a supply (exogenous) dimension, stemming from autonomous development in the basic sciences, as well as a demand (endogenous) dimension. Technical change in agriculture reflects, in addition to the effects of resource endowments and growth in demand, the progress of general science and technology. Progress in general science which lowers the "cost" of technical change may influence the direction of technical change in agriculture in a manner that is unrelated to changes in factor proportions and product demand. Similarly, advances in science and technology in the developed countries, in response to their own resource endowments, may result in a bias in the technical opportunities that become available in the developing countries. Even in these cases, the rate of adoption and the impact on productivity will be strongly influenced by the conditions of resource supply and product demand, as these forces are reflected through input and product markets.

Based on the above discussion, an attempt is made to develop and articulate a more comprehensive theory of induced institutional innovation.

An induced innovation perspective on institutional change

A distinction is often made between institutions and organisations. Institutions are usually defined as the behavioural rules that govern patterns of action and relationships. Organisations are the decision-making units — families, firms, bureaux — that exercise control of resources. There appears to be a distinction without a difference. What an organisation, a household or a firm for example, accepts as an externally given behavioural rule, is the product of tradition or decision by another organisation — a nation's court system, or the practices of organised labour for example. In our work on institutional innovation, we have found it useful to define the concept of institution broadly to include that of organisation. The term institutional innovation is used to refer to change in the actual or potential performance of existing or new organisations; in the relationship between an organisation and its environment; or in the behavioural rules that govern the patterns of action and relationships in the organisation's environment. This definition is intended to be sufficiently comprehensive, with reference to institutional innovation in agricultural development for example, to include changes in the market and non-market institutions which govern product and factor market relationships, ranging from the organised commodity market institutions to the patron-client relationships which often characterised exchange in traditional societies. It is also intended to include changes in public and private sector organisations designed to discover and disseminate new knowledge to farmers; to supply inputs such as water, fertiliser and credit; or to modify market behaviour through price support, procurement or regulation. It would encompass changes which occur as a result of the cumulative effect of the private decisions of individuals, with respect to fertility behaviour or migration for example, as well as those which occur as a result of group action designed to modify public decision-making processes.

<u>Sources of demand for institutional change</u> The demand for institutional innovation may arise out of the changes in relative factor endowments and relative factor prices associated with development. Economic historians have attempted to explain the economic growth of Western Europe between 900 and 1700 primarily in terms of changes in the institutions which govern property rights. These institutional changes were, in their view, induced by the pressure of population against increasingly scarce resource endowments. Students of economic development, focusing on more recent economic history, have identified the rising economic value of man during the process of economic development as the primary source of institutional change. The suggestion that changing resource endowments mediated through changing factor price ratios act to induce institutional innovation is, as we will show later in this chapter, consistent with considerable experience in contemporary developing countries.

The partitioning of the new income streams that result from the efficiency gains associated with technical change or improvements in institutional performance, represents a second major source of institutional innovation. In a classical or neo-classical world, unencumbered by the use of political resources to achieve economic objectives, the new income streams generated by technical change would be distributed to factors according to the Ricardian model of distribution. The gains would flow to owners of the factors that are characterised by relatively inelastic supply functions. It is readily perceived that the primary function served by the institutions which direct the new income streams to the suppliers of inelastic factors – the factors that act as a constraint on growth rather than as a source of growth – is to assure their claim on the social product.

As a result, advances in technology can be expected to set in motion attempts by factor owners, social classes and economic sectors to organise and initiate collective action for the purpose of redefining property rights or to change the behaviour of market institutions so as to modify the partitioning of the new income streams. Much of the history of farm price support legislation in the United States, from the mid-1920s to the present, can be interpreted as a struggle between agricultural producers and the rest of society to determine the partitioning of the new income streams that have resulted from technical progress in agriculture.

In the perspective outlined above, the changes in the factor endowments and factor prices arising out of economic growth and the new economic streams arising out of technical change represent important sources of demand for institutional change. The demand for institutional change may also shift as a result of changes in cultural endowments. Even under conditions of unchanging demand however, institutional change may arise out of improvements in the capacity of a society to supply institutional innovations, i.e., as a result of factors which reduce the cost of institutional change.

<u>Sources of supply of institutional change</u> The issue of the supply of institutional innovation has not been adequately addressed by either the institutionalist or analytical schools in economics. The older institutional tradition treated institutional change as primarily dependent on

technical change. Within modern analytical economics, there is a tendency either to abstract from institutional change or to treat institutional change as if it were exogenous to the economic system.

It seems reasonable to hypothesise a close analogy between the supply of institutional change and the supply of technical change. Just as the supply curve for technical change shifts to the right as a result of advances in knowledge in science and technology, the supply curve for institutional change shifts to the right as a result of advances in knowledge in the social sciences and related professions (law, administration, social service, and planning). Advances in social science knowledge should result in a reduction in the cost of institutional change just as advances in knowledge in the natural sciences and engineering have reduced the cost of technical change.

For example, research leading to quantification of commodity supply and demand relationships can be expected to contribute towards more efficient functioning of supply management, food procurement and food distribution programmes. Research on the social and psychological factors affecting the diffusion of new technology is expected to lead to more effective performance by agricultural credit and extension services, or to more effective organisation and implementation of commodity production campaigns. Research on the effects of alternative land tenure institutions or on the organisation and management of group activities in agricultural production is expected to lead to institutional innovations leading to greater equity in access to political and economic resources and to greater productivity in the generation and utilisation of resources in rural areas, provided the right choice is made in selecting from the alternative policy options.

This is not to argue that institutional change is entirely or even primarily dependent on formal research leading to new knowledge in the social sciences and professions. Technical change was not delayed until research in the natural sciences and technology became institutionalised. Similarly, institutional change may occur as a result of the

exercise of innovative effort by politicians,
bureaucrats, entrepreneurs and others, as they
conduct their normal daily activities. The timing or
pace of institutional innovation may be influenced by
external contact or internal stress. If we were
satisfied with the slow pace of technical and
institutional change which characterises most of
trial and error, there would be no need to
institutionalise research capacity in either the
natural or the social sciences.

<u>Towards a theory of induced institutional change</u> The
relationship between technical and institutional
change has represented a continuous source of concern
to economists and other social scientists interested
in the historical and institutional dimensions of
development. There has been a persistent dualism in
much of this work - with institutional change
regarded as primarily dependent on technological
change or technological change regarded as primarily
dependent on institutional change. Dissent over the
priority between technical and institutional change
is unproductive. Technical change and institutional
change are highly interdependent and must be analysed
within the context of interdependence.

The sources of <u>demand</u> for technical and
institutional change are essentially similar. A rise
in the price (or scarcity) of labour relative to
other factors induces technical changes designed to
permit the substitution of capital for labour and at
the same time induces institutional changes designed
to enhance the productive capacity of the human agent
and the control by the worker of the conditions of
his employment. A rise in the price (or scarcity) of
land (or natural resources) induces technical changes
designed to release the constraints on production
resulting from the inelastic supply of land and, at
the same time, induces institutional changes leading
to greater precision in the definition and in the
allocation of property rights in land.

The new income streams generated by technical
change and by gains in institutional efficiency
induce changes in the relative demand for products
and open up new and more profitable opportunities for

product innovations, leading to greater diversity in consumption patterns. And the new income streams generated by either technical or institutional change induce further institutional changes designed to modify the manner in which the new income streams are partitioned among factor owners and to alter the distribution of income among individuals and classes.

Shifts in the <u>supply</u> of superior techniques and institutions are also generated by similar forces. The costs of the new income streams generated by technical change are reduced by advances in knowledge in science and technology. The costs of the new income streams generated by gains in institutional efficiency, including gains in efficiency in conflict resolution, are reduced by advances in knowledge in the social sciences and related professions.

The significance of the proposed theory of institutional change is that it suggests an economic theory of induced institutional change that is capable of generating testable hypotheses regarding (a) alternative paths of institutional change over time for a particular society, and (b) divergent patterns of institutional change among countries at a particular time. It is possible to build on this model to develop a theory of induced institutional change that is not only explanatory, in the sense that the present is explained in terms of the past, but is capable of generating testable hypotheses regarding the future direction of institutional change, applicable in social science research to achieve more effective institutional performance and more rapid institutional innovation.

The induced institutional innovation hypothesis implies a strong demand for clarification of the conceptual relationships among resource endowments, cultural endowments, technological change and institutional change as they bear on the processes of development (figure 3.5). It also calls for the careful testing of those relationships against both historical and contemporary experience. In the induced innovation literature, only the relationships among resource endowments, technical change and institutional change have received significant attention.

The methodology that will be appropriate in testing the induced institutional change hypothesis is not yet as rigorous as the rather straightforward econometric tests that have confirmed the robustness of the induced technical change hypothesis. Case studies will represent an important methodological approach. Some of the implications of the interaction between technical and institutional innovation and the external environment in which the research system operates, along the lines suggested in figure 3.5, are outlined more completely for the agricultural research system in the following section.

The role of the agricultural research institute in an induced innovation model[2]

In attempting to understand the responses of the modern agricultural research institute or experiment station to the institutional and physical environment in which it finds itself, it is helpful to conceptualise the individual experiment station or research institute or the national agricultural research system along the lines suggested in figure 3.5. The figure attempts to relate the internal production processes of the station or the research system to the external environment in which it operates. The processes involve the transformation of stock and flow resources into intermediate products. The products are, in turn, transformed into outputs. The outputs can be categorised under three headings – information, capacity and influence. The most important and visible output of an experiment station or a research laboratory is the information – in a form of new knowledge or new technology – that is generated and released. In some fields, plant breeding, for example, the new technology may be embodied in higher yielding or pest resistant crop varieties (cultivars). In other fields the knowledge may be embodied in published reports on farm management, cropping practices or animal nutrition. Over the long run, the use of resources for agricultural research must be justified in terms of the economic value of the new knowledge or new technology that it produces.

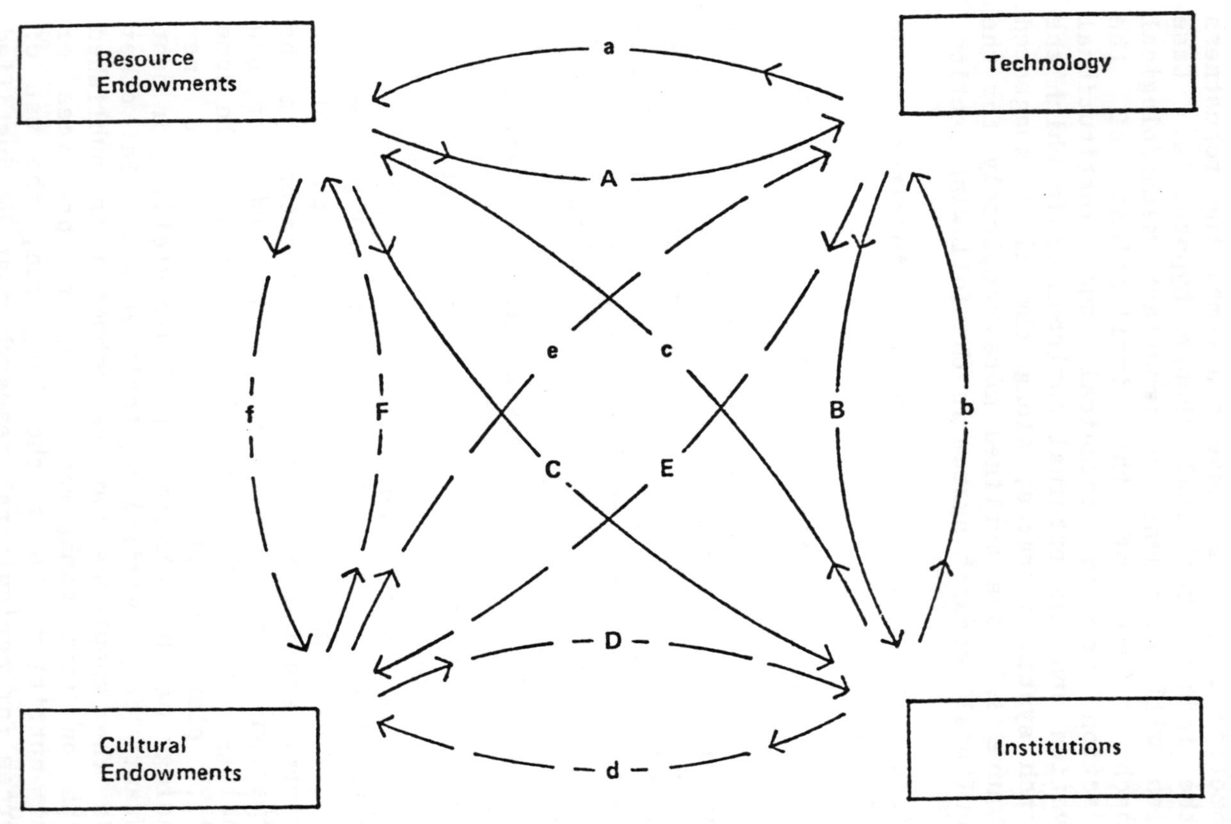

Figure 3.5: Systems model of experiment station performance and development

If a research institution is to remain a valuable social asset, it must also devote resources to reinvestment in institutional capacity - to the enlargement of its physical and intellectual capital. This means diverting some resources from the production of information that has immediate application. It means expanding the capacity of its scientific staff through time devoted to graduate education, study leaves and supporting or basic research. It also includes modernisation of the facilities, administrative structure and ideology that serves as a rationale for the research programme. The frequent arguments about the relative emphasis that should be given to the production of current services relative to the expansion of capacity - particularly to capacity embodied in the professional staff of the station or system - may also reflect the fact that there are very large spillover effects to investment in expansion of intellectual capacity. Only a small part of the benefits may be realised by the research organisation which bears on the cost of the basic research leading to capacity expansion, particularly if it leads to greater mobility of the staff member in which the expanded intellectual capacity is embodied.

Research institutions typically devote significant resources to increase their influence. In order to establish a successful claim on current and future resources, a research organisation usually finds it necessary to maintain effective relationships with funding agencies - legislative bodies, operating bureaux and divisions and private foundations. The organisation may also find it useful to devote resources to building a positive public image as a valuable institutional resource. Although the resources devoted to the production of influence may have little direct value to society, such activities are essential to the maintenance and continuity of both public and private research institutions. However, institutions which have outlived their social function as producers of information and technology frequently devote excessive resources to organisational maintenance activities.

The intermediate or transactive processes and

activities identified in figure 3.5 are for internal use. They have little direct value to society. But they are indispensable to the research institution itself. By and large these intermediate services must be produced by the research institution itself rather than be purchased from external sources. Their value is derived from their contribution to the output of the research institution.

The linkage services include the contacts and relationships with individuals and institutions outside the research station or laboratory - with other scientists, with the clients who use the services of the station and with sources of support. The linkages carry messages in both directions. It is through the linkages with the outside that a research institution influences technical and institutional change and is, in turn, influenced by the external changes in society and in science and technology. In the United States the state agricultural experiment stations are characterised by an exceedingly complex set of linkages with the external environment - with state crop improvement associations, community and regional development councils, producer and consumer interest groups, the extension service and the entire hierarchy of local, state and federal administrative and political institutions. Failure of the agricultural science community to maintain effective linkages with the general science community at times weakened the effectiveness of agricultural research. The problem of maintaining effective communication with outside groups without becoming either a captive or an adversary is exceedingly difficult.

Leadership is an extremely important intermediate product. The myth that all a research director needs to do is to hire good people and let them 'do their thing' has only minimal support at a time when the solution to many significant technical and social problems requires concerted research effort. Leadership must be sensitive to changing social goals and effectively transmit their implications to the scientific staff. Leadership also involves resource acquisition. It must be able to visualise and effectively communicate the potential contribution of

the experiment station or research system to the
solution of emerging problems. In addition, it must
be capable of mobilising and allocating both
financial and scientific resources in such a manner
as to produce the high returns to scientific and
technological effort that society has come to
expect. This means acquiring not only the necessary
human and financial resources but also the more
difficult task of creating an institutional
environment where these resources can become
productive.

The technology or the methodology of research is in
continuous flux. The research programme must be
organised in such a way that the research staff is
aware of, and contributes to, the advances in its own
and closely related fields. Resources devoted to the
production of research technology represent a capital
investment in the capacity of the individual research
worker, the research institute which makes the
investment and to the broader research system of
which it is a part. It has often been noted that
research institutes which devote a relatively high
proportion of their effort to their basic or
'frontier' type supporting research are often able to
pay lower salaries for comparable scientific manpower
because institute staff members are willing to accept
part of their 'pay' in the form of investment in
their own capacity.

Doctrine is reflected in the articulation of
institutional goals and philosophy and in the
operating style of an institution. For example,
during the 1960s the traditional production-oriented
doctrine of the American state and federal
agricultural research system experienced severe
stress under increasing pressure to give heavier
weight to the environmental 'spillover' effects of
technical change in agriculture, to the problems of
human capital formation and to the institutional
dimensions of community development in the
formulation of research priorities. The
international agricultural research institutes have
been urged to place greater weight on the geographic
and social distributional implications of the
technology they produce. The operational

manifestation of a shift in doctrine is the
reformulation of a research station's programme.
Modifications in doctrine are sterile unless
accompanied by the realisation of the revised
priorities.

Programme modifications often imply drastic
modifications in the internal structure of a research
station. Problem-oriented inter-disciplinary
'centres' or 'teams' erode the decision-making
authority of discipline- or commodity-oriented
departments. During periods of stress, the
reformulation of doctrine, the redirection of
programme and the reorganisation of internal
structure may absorb substantial resources and
seriously compete with the production of
information. These efforts must be justified
primarily in terms of their impact on the future
productivity and viability of the experiment station.

The central node of the model is a 'payoff matrix'
that identifies the gains and losses (economic,
social, political) which are expected to result from
a set of alternative technologies for the set of
particular interest groups which characterise a
particular society. Alternative technological
choices concern options among regions and commodities
(e.g., irrigated versus dry land, food versus
commercial crops, beef versus beans) and among
technological biases (land-saving biochemicals versus
labour-saving mechanical devices, modern versus
intermediate mechanical techniques, etc.). Interest
groups include, in particular, commercial farmers,
traditional landed elites, subsistence farmers,
landless agricultural workers, industrial employers,
urban workers at different income levels, private
sector suppliers of inputs and marketing services,
and government agencies, including the research
bureaucracy itself. Each relevant social group
expects to derive a specific income gain or loss from
the new income streams generated by agricultural
research.

In national research systems the supply and demand
mechanism for technical innovations in agriculture is
centred around the payoff matrix and is conditioned
by the socio-economic structure on the one hand, and

by the politico-bureaucratic structure on the other. Each economic or social group will put pressure on the politico-bureaucratic structure for research output (knowledge, technology, influence) to be (or not to be) generated, depending upon their particular expected payoffs. The relative power of different economic and social groups over the politico-bureaucratic structure is the primary determinant in getting their specific demands eventually translated into a supply of new knowledge or new technology.

The resulting supply of new knowledge and new technology creates, through the socio-economic structure, specific payoffs for each social group. The output of the research system is biased or 'screened' by the socio-economic structure or matrix. For agricultural technology the payoffs are influenced by: (1) the physical impact of the innovation in terms of yield effect and/or resource saving and resource substitution effects; (2) the diffusion effect which is conditioned by both the nature of the technological innovation and the particular position of social groups in the socio-economic structure (here, the land tenure system and the extent of access to institutions of credit, information, marketing, etc., are determinants); and (3) the relative prices which influence the economic value of the physical and diffusion effects.

Anticipated payoffs induce, again, additional demands for new knowledge and new technology. The payoff matrix thus reveals the nature of the dynamics – or lack of dynamics – of the inducement and diffusion of technological innovations. It also indicates that care must be taken to identify the specific groups who gain or lose from technical change in order to make meaningful statements about its economic and social significance. Even if the aggregate payoff is used as a criterion of net social gain from technology, care must still be taken to identify possible negative payoffs – like the loss of employment and income for some social groups – in order to deduct them from the total payoff and to determine needed economic compensations. Indeed, the commonly used net social gain measure is but a global approximation of the payoff matrix that largely hides the specific socio-economic effects of technology and also the very rationality of why technological innovations and change did or did not occur.

Some of the factors which influence the size and the distribution of the new income streams generated by agricultural research among the supplying inputs (land, labour, capital, materials) and among economic groups (labourers, landowners, suppliers of marketing services, consumers and others) are examined subsequently.

ORGANISATION AND DEVELOPMENT OF AGRICULTURAL RESEARCH SYSTEMS: SOME HYPOTHESES

An implication of the induced innovation perspective is that the demand for technical change, productivity growth and employment generation in agriculture is strongly influenced by a nation's resource endowments - by the relative scarcity of land, labour and energy resources for example. A second implication is that the capacity of a nation to respond to the demand for technical change is strongly conditioned by its cultural endowments - by its ability to organise and sustain the institutional changes conducive to research in agricultural science and the development of agricultural technology, for example. Both resource endowments and cultural endowments change over time as they interact on each other and on technical and institutional change. Together they influence both the rate and direction of technical change and the distribution of the benefits of technical change.

Among the questions which the perspective suggests and that are particularly relevant to the development of national agricultural research systems are the following: How does the economic and political environment in which a national agricultural research system evolves affect its organisation, its funding and the benefits that it generates? Does the way in which a national (or state or provincial) agricultural research system is organised and funded

affect its productivity - its effectiveness in
generating a continuous stream of new technology
capable of enhancing agricultural growth and
employment?

Stages in the development of national systems

Among the presently developed countries of Western
Europe, North America and Japan, agricultural
research has typically progressed through a series of
relatively distinct stages. The newer national
systems have drawn heavily on the experience of the
older national systems. But the pattern of evolution
of the newer systems has been conditioned in some
countries by whether the initial stages evolved under
colonial auspices and how they have been influenced,
in later stages, by particular bilateral or
multilateral assistance programmes.

<u>The first stage</u> in the development of the older
national systems was based on the innovative
activities of individual farmers and inventors. In
the United States, Jefferson's experiments with soil
fertility, Washington's efforts to introduce new
crops, and the invention of new land preparation and
harvesting equipment by Deer and McCormick, come
immediately to mind. In the United Kingdom, Arthur
Young dramatised the efforts of animals and farming
practices. In the Federal Republic of Germany, Von
Thunen's investigations into estate management
established the foundations for modern farm
management and location economics. In Japan, veteran
farmers (rono) and farming landlords (gono) were the
source of new technology such as improved methods for
seed selection and the diffusion of the best
traditional practices.
Many other examples could be cited. The
distinguishing feature of these early stage
developments was their highly personal character.
They were relatively uninformed with respect to
either scientific principles or experimental
methods. However, accumulation of agronomic practice
and craft experience did permit the gradual
advancement of agricultural technology.

The second stage is characterised by the organisation
of agricultural experiment stations staffed by
research workers with professional specialisation in
agricultural science or technology. The first
experiment stations were typically established by
private estate owners or agricultural associations.
Boussingault established an experiment station on his
estate in Alsace, France in 1834. In 1842 Scottish
farmers established the Agricultural Chemistry
Association, a chemical laboratory and a system for
farm tests and experiments. The Rothamsted
experiment station was established by Sir J.B. Lawas
on his family estate in the United Kingdom in 1843.
In the Federal Republic of Germany, the impetus for
the establishment of agricultural experiment stations
was strengthened with the publication of Liebig's
treatise on the chemical basis for plant nutrition.

The distinguishing feature of this second phase was
the research laboratory or experiment station. The
laboratory and the station represented the necessary
infrastructure to attract professionally-trained
scientists to agricultural research, and the physical
and institutional environment needed to search for
"methods of applying science to agriculture". It
permitted both the framework for assembling the
resources needed to support agricultural research and
for organising the diverse scientific and technical
skills necessary to advance knowledge. Quite
different patterns emerged during stage two for the
support of biological, chemical and mechanical
research. Support for research leading to advances
in biological technology has, in most countries, been
supported primarily by public revenues (or by
granting organised producer groups the right to tax
themselves); advances in chemical technology are
produced primarily in the laboratories of the
chemical companies; and advances in mechanical
technology continued to be due primarily to farmer
inventions but with engineering research by the
machinery companies contributing to commercial
success.

The third stage is characterised by the evolution of
integrated national agricultural research systems.

The characteristic feature of the third stage is the establishment of an agricultural research planning capacity that is capable of relating research priorities to the allocation of professional and financial resources. The third stage may evolve out of efforts by private scientific bodies, or by quasi-public advisory councils such as a national academy of sciences to draw attention to gaps in research effort - between research in crop production and in animal or human nutrition or between private sector research in plant breeding and public sector research in genetics. As the level of public sector research support expands, the national agency that is primarily responsible for the funding of agricultural research typically develops substantial "in-house" planning and resource allocation capacity.

One result of the development of a national research capacity is an attempt to rationalise the organisation of individual commodity research institutes or experiment stations, to re-examine location of branch stations, and to attempt to achieve more effective integration of national, state or provincial, and private agricultural research.

Stress in the development of research systems

The modernisation of agricultural research systems imposes substantial stress on research administrators and on research personnel. Four issues which tend to be subject to almost continuous debate in any national system are discussed in this section.

<u>Stress over linkages among research, education and extension</u> Are teaching and research complementary or competing activities? Does responsibility for disseminating the results of research to farmers conflict or complement research activities? This issue was hotly debated in the United States during the early years following the establishment of state agricultural experiment stations within the colleges of agriculture at state land grant colleges and universities.

Over time a guiding hypothesis seems to have emerged to the effect that research is highly

complementary with graduate education – but less so
with undergraduate teaching. Similarly, systematic
contact with technical and economic problems of
farmers is regarded as important in the focusing of
research effort – but the effects of responsibility
by research staff for extension programme activity is
still debated. The USDA research administration
appears to share this perspective. USDA research
staff located at or near college campuses are often
encouraged to engage in graduate training to a
limited extent.

Different perspectives have emerged in other
national systems. In systems where responsibility
for agricultural extension is located in a Ministry
of Agriculture and research is conducted by an
autonomous research institute, there may be almost
total separation between extension function and
research activities. In research systems which are
not linked administratively with extension
programmes, there is often strong pressure to develop
extension-type linkages with clientele groups in
order to assure financial support.

<u>Stress over centralisation and decentralisation</u> In
countries with a dual state (province, prefecture)
and national system, there is typically continuous
effort by the national system to achieve effective
co-ordination of the combined efforts of the state
and national systems. In systems which are highly
centralised, there is a continuous struggle to assure
that the research programme is relevant to the needs
of the less advantaged regions of the country and the
less well organised agricultural constituencies. In
the Brazilian system, for example, reform efforts
have been directed at achieving greater
centralisation of planning and management. In the
Indian system, reform efforts are being directed
towards removing the excessive bureaucratic
restraints on research activity in a system with
strong central direction.

Hayami and Ruttan have suggested that a system that
is decentralised with respect to location, management
and funding tends to be more responsive to resource
and cultural endowments, and hence more productive in

supplying new technology to farmers, than a highly centralised system. This hypothesis has been disputed by Pastore and Alves.

<u>Conflict between basic and applied research</u> In every productive agricultural research institution, there is continuous tension between the commitment of professional and financial resources between mission-oriented and more fundamental or basic research. The problem of how to achieve convergence between the professional objectives of the individual research scientists and the social objectives or mission of the research institution is subject to continuous debate between research staff and administration. The solution is to link support for fundamental research to the demand for knowledge on the part of the mission-oriented applied research. Fundamental research on nutrient uptake may, for example, have a high payoff for applied research on varietal development.

The issue of the appropriate mix of basic, supporting and applied research is particularly difficult for many of the smaller national research systems. The agricultural sector in many of the smaller countries is no larger, in economic terms, than the areas served by the smaller states in India or the United States or even by the prefectural stations in Japan. The cost of developing substantial domestic capacity in the basic and supporting sciences would be very expensive relative to the value of commodity production. The development of regional research networks such as the West African Rice Development Association and the establishment of effective links with the international agricultural research system represent one way of dealing with the limitations imposed by the small size of many national systems. The problem of the appropriate mix of research and development activities for the smaller national systems remains difficult to resolve.

<u>Conflict over who should pay for research</u> In countries organised along federal lines, the question of what share of the costs of research should be

borne by the local government and what cost should be borne by the central government is of continuous concern. Central funds are often accompanied by central control. Yet the benefits of research done by one state or province spill over into other states or provinces. What part of the total agricultural research effort is appropriately within the public sector and what part is most appropriately conducted by the private sector? Of that part that is conducted in the public sector, what part should be paid for by consumers in the form of general revenue measures and what part should be paid for by producers in the form of a marketing cess or tax? The answers to these questions must be sought in terms of how the benefits of research are partitioned between producers and consumers, the economic organisation or structure of the agricultural sector, and the fiscal and administrative capacity of governments.

A major gap in our knowledge about the development of national agricultural research systems is the lack of a body of research on the mobilisation of political resources for the support of agricultural research. Why do some societies find it so difficult to sustain support for an activity which pays such high growth dividends? It has been hypothesised by de Janvry that an agricultural economy characterised by great disparity in size (or income) in its farm structure will be less effective in generating a sustained demand for a productive national agricultural research system than an agricultural system characterised by reasonable equity in farm size distribution. This hypothesis has been challenged by Huffman and Miranowski.

<u>Stress over the commodity and factor focus of research</u> The history of most agricultural research systems indicates continuous stress over the relative emphasis that should be placed on alternative commodities and on the several factors of production. These stresses reflect the political and economic power of commodity constituencies and of the owners of factors of production.

The great agricultural research institutions

developed under colonial administrations were directed primarily towards improvements in the productivity of export commodities - typically produced under a plantation form of organisation and expatriate management and ownership. In many post-colonial economies, particularly where agriculture remains organised along highly dualistic lines, this pattern continues. In other economies changes in political organisation have been accompanied by a shift in research priorities. In Malaysia, for example, the Rubber Research Institute is now much more strongly oriented toward the problems of smallholder rubber producers than a decade ago.

The choice of factor-oriented research strategies is also strongly influenced by economic and political organisation. Public investment in irrigation development may result in a focus of research effort on crops that are grown in irrigated rather than upland areas. Research designed to improve the opportunities of landless labour is rarely pursued with the same intensity as research designed to improve land productivity.

EVALUATING THE COSTS AND BENEFITS OF AGRICULTURAL RESEARCH[3]

To the extent feasible the several national studies should develop historical data on the sources and uses of research resources. An attempt should be made to estimate the impact of research on production and productivity; and attempts should be made to assess the distributional implications of research. Some of the considerations involved in each of these efforts are outlined in this section.

Sources and uses of agricultural research resources

Where do the funds that support a national (or state) research programme originate? Are they generated by a tax or a cess on commodity production on exports? From public revenues? From international assistance agencies? What agencies actually conduct the research? Is it spent by a ministry of agriculture research service; by a college of agriculture; by free standing research institutions, or by some combination of these?

<u>A first step</u> in any attempt to analyse the development of a national or provincial research system is a careful historical review of the sources and uses of research sources - including both financial and scientific resources.
   Identification of the sources and uses of research funds is itself often a rather complex task since the institutions that generate the funds for research are not the same institutions that perform the research.

<u>A second step</u> is a careful historical analysis of the use of the resources that are available for agricultural research. Little can be said about either the productivity or the distributional implications of agricultural research without a careful identification of how the financial and scientific resources available to a national research system have been used. In the United States, for example, the USDA has been criticised for excessive attention to mechanisation research. Yet a review of the Agricultural Research Service and the State Agricultural Research Service budgets indicates that, with the exception of a few states such as California, very little effort has been devoted to mechanisation research. Most research on mechanisation has been conducted in the private sector.
   The data on the allocation of financial and scientific resources that will be available in the several national studies will rarely be as complete as might be desired. In an ideal system it would be possible to obtain data on expenditures and on scientific manpower classified by the following five criteria:
   - among commodities - such as wheat, cotton, and beef;
   - among resource (or factor) categories - such as soil and water, or labour and management;
   - among stages or levels - industrial inputs (fertiliser, pesticides, mechanisation); farm

production activities; post-harvest technology and marketing;
- among geographic regions;
- among disciplines - such as plant breeding, agronomy, human nutrition, agricultural economics.

In attempting to develop historical data on the allocation of research resources, compromises will need to be made with existing data systems. In making these compromises it should be kept in mind that the one analytical use for the data will be simple comparisons of the way resources are allocated with the importance of the several commodities, resource categories, stages, regions and disciplines. For example, it will be extremely useful to be able to make the following comparisons:
- a comparison of the ratio of research expenditure by commodity to the value added in farm production for each commodity by region;
- a comparison of the ratio of research expenditure by factor (or resource) input to cost or economic value of the factor (or resource) in production;
- a comparison of the ratio of research expenditure to the value added at each stage in the food production chain from purchased inputs to the consumer.

Capital investment in facilities and equipment should be analysed separately from annual personnel and operating costs.

When comparisons of the type suggested in this section are made, there is often an implicit assumption that research resources should be allocated equitably among commodities, factors and regions. This has been referred to as the "parity" or "consequence" model of research resource allocation. In the United States, for example, soya bean research expenditures amount to about US$2.00 per US$1,000 of product. Cotton research expenditures amount to about US$15 per US$1,000 of product. Are these ratios efficient? Are expenditures on cotton too high? Are expenditures on wheat and soya beans too low? When such figures are cited there is often a presumption that the commodity characterised by a low research/output ratio is not getting its fair share of the research dollar. At a more sophisticated level there is usually an implication that the return from an additional dollar would be highest if invested in research on the commodity with the lower research/output ratio. In either case the critic usually has in mind, at least implicitly, what might be termed a "parity model" of research resource allocation.

The parity perspective is a useful first step in any analysis of research resource allocation. There are, however, two assumptions that are usually implicit in its application. The first assumption is that the opportunities for productive scientific effort or productivity-enhancing technical change are equivalent in each commodity and resource category. The second is that the value of a scientific or technical innovation is proportional to the value of the commodity or the value of the contribution of a particular resource to production. The compilation of a set of research parity accounts should not imply a judgment that resources should be allocated by a parity rule. It does not suggest that an explicit rationale should be developed for any departures from a parity rationale.

The next sections of this chapter will be devoted to a more systematic review of some of the criteria that might be used as a basis for evaluating departures from the parity model of research resource allocation.

Measuring the contribution of research

The beginning of successful modernisation of agricultural production, as suggested in the above section, is signalled by the emergence of sustained growth in productivity. During the initial stages of development, productivity growth is usually accounted for by improvement in a single partial productivity ratio such as output per unit of labour or output per unit of land. In the United States, and in other countries of recent settlement such as Canada, Australia, New Zealand and Argentina, increases in labour productivity initially carried the burden of growth in total productivity. In countries which entered the development process with relatively high man-land ratios, such as Japan, Taiwan (China), Denmark and the Federal Republic of Germany, increases in land productivity were initially the primary source of productivity growth.

In tables 3.1 and 3.2, changes in total productivity and in the two partial productivity growth rates are presented for the United States and Japan for the period since 1870. The figures illustrate the point made earlier in this section. Prior to the mid-1950s, productivity growth in Japanese agriculture was dominated by growth in land productivity. Prior to the 1940s, productivity growth in American agriculture was dominated by growth in labour productivity. Note also that both countries experienced periods of relatively slow productivity growth. During the first quarter of the 20th century, the rate of growth in labour productivity declined in the United States. Total inputs grew more rapidly than output. Total productivity declined.

Growth in total productivity has been influenced by a number of factors. Research leading to new knowledge and new technology has clearly been important. The education of farm people through formal schooling, through organised and extension activity and through agricultural publications has contributed to the rapid diffusion and efficient use of new technology. Transportation improvements have reduced the costs of industrial inputs and the costs of marketing. Rural mail and telephone services have exerted a pervasive impact on productivity.

Considerable evidence has, however, been accumulated on the contribution of research. The results of a large number of studies of the contribution of research to productivity growth and employment generation have been assembled in table 3.3.

Table 3.1
Annual average rates of change (per cent) in total
outputs, inputs, and productivity in United
States agriculture, 1870–1970[a]

Item	1870–1900	1900–25	1925–50	1950–65	1965–79
Farm output	2.9	0.9	1.6	1.7	2.1
Total inputs	1.9	1.1	0.2	-0.4	0.3
Total productivity	1.0	-0.2	1.3	2.2	1.8
Labour inputs[b]	1.6	0.5	-1.7	-4.8	-3.8
Labour productivity	-1.3	0.4	3.3	6.6	6.0
Land inputs[c]	3.1	0.8	0.1	-0.9	0.9
Land productivity	-0.2	0.0	1.4	2.6	1.2

[a] Sources: USDA (1979) and Durost and Barton (1960).
[b] Number of workers, 1870–1910; man-hour basis, 1910–71.
[c] Cropland used for crops, including crop failure and cultivated
summer fallow.

Table 3.2
Average annual change in total outputs,
inputs and productivity in Japanese agriculture,
1880–1965*

Item	1880–1920	1900–25	1935–55	1955–65	1965–75
Farm output	1.8	0.9	0.6	3.6	1.4
Total inputs	0.5	0.5	1.2	0.7	–
Total productivity	1.3	0.4	–0.6	2.9	–
Labour inputs	–0.3	–0.2	0.6	–3.0	–3.6
Labour productivity	2.1	1.1	0.0	6.6	5.0
Land inputs	0.6	0.1	–0.1	0.1	–0.7
Land productivity	1.2	0.8	0.7	3.5	2.1

*Source: Yamada and Hayami (1979), pp. 33–58 and extended to 1975 by
S. Yamada, "Letter", Dec. 17, 1979.

Table 3.3
Summary of studies of agricultural research
productivity[a]

Study	Country	Commodity	Time period	Annual internal rate of return %
Index number				
Griliches 1958	USA	Hybrid corn	1940–55	35–40
Griliches 1958	USA	Hybrid sorghum	1940–57	20
Peterson 1967	USA	Poultry	1915–60	21–25
Evenson, 1969	South Africa	Sugarcane	1945–62	40
Ardito Barletta 1970	Mexico	Wheat	1943–63	90
Ardito Barletta 1970	Mexico	Maize	1943–63	35
Ayer, 1970	Brazil	Cotton	1924–67	77+
Schmitz & Seckler 1970	USA	Tomato harvester	1958–69	
	with no compensation to displaced workers			37–46
	assuming compensation of displaced workers for 50% of earnings loss			16–28

Ayer & Schuh 1972	Brazil	Cotton	1924–67	77–110
Hines 1972	Peru	Maize	1954–67	35–40[b]
				50–55[c]
Hayami & Akino 1977	Japan	Rice	1915–50	25–27
Hayami & Akino 1977	Japan	Rice	1930–61	73–75
Hertford, Ardila,	Colombia	Rice	1957–72	60–82
Rocha & Trujillo	Colombia	Soya beans	1960–71	79–96
1977	Colombia	Wheat	1953–73	11–12
	Colombia	Cotton	1953–72	none
Pee 1977	Malaysia	Rubber	1932–73	24
Peterson &	USA	Aggregate	1937–42	50
Fitzharris 1977			1947–52	51
			1957–62	49
			1957–72	34
Wennergren &	Bolivia	Sheep	1966–75	44.1
Whitaker 1977		Wheat	1966–75	–47.5
Pray 1978	Punjab (British India)	Agricultural research and extension	1906–56	34–44
	Punjab (Pakistan)	Agricultural research and extension	1948–63	23–37
Scobie & Posada 1978	Bolivia	Rice	1957–64	79–96

Table 3.3 (Continued)

Study	Country	Commodity	Time period	Annual internal rate of return %
Regression analysis				
Tang 1963	Japan	Aggregate	1880–1938	35
Griliches 1964	USA	Aggregate	1949–59	35–40
Latimer 1964	USA	Aggregate	1949–59	not sig.
Peterson 1967	USA	Poultry	1915–60	21
Evenson 1968	USA	Aggregate	1949–59	47
Evenson 1969	South Africa	Sugarcane	1945–58	40
Ardito Barletta 1970	Mexico	Crops	1943–63	45–93
Duncan 1972	Australia	Pasture improvement	1948–69	58–68
Evenson & Jha 1973	India	Aggregate	1953–71	40
Cline 1975	USA	Aggregate	1939–48	41–50[d]
(revised by Knutson and Tweeten 1979)		Research and extension	1949–58	39–47[d]
			1959–68	32–39[d]
			1969–72	28–35[d]

Bredahl & Peterson 1976	USA	Cash grains	1969	
		Poultry	1969	
		Dairy	1969	
		Livestock	1969	
Kahlon, Bal, Saxena & Jha 1977	India	Aggregate	1960–61	63
Evenson & Flores 1978	Asia national	Rice	1950–65	32–39
			1966–75	73–78
	Asia international	Rice	1966–75	74–102
Flores, Evenson & Hayami 1978	Tropics	Rice	1966–75	46–71
	Philippines	Rice	1966–75	75
Nagy & Furtan 1978	Canada	Rapeseed	1960–75	95–110
Davis 1979	USA	Aggregate	1949–59	66–100
			1964–74	37
Evenson 1979	USA	Aggregate	1968–76	65
	USA	Technology oriented	1927–50	95
	USA	Science oriented	1927–50	110
		Science oriented	1948–71	45
	USA – South	Technology oriented	1948–71	130
	USA – North	Technology oriented	1948–71	93
	USA – West	Technology oriented	1948–71	95
	USA	Farm management research & agricultural extension	1948–71	110

Footnotes to table 3.3

---

a Sources: See names corresponding to each study listed in the bibliography.

b Returns to maize research only.

c Returns to maize research plus cultivation "package".

d Lower estimate for 13, higher for 16 year time lag between beginning and end of output impact.

e Lagged marginal product of 1969 research on output discounted for an estimated mean lag of five years for cash grains, six years for poultry and dairy, and seven years for livestock.

The contributions of research to increased agricultural productivity have been studied primarily by two methods. The estimates listed under the _index number_ heading were computed directly from the costs and benefits of research on, for example, hybrid corn. Benefits were estimated using accounting methods to measure the increase in production attributed to hybrid corn. The contribution of research was usually measured as the residual after all other factors that contributed to increased production were accounted for. The calculated returns represent the average rate of return per dollar invested over the period studied with the benefits of past research assumed to continue indefinitely. Benefits are defined as the benefits retained in the form of higher incomes to producers or passed on to consumers in the form of lower food prices.

The estimates listed under the _regression analysis_ heading are computed by a different method, which permits estimation of the incremental return from increased investment rather than the average return from all investment. Further, this method can assign parts of the return to different sources, such as scientific research and extension advice. Because regression methods are used, the significance of the estimated returns from research can be tested statistically.

The regression analysis studies build directly on the concept of an aggregate production function. In one variant changes in production over time or cross sectional differences in output among geographic units (states, districts, farms) form the dependent variable. The independent variables include (a) the conventional inputs such as labour, land, capital (structures and equipment), and operating inputs (fuel, maintenance, fertiliser, pesticides) and (b) factors which shift the production function such as research, extension and schooling. In a second variant the dependent variable is a measure of changes on differences in total productivity, measured by changes or differences in the ratio of output to conventional inputs. The independent variables are the shift factors such as research, extension and schooling. In both cases the purpose of including the conventional inputs is to "account" for their effects so that the impact of increases in conventional inputs on productivity is not attributed to research or education.

Analysing the distributional effects of research

Research administrators and policy makers have traditionally attempted to avoid overly precise specification of research objectives. They have preferred to be able to mobilise support from farmer clientele and clientele representatives by emphasising the contributions of research to the reduction of production costs, to increases in yields, or to the adaptation of crops to different environments. Administrators have preferred, at the same time, to emphasise to other constituencies the gains to consumers in the form of lower food costs or to the contribution of agricultural exports to the solution of balance of payments difficulties.

In nations characterised by strong farmer organisation, benefits to producers tend to be emphasised and benefits to consumers muted. In countries where farmers are poorly organised, consumer benefits tend to be emphasised. Thus in many developing countries with large, but politically inert rural populations, the primary emphasis in establishing a claim to research resources is often placed on meeting national food needs. In many developed countries, with relatively small but politically articulate agricultural constituencies, discussion of research benefits tends to emphasise the gains to agricultural producers.

However, research administrators are generally very uneasy about attempts to implement "demand-oriented" research programmes – that is, to direct research efforts towards specific social or economic objectives. They are, in contrast, much more comfortable with a "supply orientation" – that is, with research efforts which attempt to take advantage of perceived opportunities for scientific or technical advance. This orientation towards the supply or opportunity side of the equation rather than towards the demand or value side has, in our judgment, often led to a lack of effectiveness on the part of research administrators in dealing with budget offices, legislative committees, and special interests. There is an infinite number of interesting scientific problems, but not all of them are important. The effective research administrator must be able to reconcile the interest of scientists in exploring the endless frontier of interesting problems with the legislative demand that funds be allocated to those areas that are most important.

In this section, we give particular attention to the factors which determine the distribution of the gains, and the losses, from research and which influence the incentives to support research.

If the objective of research is to increase income to producers, it should be committed first to those commodities which are currently exported and secondly to those commodities where research would enable the country to achieve a substantial export market. If the object is to transfer the gains to consumers, research should be focused on commodities in which the country imports a small share of its consumption. A small increase in production, relative to consumption, will, in this situation, result in a transition to a net export position, or at least a potential export position, and push domestic prices below world market prices.

<u>Saving land, labour, and energy</u> The allocation of resources to research also involves choices about the importance of releasing the constraints imposed by resource supplies – of saving land and saving labour, for example. These choices are implicit in decisions about expanding or contracting research on problems such as soil and water conservation, soil fertility, and photosynthetic efficiency, mechanisation and energy use, and labour efficiency and management. What criteria are available to determine the relative balance between agricultural research directed primarily to decreasing or increasing labour requirements per hectare or to increasing the amount of product that can be produced per hectare?

Historically the answer to this question has been quite clear. In countries such as the United States, Canada and Brazil – with relatively abundant land resources and a strong demand for labour in industry – the primary thrust was towards improvements in mechanical technology that would enhance labour productivity. Only after expansion of the area available for cultivation became limited was attention turned to the development of technologies to expand output per hectare. The mechanical revolution in American agriculture began in the middle of the 19th century. The biological and chemical revolution did not begin until after the first quarter of the 20th century.

In countries such as Japan and Denmark, with abundant labour and relatively limited or poor land resources, the primary thrust in agricultural technology was to achieve increased output per hectare. In Japan the emphasis was placed on increases in crop yields. In Denmark the emphasis was placed on increased crop yields per hectare and on technologies to facilitate intensification in the production of livestock and livestock products. In both countries research designed to enhance labour productivity was delayed until the agricultural labour force began to decline in response to the rising demand for labour in the non-agricultural sectors of the economy.

In addition to the changing prices of land and labour, declining real prices of energy gave a further impetus to the invention of technologies which permitted the substitution of mineral fuels for organic sources of energy - tractors that utilised petroleum-based fuels were substituted for animal power that utilised farm-produced feed. The declining price of energy, embodied in chemical fertilisers, encouraged the development of crop varieties capable of responding to higher levels of nutrition.

Since the early 1970s the world has entered into a period of great uncertainty with respect to changes in the relative prices of labour, land and energy. The end of the era of cheap energy, like the end of the era of cheap land, is inducing a reallocation of research effort. But the new sources of productivity growth have not yet been clearly identified. Until the new trends become more evident than they have been during the 1970s, the appropriate allocation of research effort among land, labour and energy-saving alternatives will remain uncertain. In this environment of great uncertainty, an efficient research portfolio will include a wide range of options. It should avoid becoming locked into a commitment or a "fix" on any single option. It is, for example, too early to be able to make firm judgments about the proportion of future nitrogen that can be expected from advances in biological nitrogen fixation, low pressure nitrogen systems, or improvements in the efficiency of conventional high pressure nitrogen technology. As long as this uncertainty remains, resources should be allocated to the exploration of the possibilities for each of the several potential options.

<u>Research for large farms and small farms</u> During the 1960s and 1970s agricultural research institutions, in both developed and developing countries, have been widely criticised for focusing their research efforts on the problems of large farmers and of neglecting

research that would be beneficial to small farmers. Research designed to improve labour productivity has been criticised on the grounds that it leads to displacement of workers by machines. An attempt has been made in the state of California to legislate restrictions on mechanisation research at the University of California.

Some of this criticism is valid. Some of it is ideologically motivated. Much of it is confused. The long-term thrust of research and development on mechanical technology in American agriculture is a response to the rising real prices of labour. It has been primarily a response to labour shortage rather than a source of labour displacement. But this does not mean that the concern with a premature mechanisation may not be valid in specific situations. Subsidies to mechanisation have occurred in several forms. In the United States tax rules that provide for investment tax credits and accelerated depreciation schedules have driven a wedge between private profitability and economic efficiency. In Brazil and India access to foreign exchange on excessively favourable terms has at times biased the choice of technology in favour of mechanisation, and often in favour of large-scale rather than smaller, intermediate-scale equipment. When the choice of technology is subsidised in this manner, it has the effect of inducing related research, in fields such as agronomy and farm management, designed to improve the efficiency and speed the diffusion of a technology that is itself not appropriate.

But this is only part of the issue of technology for small farms. Is it possible to design technologies that are specifically suited to the needs, or the factor endowments of small farms? It is very difficult to think of examples of technology that would have greater benefits – that is, result in greater unit cost savings – for small farms than for large farms. But it is not too difficult to think of technologies that are roughly neutral in their impact. Indeed, much of the yield-increasing biological and chemical technology is roughly neutral with respect to scale. The effect of a new pesticide on the yield of rice may be no different if it is applied by aerial spraying or by a backpack sprayer. The choice of appropriate technology in this case would depend on the price of labour relative to the price of capital equipment.

There is, however, one way in which research benefits may, in some cases, be biased in favour of small farms. That is in the choice of commodity emphasis. In many countries in Latin America beans are produced on small farms and beef is produced on large farms. A decision to improve the productivity of bean production does, therefore, have the effect of biasing the direct impact of the gains from productivity growth in favour of small farms. A decision to conduct research on beef does bias the direct impact in favour of large producers.

A choice in favour of beans relative to beef also directs the gains that get transferred to consumers to low-income consumers – many of whom are small farmers or hired labourers. How much of the gains from productivity growth will be retained by the farmers who initially adopt the new technology will depend on the relationship between growth of productivity and growth of demand, outlined in the section on closed and open economies.

<u>Public research and private research</u> In all countries a very large share of agricultural research is supported with public funds and conducted in public institutions. In the United States somewhere in the neighbourhood of two-thirds of agricultural research activity is conducted in the public sector. In most countries the share of private research is much smaller. And in some countries private sector research, in plant breeding for example, is actively discouraged. What are the principles that can be utilised in deciding on the appropriate mix of public and private research?

The traditional rationale for public sector agricultural research is that individual farmers had but little capacity to conduct research and little possibility of realising a significant share of any gains from the results of research. The difficulty of establishing effective proprietary control over the results of research directed to the development of open pollinated crop varieties, the development of agronomic practices or the production of fundamental knowledge has also been important in limiting private incentives to conduct agricultural research. It was widely recognised that under these conditions the gains from technical progress tended to be rapidly externalised with consumers rather than farm producers or innovating firms realising the major share of the growth dividends generated by technical change.

This rationale applied more accurately to research leading to new crop varieties and agronomic or farm management practices than to research leading to new machines or new chemicals. In the area of mechanical technology the private sector has been remarkably successful in generating a continuous stream of technology. The classic pattern has been the invention of new machines or equipment by farmers, blacksmiths, and mechanics, followed by improvements in design and performance by the research and development units of the farm equipment companies.

The argument is often made that this pattern cannot be expected to work effectively in less developed countries where farmers and village blacksmiths are less sophisticated in their knowledge of the principles of mechanical technology and where a tradition of entrepreneurship in farm equipment does not exist. Yet we recall visiting the shop of a blacksmith in a village in Bihar, India, where rice threshers, remarkably similar to those being developed at the International Rice Research Institute, were being fabricated.

It is difficult to find public sector research units in either the developed or developing economies that have exerted a significant impact on the progress of mechanical technology in agriculture. A similar comment seems to apply, though perhaps with somewhat less force, in the area of post-harvest technology.

In the area of biological technology – the development of new crop varieties, new production practices in crop and animal physiology and pathology, and in entomology – the public sector has and continues to be the major source of new knowledge and new technology. The private sector has become a dominant source of improved seed in those areas, primarily maize, where hybridisation has given the inventor control over genetic materials. But even in the case of hybrid maize the public sector is still the dominant source of inbred (i.e. parent) lines for the commercial hybrids in the United States. The advent of "breeders'rights" legislation, which provides patent-like protection for the developers of new seed varieties, may result in greater incentive for the private sector to participate in the development of new varieties of open pollinated crops. As yet, however, the private sector is primarily engaged in only the final stage of genetic engineering involved in varietal development.

The role of public sector research and development in the area of chemical technology falls somewhere between the patterns noted for mechanical and biological technology. Most new plant protection chemicals are invented by the private sector. The public sector, however, plays an important role in the evolution of alternative materials and in the design of practices. Efforts made to minimise environmental impacts through integrated pest control have resulted in an expansion of the public sector contribution. In the fertiliser sector, private R & D on materials and processes in the United States is conducted at the National Fertiliser Development Centre operated by the Tennessee Valley Authority (TVA). Related research on soil fertility and fertilisation practices is, however, conducted primarily at the state agricultural experiment stations.

Who captures the gains from private sector research and development — or from public sector research embodied in private sector products? Great concern has been expressed by opponents of "breeders'rights" legislation, for example, that the gains that otherwise would be captured by farmers and consumers will be captured by the private suppliers of industrial inputs to agriculture. No careful studies are available for the agricultural input industries. Studies by Edwin Mansfield and his associates of the distribution of gains from industrial research generally indicate that the social returns tend to be roughly double the private returns. This means that the suppliers of the new technology are able to capture about half of the gains while market processes transfer another half to consumers. Similar studies are needed for the agricultural input, processing, and marketing industries.

In spite of our lack of knowledge, however, it seems clear that two general principles should be involved in the decision regarding the appropriate role of public and private sector research. The _first_ principle is that where adequate incentives to private sector research are present, intensive private sector research effort is readily induced. Indeed, it is difficult to imagine that the public sector would be willing to invest the level of resources in the field of maize improvement that has been profitable for the private sector. In some cases the private sector incentives have occurred as a result of institutional innovations, patent and breeders rights legislation, for example. In other cases, as in the case of hybrid maize, controls over inbred lines have provided a natural control over the product of research.

The _second_ principle is that when private incentives have been difficult to institutionalise, public sector support for agricultural research remains essential. In the absence of strong public sector support, the direction of research effort would tend to be strongly biased in the direction of proprietary mechanical and chemical technologies. Even in those areas where there are substantial inducements for product development by the private sector, investment in the supporting sciences necessary to sustain development tends to be limited. Few private seed companies, for example, devote significant resources to farm management, production economics, and cropping systems research or to plant pathology, plant physiology, or even genetics.

<u>A note of caution on the incidence of benefits and costs</u> In thinking about the incidence of benefits

125

and burdens of technical change, it is important to distinguish between embodiment, augmentation and incidence. Technology embodied in one factor of production has the capacity to augment the productivity of other factors, while the incidence of benefits and burdens may be experienced by other factors or even in other sectors. Embodiment is a characteristic of the technology itself. Augmentation and incidence are influenced by the institutional environment into which the technology is introduced.

Let us illustrate this point. Research embodied in a highly divisible technology such as improved seed, available to both small and large farmers, may have a very different impact on income distribution depending on the economic or institutional environment into which it is introduced. If introduced into an environment in which the supply of labour is more elastic than the supply of land, it will augment the returns per unit of land more than per unit of labour — even though it may greatly expand the employment of labour per unit of land. If introduced into an environment in which the supply of labour is inelastic, it will augment the returns per unit of labour more than per unit of land — even though there is little increase in employment per unit of land.

The incidence of benefits is also affected by institutional endowments and changes. If a divisible technology is introduced into an economy in which agriculture is carried out under a share tenure system, the incidence of benefits will be different than if introduced in a system in which farming is carried out primarily by owner-cultivated families.

CONCLUDING REMARKS

The development of effective national agricultural research systems has, during the last half of the 20th century, become a major focus of both national economic policy and of bilateral/multilateral assistance programmes. The forces which affect the capacity of a nation to effectively institutionalise agricultural research capacity; that determine the rate and direction of technical change that is most appropriate for a particular economy; and that determine the size of the new circumstances and the distribution of the benefits generated by technical changes are, however, not well understood.

In this chapter an attempt has been made to develop a conceptual framework for research studies designed to achieve a better understanding of forces that influence the rate and direction of technical change and the level and distribution of the benefit from research. An induced innovation model has been elaborated which provides a framework for the analyses of the impact of resource and cultural endowments on the rate and direction of technical change; that explores the role of resource and cultural endowments on the institutional changes that effect a nation's capacity to generate technical change in agriculture and that condition the distribution of the benefits of technical change among factor owners and social classes; and that suggests the need for a more adequate understanding of the complex interaction between technical and institutional change.

NOTES

[1] This section draws principally on materials developed more fully in Hayami and Ruttan 1985 and Binswanger and Ruttan 1979.
[2] This section draws very heavily on Blase and Paulson 1977, pp. 11-16 and on de Janvry 1977 and Ruttan 1978, pp. 1-18.
[3] This section draws very heavily on Ruttan 1982.

4 Factors influencing agricultural research and technology: A case study of India

SUDHIN K. MUKHOPADHYAY

INTRODUCTION

It is being increasingly realised that sustained agricultural development requires much more than just the provision of physical inputs or the knowledge of how to use them. An effective explanation of agricultural development is in terms of inputs and technologies as endogenous to the process of development itself and it is through such models that one could understand why the same technologies or inputs might meet with differential successes in different countries or regions. Experience in several developing countries during the last two decades in particular has helped to highlight the importance of seeking an explanation of the varying degrees of success of agricultural policies through the extension of the horizon of the traditional development models that generally used to treat technological changes as exogenous.

In particular, the basic question one has to answer in dealing with the generation and transfer of technological changes in agriculture is how far such changes are consistent with the physical and the socio-institutional endowments of the given country or region. What is more important, however, for an understanding of the disparate successes of agricultural technology is to go beyond the consideration of their appropriateness in given situations, and to examine the nature and working of institutions that lie behind the generation and management of the technology. In other words, the mechanism of appropriate institutional changes should be ensured along with that of appropriate technological changes in order for the process of development to be sustained. Significant contributions have recently been made towards the construction of such a theoretical model incorporating the mechanism of technological as well as institutional innovations (Arndt, et al. 1977; Binswanger, Ruttan, et al. 1978); (Ruttan, Chapter 3 of the volume).

OBJECTIVES

An attempt is made in this chapter to trace the institutional and technological development in Indian agriculture and to examine how far this can serve as a satisfactory explanation of the agricultural development of the country. The basic objective of this study is to examine the extent to which agricultural development in India has been affected by the institutional framework for appropriate technology generation and diffusion, and how far this institutional development has been consistent with the fundamental characteristics and regional distribution of the resource endowments of the country in its economic, demographic, historical and politico-socio-institutional spheres. The study should help throw some light upon the extent of sustained agricultural development of the national economy as well as its varying rates across regions.

OUTLINE OF STUDY

In the next section the elements of an integrated

model are presented to indicate how agricultural research may be viewed in the context of such a model. The third section traces the organisational development and structural pattern of agricultural research in India. The fourth section points out some potential sources of complementarities and conflicts that emerge in the process of development of agricultural research and indicates their importance in India. The fifth section deals with some evidence on the growth of income and employment in Indian agriculture during the recent period of technological change. The sixth section briefly discusses estimates of productivity of investment in agricultural research and attempts a fresh estimate. The seventh section is an attempt at explaining the allocation of research in Indian agriculture with the help of economic, socio-political and demographic variables. The final section presents the conclusions. The discussion and the quantitative analysis presented here have been constrained severely by the lack of appropriate detailed data. The problem was especially critical for data on state-wise breakdown of expenditure on agricultural research, education and extension. It was, therefore, not possible to consider the problem of extension separately in the quantitative analysis.

ANALYTICAL MODEL

Most traditional models and analytical designs dealing with agricultural development treat as exogenous the basic forces in an economy generating and governing a successful process of technological and institutional transformation of agriculture. Such analysis, however, would remain largely incomplete unless an effective link can be established between the overall resource environment in which an economy finds itself at a given point in time and its developmental dynamics. The emerging theory of induced technical and institutional innovation seeks to meet this requirement by internalising the mainsprings of agricultural growth into the overall model (Chapter 3).

Agricultural research-education-extension

In attempts at the application of the induced innovation model, a key position is held by the institutions of agricultural research, education and extension. Retarded agricultural development is often related to deficiencies in this research-education-extension system and inadequate investments in the generation of technology specific to the resource and cultural endowments of the region. The research-education-extension system is basically instrumental in generating and disseminating technology, while in its turn, this system is subject to a set of forces emanating from the social-material environment lying behind it. The research system's effectiveness is subject to a variety of interactive influences - both internal and external to the organisation. The "conflicts and complementarities" in the internal organisation and development of agricultural research systems, e.g. "centralisation against decentralisation" in research management, links between research, education and extension, and the choices between "science orientation" and "clientele orientation" - all these critically determine the sensitivity and success of the research system (Evenson, Waggoner and Ruttan 1979). A test of this performance of the research system can be attempted with the help of examination of the productivity of agricultural research in given countries and regions. It may appear that although agricultural research is highly productive in terms of resources utilised, and the rates of return to this investment are higher than in other alternative channels, actual investments in agriculture do remain low and biased against the needs of the resource structure. The question would require one to go behind the operation of the agricultural research system and would call for an exploration into the factors that determine the nature of decisions taken there. Some of the factors are: the existing resource pattern reflected through its relative price-cost structure perceived at the policy level, the heritage of past policies and the ease with which the political system can "articulate", "decentralise" and "evaluate" in the process of technology generation through adoption, etc. (Anderson 1979 and Ruttan 1980).

A hypothesised framework

The critical position held by the agricultural research system and the linkages that determine its performance can be hypothesised in terms of directions of influences emerging from different agents in an inter-connected way. The environment of historical, social, political and physical resource endowments exert its influences, through government policies and market factors, upon the extent and nature of agricultural research, which in its turn would work through technology generation, extension, and adoption to increased agricultural income and employment leading to changes in demand calling for sustained research. A hypothesised schematic framework (figure 4.1) of this mechanism could be used to understand the integrated nature of the institutional-technological nexus in a continuous process of agricultural development.

Keeping this broad framework in mind, an attempt is made in the following pages to construct the elements of a case study of the growth and structure of the organisation of agricultural research in India, a major developing country. Efforts would then be made to view this organisational development as a source of influence determining the performance of research leading through technological development and sustained growth.

AGRICULTURAL RESEARCH ORGANISATION IN INDIA

India in the developing world

In a number of ways India would occupy an important position in any study of agricultural development. It ranks first among 15 developing countries in total population (40 per cent of total), total agricultural product (24.2 per cent), and in the number of agricultural scientists (24.1 per cent) (table 4.1).

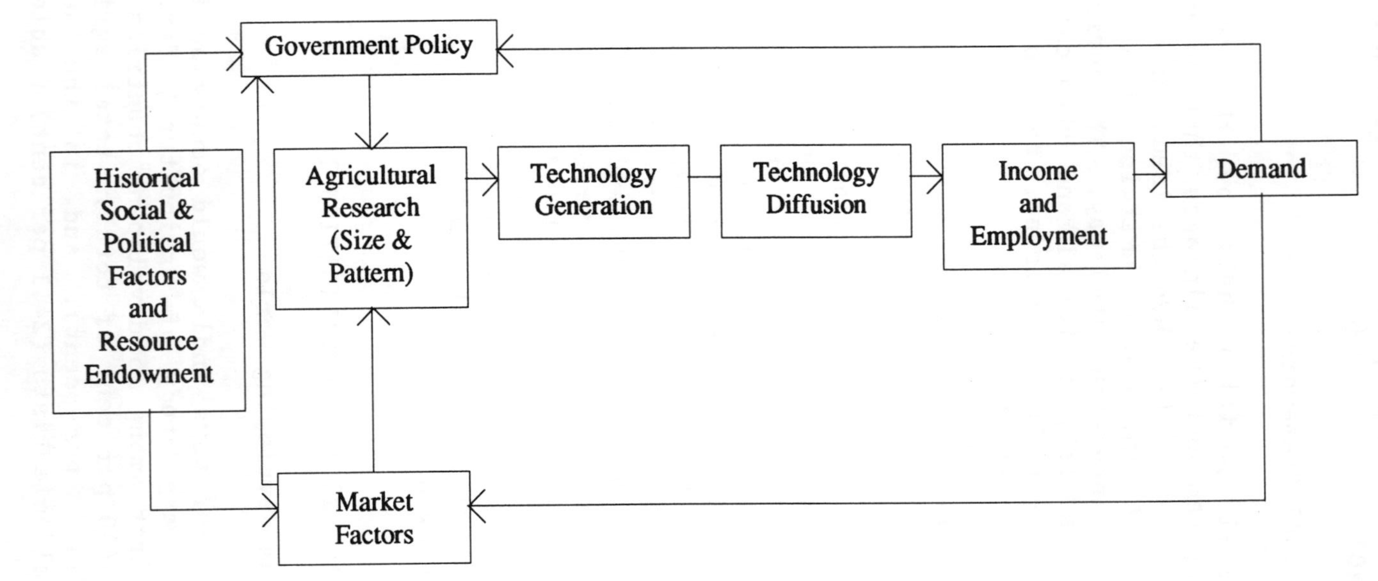

Figure 4.1: A schematic framework of the interrelationships between institutional and technological change

In terms of total expenditure on agricultural research also, India's position is third (12.5 per cent of total). However, in both percentage of agricultural product spent on agricultural research (only 0.3 per cent) and the rate of growth of agricultural output (about 2.6 per cent) India's position is quite low (Oram and Bindlish 1981). It would evidently be an important exercise to examine the nature of organisation of agricultural research and development in India that could shed some light upon the complex mechanism of agricultural and economic development. The focus in this section is upon tracing briefly the history of the organisation of agricultural research in India, its structural and functional characteristics, and its relationship with the requirements of the agricultural sector.

Phases of growth in research

The Indian agricultural research system has come a long and varied way since its formal beginning in the setting up of the Imperial Council of Agricultural Research in 1929 (India, National Commission on Agriculture 1976). This period of more than half a century of growth in agricultural research in India can be broadly classified into three phases. First, from 1919 to 1947 was the colonial period characterised by the laying of research infrastructures with a view mainly to the imperial trade interest that induced priority for commercial as against food crops and led to the relative growth of port-linked transport and communication as against irrigation systems. The total volume of research investment also was small and inadequate considering the need and potential of the nation's agriculture. The second phase commenced in 1947 with the attainment of independence and adoption of national economic reconstruction and development programmes, and it continued through the mid-1960s. Till then agricultural productivity had changed very little and the terms of trade also had remained against the agricultural sector. During this period the volume of investment in agricultural research first showed an upward thrust and then maintained a slowly rising trend in real terms (table 4.1). However, in the late 1950s with the undertaking of the second Five Year Plan, attention of the planning authorities turned increasingly to the need for bringing about technological improvement in the nation's biggest sector and various avenues started being explored for promoting agricultural research and education. The third phase began in the mid-1960s. This is the period that witnessed the emergence of the high-yielding varieties of some major crops and brought about important changes in the outlook of national planners leading to expectations of higher returns from research in agriculture. In 1964, major changes were introduced in agricultural research promotion, coordination and direction in the country through reorganisation of the Indian Council of Agricultural Research in accordance with the report of the Agricultural Research Review Team.

The Indian Council of Agricultural Research (ICAR)

The entire agricultural research system in India has evolved around one pivotal central organisation – the ICAR – and in that sense it can be said to have set an organisational model for agricultural research. The ICAR, however, did not always have its present power and position. It was only in 1964, following the recommendation of the Agricultural Research Review Team, that the pre-eminent position in agricultural research was given to the ICAR. The present objectives of the ICAR are as follows:

(i) "to undertake, aid, promote and coordinate agricultural and animal husbandry education, research and its application, development and marketing to increase scientific knowledge of the subjects and to secure its adoption in everyday practice";

(ii) "to act as a clearing house of information not only in regard to research but also in regard to agricultural and veterinary matters generally";

(iii) "to establish a research reference library with reading and writing rooms and to furnish the same with books, reviews, magazines,

Table 4.1
Population, agricultural output and agricultural research: selected developing countries, 1980*

Country	Share of total population		Share of total agricultural GDP		Rate of growth of agricultural GDP (1970–80)		Share of total agricultural scientists		Share of total agricultural research expenditure		Agricultural research expenditure as % of agricultural GDP	
	%	Rank	%	Rank	Rate	Rank	%	Rank	%	Rank	%	Rank
Argentina	1.6	11	4.6	6	2.3	11	3.6	9	13.4	2	1.64	1
Bangladesh	5.0	4	2.4	13	1.6	14	5.6	4	2.1	11	0.48	9
Brazil	7.2	3	9.6	2	5.3	3	10.1	2	19.8	1	1.15	3
Colombia	1.5	12	3.2	10	4.9	6	1.1	15	3.9	6	0.67	7
India	39.5	1	24.2	1	2.6	10	24.1	1	12.5	3	0.29	12
Indonesia	8.6	2	4.7	7	4.0	7	5.	6	3.6	7	0.44	10
Kenya	0.9	13	0.9	15	5.5	2	1.4	13	1.8	14	1.08	4
Korea,Rep.of	2.2	10	5.7	5	4.0	8	3.3	11	2.3	10	0.23	14
Malaysia	0.7	15	2.5	12	5.0	4	2.8	12	3.5	8	0.81	5
Mexico	4.0	7	5.8	4	2.1	12	4.3	7	6.8	5	0.65	8
Nigeria	4.4	6	7.9	3	-1.5	15	3.7	8	9.8	4	0.70	6
Pakistan	4.7	5	2.8	11	1.9	13	9.9	3	2.0	12	0.41	11
Philippines	2.9	8	3.8	9	4.9	5	3.5	10	1.0	15	0.16	15
Thailand	2.7	9	4.0	8	5.6	1	5.2	5	1.9	13	0.26	13
Venezuela	0.8	14	1.3	14	3.5	9	1.2	14	3.2	9	1.32	2

*<u>Source:</u> Oram, Peter, A. and Bindlish, Vishva (1981).

newspapers and other publications"; and

(iv) "to do all other things as the Society may consider necessary, incidental or conducive to the attainment of the above objectives". (India 1976).

Through the recent revisions (1975) in the Government of India policy toward the ICAR, this organisation has now the authority and obligation to seek to fulfil the above objectives by means of the institutional network for agricultural research, education and extension in the country. The ICAR can exercise this authority both directly through the operation of its own institutes and indirectly through the funding mechanism for research at other organisations under the financial control of the ICAR. A brief discussion is presented below of the organisational set up of agricultural research in India and the major control mechanism used by the ICAR follows.

ORGANISATIONAL STRUCTURE FOR AGRICULTURAL RESEARCH IN INDIA

The ICAR, a central Government organisation, derives its statutory power mainly from a general clause in the Constitution of India that specifies that the central Government would be responsible for "co-ordination and determination of standards in institutions for higher education or research and scientific and technical institutions". By virtue of this position the ICAR conducts its operation in agricultural research in the country through its control and linkages with a big network of research institutes, universities, state departments and other organisations. A schematic presentation of the different organisations engaged in agricultural research in India and how they are linked to each other is given in figure 4.2.

Central institutes

At present the ICAR has directly under its own control 33 research institutes. Of these, ten are concerned with specific crops, three with animal sciences, four with fishery and three with soils. The others are either of general type or concerned with crop/fish technology, engineering and statistics. Some of these institutes are descendants of the earlier commodity boards, some have been transferred from the Central Ministry of Food and Agriculture and some have been set up recently under the ICAR. The size, scope and efficiency of these institutes vary widely. The range of variation in terms of staff strength is from 12 in the Central Agricultural Research Institute of Andaman and Nicobar to about 1,000 in the Indian Agricultural Research Institute, New Delhi, and in terms of expenditure (in 1973-74) per scientist from 15 to above 60 thousand rupees per year.

Reviewing the efficiency of the major institutes in terms of growth in production, yield and acreage under the modern varieties of crops relevant for the institutes, the National Commission on Agriculture showed that no systematic impact of the research institutes could be identified either upon the crops or in the regions of location of the institutes. The National Commission concluded that there might be three possible reasons for this:

(i)   the research recommendations of the institutes were valid and effective but these had not reached farmers through adequate extension;

(ii) the research recommendations were broadly valid and effective but not suitable for the agro-climatic regions concerned; and

(iii) the research recommendations were not valid and effective and so not acceptable to the farmers.

Commenting on the need for change in the attitude governing the working of these institutes, the Commission observed that there was an "unhealthy competition" among the institutes to raise their status and expand their size and scope which tended to affect their efficiency adversely.

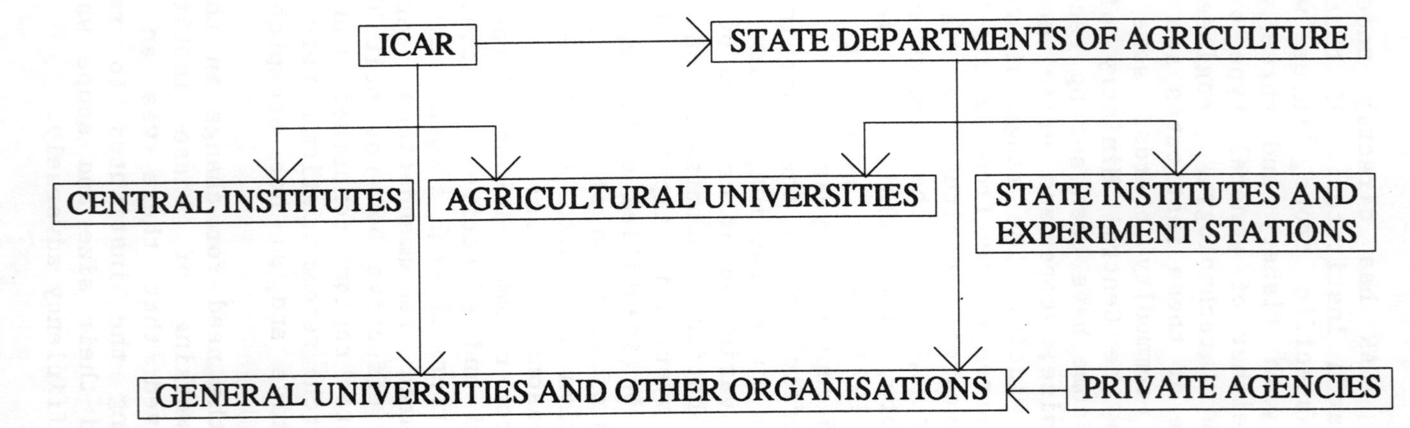

Figure 4.2: Organisations engaged in agricultural research in India

Agricultural universities

In the early years of promoting agricultural research in the country during the post-independence era it was being increasingly realised by the ICAR that a distinction was necessary to be made between research and education in general and agricultural research in particular. The need for integrating agricultural education, research and extension was also being felt. The joint Indo-American teams set up to review agricultural research in India stressed this requirement and recommended the establishment of agricultural universities following the model of the Land Grant Colleges of the United States where a balanced integration of agricultural education, research and extension is pursued. The result was a steady growth in the number of such universities in India, with all the major states having at least one and some even more of such universities, the total being 22 at present. The agricultural universities are under a dual funding system, the ICAR and the state governments, the share of each being about half. The basic objective of the agricultural universities is to undertake state-wide responsibility for agricultural education, research and extension, with a focus on location-specific applied research. The agricultural universities model also envisaged the transfer of research responsibilities in the state from the state government to these universities. By and large, the performance of the agricultural universities in India so far has been judged to be satisfactory. However, the process of transfer of powers and responsibilities from the state governments to the universities has not been smooth, costless or regionally even. Similarly, problems have often been encountered in the assignment of priority to basic, applied and adaptive research. Moreover, although generally the quality of research and education above the master's level has been considered satisfactory, the same cannot be said about undergraduate education.

As a result, different states can be found to have attained different levels of progress in education and research through their agricultural universities

- partly an expression of the extent of friction between the state agricultural departments and universities.

State governments

The supremacy of the ICAR in the Indian agricultural research system has to be viewed in the context of the Constitution of India which declares agriculture to be a state subject. The state responsibility according to the Constitution includes: (a) agricultural education and research, protection against pests and plant diseases; (b) preservation, protection and improvement of stock and prevention of animal diseases; and (c) veterinary training and practice. Thus, the constitution gives to the states the basic and primary responsibility for agricultural education, research and policy. The role of the ICAR, therefore, may be considered as a measure of reinforcement of agricultural education and research collateral with the state's role.

The state departments of agriculture have traditionally been entrusted with agricultural research in their respective areas and they still retain some of these authorities, though in varying degrees in different states. The state departments provide part of the financial and organisational support to the agricultural universities which ensures the location-specific character of the research done at the universities. In some states, however, the state departments still do retain a major share of the applied and adaptive research in agriculture. Apart from this, many state agricultural departments have research centres and farms under their own control. These are, by and large, engaged in problems of local importance assigned by the state government from time to time.

Other organisations

The last category of organisations engaged in agricultural research consists of some general universities, some general research institutes, some crop-specific institutes and various other centres of local regional and national importance. These organisations receive their support and sustenance from the ICAR, the state governments and also from private agencies. Their research programmes range from the basic and fundamental to the applied and adaptive areas of agricultural research.

All India Coordinated Research Projects (AICRP)

In carrying out its role in promoting and conducting agricultural research in India, the ICAR had been traditionally supporting individual research projects at different centres in the country. Such a method obviously led to isolated research with the possibility of duplication and overlap as well as a narrowing down of the scope of research. This problem was overcome by an administrative innovation by the name of All India Coordinated Research Projects (AICRP). The innovation was introduced in 1957 through the All India Coordinated Maize Improvement Project initiated with the collaboration of the Rockefeller Foundation. In this project, a departure was made from the old method of financing research based on political boundaries of the states, and instead the country was divided into agro-climatic zones for the purpose of conducting research on maize. Scientists working on the crop in various institutes and centres were called upon to meet and prepare a co-ordinated research outline to be followed at the respective locations. The results of this successful coordinated experiment in research encouraged the ICAR to introduce this method in larger numbers. The All India Co-ordinated Rice Improvement Project was initiated in 1965 and research programmes on rice at more than 100 centres around the country were brought under its purview. The basic criteria for initiating an AICRP at a given location were:
   (i)   necessary physical facilities and technical
           expertise should be made available;
   (ii)   the programme had to be multi-disciplinary
           and a group of scientists from relevant
           disciplines should be involved in it; and
(iii)   the location should satisfy the agro-climatic

characteristics typical for the experiment.

Beginning in the early sixties during the fourth plan period, the number of projects under this coordinated programme increased steadily to 70. The figure came down to 51 during the fifth plan period and in 1982 the number stands at 53. The programme has some outstanding merits in ensuring coordination among leading centres and researchers working on the same or similar problem and thereby avoiding waste, enhancing positive interaction among workers and increasing the probability of success, and in building into the research programme the ingredients necessary for adoption by farmers. However, several problems also arose in the process of the programme's operation. Following are some of them suggested by the National Commission on Agriculture.

The AICRP being a separate category of research programme with special funding facilities, there has grown a tendency among workers in these programmes to be isolated from other research workers in the country as well as in the same centre. This has led to a dichotomy among research workers — with undesirable effects on their morale. Secondly, some of the research projects have developed into mere data collecting units in a routine manner without much consideration for the basic objective of the AICRP. Thirdly, co-ordinated projects have often been initiated at some centres simply by replacing some ongoing research on similar problems although it might have been of good quality. The gradual expansion of the AICRP has a tendency to leave very few resources for individual research projects outside this co-ordinated scheme. This has a negative impact upon good quality research, especially that coming from universities. The National Commission on Agriculture recommended that the ICAR should try to resist this tendency and should sponsor carefully drawn research programmes to be carried out at universities and research centres, both central and state. These programmes should be funded by the ICAR particularly for the universities with a view to developing a strong base there for fundamental and applied research.

Ad hoc projects

A substantial amount of research is undertaken at universities and institutes outside the scope of the AICRP and under the ICAR Cess Fund Scheme. In this scheme, projects are prepared and submitted by individual scientists and centres and are supported on an ad hoc basis. The number of such projects approved during the seventies was about 700. During the 50 years from 1929 onwards, the number of projects supported was 2,531 and the amount of funds involved was Rs.203.7 million (ICAR 1979). This grant-in-aid method of financing agricultural research often runs against the difficulty of paucity of funds and there is also a considerable bureaucratic delay involved in securing clearance. Stressing the importance of this method of research support, the National Commission on Agriculture suggested that while AICRP should be restricted to only problem-oriented applied research, no such restriction should be imposed on individual research projects which should be judged on the basis of their merit alone.

COMPLEMENTARITIES AND CONFLICTS IN RESEARCH ORGANISATION

Research in general, and agricultural research in particular, is characterised by strong linkages and complementarities among different organisational components – both vertical and horizontal. In a developmental phase such complementarities are subjected to a certain degree of stress and often conflicts of various nature arise. These conflicts have a direct bearing upon the generation of technology and employment, and if these are not adequately taken care of in time, the performance of research organisations and productivity of research investment might be affected substantially and thus agricultural growth potential might be impaired. Several such conflicts could be identified in the process of the organisational development of agricultural research in India, and some of them are briefly mentioned below.

Research categories

Although research can be viewed as a continuum from its conception at the scientist or laboratory level to its widespread adoption in real life, for analytical purposes agricultural research can be broadly categorised into three types: (i) basic or fundamental research, (ii) applied research, and (iii) adaptive research. Basic or fundamental research in agriculture is carried out by scientists mostly in their laboratories or experiment stations under controlled conditions in order to bring about new breakthroughs or inventions and thereby to increase the stock of knowledge. Applied research is one step forward from basic research towards real life conditions, and it is concerned with the application of the new ideas derived in basic research to farms under different agro-climatic conditions. Adaptive research goes still further and deals with the feasibility and viability of the ideas generated by basic research through trials, verifications and demonstrations. Adaptive research thus would be concerned with the economic and social conditions that farmers face and would thus make it possible to examine how far the preceding applied and basic research on the problem would be of use for farmers under different agro-climatic as well as socio-economic environments.

The close complementarity among the three categories of research mentioned above is evident through the fact that there can be hardly any adaptive research unless there are ideas and technologies that have been tested through empirical trials, and similarly applied research would have an empty basket unless there is a continuous stream of new technologies. It should be noted that the three categories of research suggested above are not necessarily in that sequence in all cases. Adaptive and applied research do often throw up problems that basic research would be called upon to solve through laboratory experiments which can then be transmitted back to applied and adaptive research for verification and adoption trials. A balanced distribution of research resources among these categories is a prerequisite for sustained technological progress.

Such a balance seems yet to be attained in the agricultural research system in India. Examination of the research projects approved and undertaken in the country shows a very high proportion going to applied research, while basic research trails behind even adaptive research. The most important reason for this is the fact that basic research is time consuming and its results are uncertain, while most applied and adaptive research is time bound and result oriented. The system of incentive, award and recognition for scientists being based upon immediate results obtained, their preference is often for the latter two categories of research. The Indian agricultural research system is thus characterised by relative scarcity of scientific talent to undertake basic research as well as a relative lack of balance among the categories of research. This problem is also evident in the development of the organisational components that are entrusted with the responsibility to carry out the respective types of research.

Universities, institutes and State Departments of Agriculture

The National Commission on Agriculture (1976) discussed the above three categories of agricultural research in their various aspects and observed that in India there is a great deal of overlap and duplication of research responsibilities among institutions causing substantial waste of resources. It was noticed that although general universities and agricultural universities had the resources – both intellectual and financial – for basic research, there was a strong inclination among most of them to take up applied research. Departments of basic sciences in agricultural universities have a low priority and scientists engaged there have relatively less potential for future development. As a result there is a dearth of qualified personnel for such departments and so basic research remains at a relative disadvantage. State departments of agriculture have got the benefit of experiment stations and farms under different agro-climatic conditions and are, therefore, equipped to take up adaptive researches. However, in reality even now there are state departments working in competition rather than co-operation with agricultural universities and institutes in carrying out applied and even basic researches. Considering the need for a clear demarcation of responsibilities for research among these major participant institutions, the National Commission on Agriculture suggested a pattern which is presented in table 4.2 by categories of research.

What is critical in the allocation of responsibility of agricultural research among institutions is the need for close collaboration and continuous contact and feedback among them for successful generation of the right technology, their scientific experimentation, and widespread and economic adoption. A way of looking at organisational pattern of research resources is to examine the allocation of funds to them. If it can be assumed that general universities are mostly engaged in basic research while agricultural universities carry out applied research it would appear from table 4.3 that in the allotment of ad hoc research grants, the position of central institutes, which was the highest in 1974-75, has gone down substantially after yielding place to agricultural universities. This would suggest that there has been a rise in the importance of applied research in Indian agriculture. However, it must be mentioned here that these figures refer only to a part of the organisational structure and to the ad hoc projects alone, and the total picture may be somewhat different.

Table 4.2
Institutional allocation of research responsibility

Institution	Recommended responsibility in		
	Basic Research	Applied Research	Adaptive Research
Agricultural Universities	Partly	Mainly	None (except through liaison with State Governments)
General Universities	Mainly	Partly	None (except through liaison with State Governments)
Central Institutes	Partly	Mainly	None (except through liaison with State Governments)
State Governments			Mainly (in close collaboration with Universities and Institutes)

Table 4.3
Percentage allocation of funds for ad hoc research projects*

	1974–75			1981–82		
	Animal Science	Crop Science	Total	Animal Science	Crop Science	Total
Institutes	31.27	51.87	44.11	22.17	13.24	14.46
Agricultural Universities	50.18	35.60	41.09	61.32	74.27	72.50
General Universities and other Organisations	18.54	12.53	14.79	16.51	12.49	13.04

* Source: ICAR Budget proposals, 1982–83

Research, education and extension

Like research categories and implementing agencies, another source of complementarity and conflict in agricultural research organisation lies in the relative importance of research, education and extension. It is well recognised that the three are integrated in their effect upon successful agricultural development and this recognition had been accepted in India as the basic principle of the agricultural universities. However, in practice stresses have often arisen in the importance to be attached by different organisations to these three areas. Agricultural research has been found to be complementary with higher education in agriculture while it is not found to be very closely related to the undergraduate education (Ruttan 1980). But obviously higher education would be related closely with undergraduate education and, therefore, it would be vital to the interest of agricultural research to have a strong educational programme. Similarly, research and technology generation provides the input for extension programmes while the latter is essential for carrying the fruits of research to the community through the farmers.

The National Commission on Agriculture noted more than five years ago, that although there has been an understanding in India of the importance of all these three facets of agricultural development programmes, organisations and agencies often do vie with each other in taking up programmes irrespective of their functional jurisdiction. For example, the National Commission observed that although education and training programmes were supposed to be undertaken primarily at agricultural universities and only at a very few selected other centres, many institutes were building training and teaching programmes and were also asking for the authority to award degrees and diplomas. Such overlap in individual functions tends to dissipate organisational responses. The institutes have a very important role to play in conducting adaptive research and undertaking extension programmes. This must necessarily be done in close and continuous collaboration with agricultural universities and state departments of agriculture. Such harmonious collaboration has not been very common so far in India, except in a few states, e.g. Punjab, where sources of conflict and overlap seem to have been minimised. Competition and conflict also exist in the allocation of research funds among the three functions. Data are hard to find on such allocation in the states and whatever could be gathered suggest that there is still a relative neglect of extension as compared to other activities (table 4.4). However, very recently extension programmes have been substantially strengthened at the national level through the World Bank-aided National Agricultural Research Project, the Lab-to-Land Programmes and the National Demonstration Schemes. Individual states also are trying to develop their extension programmes. For example, West Bengal has undertaken an Agricultural Extension and Research Project with the help of the World Bank. These have helped improve the relative position of extension in agricultural development programmes although the results achieved leave much to be desired.

Research allocation among crops

It has been observed that the volume of agricultural research in India has experienced a substantial increase over time especially since Independence. However, volume alone could not speak adequately about the relevance of the type of research that has been undertaken. In the imperial days the priority of policy was upon commercial and plantation crops as against food crops. The pattern was expected to change with the national research system giving increasing importance to food crops. However, as table 4.5 shows, the percentage distribution of grant-in-aid by the ICAR for major crops and groups of crops is quite at variance with the importance these crops enjoy in the share of national output. For example, in 1981-82, rice, contributing one-third of the total agricultural output of the country, got only 7.37 per cent of the ICAR grant of research funds while fruits, vegetables and other crops

Table 4.4

Allocation of funds to agricultural research, education and extension: selected states[a]
(Millions of rupees)[b]

State	Research	Education and Administration	Extension
Gujarat	36.39	53.92	11.20
(1977–78)	(35.85)	(53.12)	(11.03)
Karnataka	61.06	186.19	36.61
(1978–79)	(21.51)	(65.59)	(12.90)
Madhya Pradesh	25.83	23.71	0.76
(1978–79)	(51.29)	(47.12)	(1.59)
Orissa	17.78	82.16[c]	2.85
(1979–80)	(17.79)	(79.92)	(2.78)
Punjab	45.03	41.82	7.26
	(49.00)	(43.00)	(8.00)

[a] Source: State reports for <u>National Agricultural Research Project</u>, ICAR 1982.

[b] Figures in parentheses denote percentages.

[c] Includes campus development.

Table 4.5

Cropwise percentage allocation of ICAR grant-in-aid for research in India, 1981-82

Crop	Percent of total expenditure approved as grant-in-aid by ICAR (1981-82)	Weight of crop in national aggregate agricultural output
Rice	7.37	33.94
Wheat and other cereals	14.07	26.07
Millet and pulses	16.94	8.07
Fruits, vegetables and other crops	61.61	31.88

(including plantation crops) received as high as 62
per cent. Such discrepancy between research priority
and product importance might perhaps be able to
explain part of the sluggish growth of aggregate
agricultural output in spite of steep increases in
research expenditure (of course, research results may
not be proportional to research expenditure).

Research allocation among disciplines

Success in research for technological progress in
agriculture consistent with a given resource
situation is dependent upon how far the pattern of
research is optimal with respect to the various
disciplines relevant for augmenting the resources of
the agricultural economy. For example, one would
expect a consistency between research for technology
generation on the one hand and technology diffusion
on the other, between soils and manure on the one
hand and agricultural engineering on the other. The
allocation of the total amount of grants-in-aid for
research by the ICAR for projects already approved
and in progress in 1981-82 is given in table 4.6.
Although the figures are partial and relate only to
one year, they indicate the kind of priority given to
the different disciplines in agricultural research.
Agronomy and agricultural education here seem to have
been accorded relatively low shares of the
grant-in-aid. More data and research on this aspect
would be useful.

Allocation among regions

A rational allocation of research funds among regions
may also be considered a prerequisite for successful
and balanced generation and diffusion of technology
in agriculture. Regional disparity being often a
concomitant of agricultural growth, allocation of
research resources might act as a policy variable to
keep such disparity under control so that growth
could be balanced. In India such increase in
disparity in the recent past has been particularly
noticeable (Mukhopadhyay 1976, 1980). As table 4.7
shows, the annual compound growth rate of agriculture

Table 4.6
Allocation of ICAR grant-in-aid according to discipline

Discipline	Allocated fund (rupees)	Percentage of total
Soils and manures	83,38,966	7.58
Agronomy	25,04,985	2.28
Physiology, entomology mycology and floriculture	1,93,73,076	17.61
General	60,49,646	5.50
Engineering	50,71,711	4.61
Agricultural education	7,97,926	0.72
Lab to land (extension)	6,78,63,830	61.69
TOTAL	11,00,00,140	100.00

during 1960-61 to 1978-79 varied from 1.19 in Orissa to 8.10 in Punjab. What is important, however, is that the allocation of expenditure on research and education on agriculture during 1979-82 has practically very little relationship with the percentage contribution of the states to the national output. Out of the 14 states shown, seven have their share of agricultural research funds exceeding their share in national agricultural output. Kerala, with the lowest share in the national agricultural output (1.2 per cent) had 6.82 per cent of the research funds, while Uttar Pradesh with 15.8 per cent of the country's output had only 11.15 per cent of the research funds.

Conclusion

The discussion and evidence presented in this section suggest that during the significant rise in the nation's level of expenditure on agricultural research and education and the expansion and proliferation of its research organisation, India has had to face several sources of complementarities and conflicts. Not all of these have been very easy to deal with and indications are that these conflicts might have substantially eroded the efficiency of the agricultural research system in India. It appears that while the size of investment in research may be important in the generation of agricultural technology and creation of employment in a labour abundant economy like India, volume alone does not provide a sufficient condition for the success of research. Appropriate balance between each of the components of the research-technology-employment programme is a critical requirement for sustained growth.

TECHNOLOGY, INCOME AND EMPLOYMENT

It has been hypothesised in this study (see the second section) that a key position is held by the agricultural research system of a nation in its process of development. While the nature of this

Table 4.7

Allocation of research expenditure and agricultural growth in states of India

State	Percentage share of state in all India output (1960–61 to 1978–79)	Growth rate in agriculture output and education (per cent)	State as percentage of total expenditure on agricultural research
Andra Pradesh	8.4	1.69	1.86
Bihar	7.7	1.92	3.42
Gujarat	3.3	3.56	8.74
Haryana	3.8	5.33	4.29
Karnataka	6.2	3.40	5.02
Kerala	1.2	1.39	6.82
Madhya Pradesh	10.0	1.67	5.63
Maharashtra	7.4	1.77	6.72
Orissa	4.8	1.19	5.15
Punjab	7.5	8.01	10.21
Rajasthan	6.0	2.97	4.36
Tamil Nadu	5.9	1.83	14.31
Uttar Pradesh	15.8	2.79	11.15
West Bengal	7.2	2.72	6.88

system is determined by the prevailing historical, social, political and physical resource endowment, the research system, in its turn, is sustained by its ability to effectively generate and disseminate appropriate technology leading to the creation of employment and income that provide the demand for continued growth. In the preceding sections an outline of the growth and organisational structure of the agricultural research system of India has been presented, and some of the critical problems that emerge in the process of development and largely determine the performance of agricultural research have been identified. What is important now is to explore how far the Indian agricultural research system has succeeded in generating appropriate technology leading to increases in incomes and employment in the country.

In this section an attempt is made to put together some pieces of information to examine the nature and extent of changes in income and employment that have occurred in the different regions of India during the last two decades. It must be mentioned here that systematic and comprehensive time-series or cross-section data on income and especially employment in agriculture in India is extremely difficult to come by. We have, therefore, tried to collect data of different types from different sources with a view to obtaining whatever meaningful insight might be available.

Technology and income change

The basic question that has been asked time and time again in the context of the recent technological changes in Indian agriculture is with regard to their income generation ability and the regional pattern of such income generation. Evidence presented in this section is strongly in support of the contention that there has been a substantial increase in farm income generated by the new technology. The growth was substantial during the 1960s, and in the following decade it continued, but at a lower rate. Table 4.8 presents the value of average agricultural productivity per hectare in the different

Table 4.8
Regionwise agricultural production for
1962–65 and 1970–73*

State/Region	Value of agricultural output per hectare (Rs.)		
	1962–65	1970–73	1970–73 (–) 1962–65 (Difference)
1. Andhra Pradesh			
Coastal	1,380.2	1,475.9	95.7
Inland Northern	744.5	652.6	–91.7
Inland Southern	1,028.7	1,192.2	163.5
2. Bihar			
Southern	892.2	865.6	–26.6
Northern	906.4	950.5	44.1
Central	983.1	1,135.5	152.4
3. Gujarat			
Eastern	1,004.4	987.7	–16.7
Plains, Northern	891.4	1,175.1	283.7
Plains, Southern	956.9	1,098.4	141.5
Dry Areas	414.5	563.2	148.7
Saurashtra	760.1	911.9	151.8
4. Haryana			
Eastern	905.0	1,372.8	467.8
Western	706.2	913.7	207.5
5. Karnataka			
Coastal and Ghata	1,515.8	1,665.8	150.0
Inland Eastern	1,257.1	1,582.3	325.2
Inland Southern	1,018.6	1,412.3	393.7
Inland Northern	539.1	719.5	180.4

160

6. Kerala

Northern	1,631.2	1,751.9	120.7
Southern	1,601.0	1,800.0	199.0

7. Madhya Pradesh

Eastern	818.1	897.4	79.3
Inland Eastern	519.3	592.5	73.2
Inland Western	570.7	617.7	47.0
Western	621.6	629.2	7.6
Northern	626.6	674.3	47.7

8. Maharashtra

Coastal	1,370.3	1,345.6	−24.7
Inland Western	109.4	632.3	522.9
Inland Northern	655.2	497.2	−158.0
Inland Central	475.7	293.8	−101.9
Inland Eastern	511.2	399.7	−111.5
Eastern	706.1	657.1	−49.0

9. Orissa

Coastal	1,169.0	1,067.5	−101.5
Southern	1,065.8	975.3	−90.5
Northern	1,067.9	1,011.9	−56.0

10. Punjab

Northern	1,194.2	1,793.7	599.5
Southern	1,124.1	1,738.8	614.7

11. Rajasthan

Western	159.2	225.7	66.5
North Eastern	484.2	720.5	236.3
Southern	746.7	775.1	28.4
South Eastern	537.9	705.3	167.4

Table 4.8 (cont'd)
Regionwise agricultural production for
1962–65 and 1970–75*

State/region	Value of agricultural output per hectare (Rs.)		
	1962–65	1970–73	1970–73 (–) 1962–65 (Difference)
12. Tamil Nadu			
Coastal Northern	1,588.4	2,030.2	441.8
Coastal Southern	1,488.4	1,821.5	333.1
Inland	1,383.1	1,564.0	180.9
13. Uttar Pradesh			
Himalayan	880.2	1,034.7	154.5
Western	1,115.5	1,344.9	229.4
Central	897.2	1,014.9	117.7
Eastern	858.7	928.3	69.6
Southern	596.9	721.9	125.0
14. West Bengal			
Himalayan	1,264.9	1,320.8	55.9
Eastern Plains	1,231.8	1,371.0	139.2
Central Plains	1,532.5	1,608.3	75.8
West Plains	1,314.2	1,433.0	118.8

* Source: Ranade, C.G. 1980.

geo-climatic regions of states in the early and late 1960s (Ranade 1980). Practically all states had increases in value productivity of land, though the extent of increase in value varied substantially among states and also among regions within given states. The states leading in the extent of increase were Punjab and Haryana in the north and Karnataka and Tamil Nadu in the south. These are also the states which were early adopters of the new technology. The differences in average value productivity, however, include the effects of changes in price level. Nonetheless, substantial differences would remain even at constant prices.

The relative position of agricultural income as against non-farm income of the states is indicated in table 4.9. It appears that the share of the primary sector (agriculture, animal husbandry and mining) declined in most of the states (all except for Maharashtra) over the seventies. The rate of growth of SDP during the 1970-71 to 1977-78 ranged from 1.1 in Rajasthan to 5.3 in Maharashtra, all states recording a growth rate above 1 per cent per year. The table also shows that barring Kerala and Rajasthan, the states had a low to moderate increase in income per capita during the 1970s. Kerala had a decline in agricultural output accompanied by a slow increase in overall SDP while agricultural production in Rajasthan increased slightly with a low overall SDP that fell below the growth in population. The per capita income growth of states again points out the leading position of Punjab, Haryana and Maharashtra, while Tamil Nadu and Karnataka slid down to lower levels.

The argument of increase in agricultural income at the state level is also supported by micro-level evidence of farm business income. Looking at an illustrative set of figures for farm business income, computed on the basis of Farm Management Survey data, it appears that income per farm increased substantially in all the six districts of West Bengal during 1968-78 (Mukhopadhyay 1983). Data for Muzaffarnagar (Uttar Pradesh), Ferozepur (Punjab) and Coimbatore (Tamil Nadu) also suggest large increases in both farm business income per farm and income per

Table 4.9
Growth and pattern of agricultural income in states of India*

State	State share in national domestic product at current prices		Share of primary sector in state domestic product (SDP)		Average annual rate of growth of SDP	Average annual growth of SDP per capita
	1970–71 to 1972–73	1975–76 to 1977–78	1970–71 to 1972–73	1975–76 to 1977–78	1970–71 to 1977–78	1970–71 to 1977–78
Andhra Pradesh	7.3	6.7	52.1	47.2	3.1	0.8
Bihar	6.6	6.4	62.4	58.1	2.9	1.0
Gujarat	5.8	5.9	44.2	38.8	3.7	1.4
Haryana	2.5	2.4	56.0	52.7	4.8	2.1
Karnataka	5.5	5.0	58.0	55.3	3.2	1.2
Kerala	3.3	3.3	53.8	49.0	2.0	–1.4
Madhya Pradesh	6.0	5.7	60.6	55.6	3.0	0.4
Maharashtra	11.5	12.4	24.9	28.6	5.3	3.1
Orissa	3.0	2.6	68.4	65.7	3.0	0.9
Punjab	4.0	3.9	59.5	57.1	4.6	2.7
Rajasthan	4.1	3.9	59.3	57.1	1.1	–1.5
Tamil Nadu	7.2	6.1	41.0	37.3	3.0	1.4
Uttar Pradesh	12.6	11.6	58.7	57.3	2.4	0.6
West Bengal	9.2	8.7	44.3	45.1	4.4	1.0

* Source: <u>Reserve Bank of India Bulletin</u>, Sept. 1961; and <u>Indian Journal of Agricultural Economics</u>, April–June, 1980.

work day for farm family labour (Vaidyanathan 1978).

Technology and employment

The implication of the new technology for employment
is a complex question and no unique answer has been
obtained on this issue so far. The reason for this
complexity lies in the inter-related effects of the
components of the technology packages as well as the
physical and institutional factors governing the
environment where the technology is adopted. Effects
of the HYV seed, fertiliser, and irrigation are
invariably intermingled with those of cropping
intensity, cropping pattern, farm size, land tenure,
etc. However, evidence is available to suggest that
regions with a higher rate of adoption of the new
technology also had a higher rate of labour
absorption (Bardhan 1978, Vaidyanathan 1978).
  With the help of farm level data available from the
farm management surveys in West Bengal, 1972-73, an
attempt has been made here to examine the effect of
technology upon labour demand. The hypothesis is
that total labour days used are dependent upon the
amount of fertiliser, irrigation, machinery and the
scale of operation. Normalised labour demand
functions of the Yotopoulos-Lau type (Yotopoulos and
Lau 1979) have been estimated for aman (local) rice,
boro (HYV) rice and wheat (HYV) for the districts of
West Bengal. The results (table 4.10) show that the
four factors explain significant variations in the
total labour days used at the farm level. For boro
rice labour demand seems to be positively affected by
the level of output and the amount of fertiliser
used, whereas machinery and irrigation seem to have a
negative but statistically insignificant effect on
labour demand. For wheat, irrigation continues to
have a negative effect whereas scale of output,
fertiliser and machinery seem to generate a positive
demand for labour. It thus seems, in view of the
input demand function estimated here, that the HYV
rice and wheat have been by and large employment
generating. The effect would of course be
strengthened if the right type of
irrigation-water-control and crop planning could be

introduced. Mechanisation, to the extent it has been introduced, might have suppressed aggregate employment generation in the farm sector (Binswanger 1978, Mukhopadhyay 1982).

In view of the absence of consistent data on the regional distribution of farm employment it is difficult to judge the impact of the new technology on employment in the different parts of the country. The Census of India provides data on workers in agriculture that consist of cultivators and agricultural labourers. Comparability over time, however, is marred by significant changes in the definitions of worker from census to census. For example, the 1971 Census defined a person as worker depending upon his/her <u>main</u> occupation, whereas in 1981 both <u>main and marginal occupations</u> were used to define a worker. If it can be assumed that the proportion of marginal to main workers does not vary too widely among states, the cross-sectional data on the proportion of each state in the total agricultural workers of the country could be compared between 1971 and 1981 to see whether there was any significant change in the inter-state distribution of agricultural workforce. This would, then, suggest whether the impact of the new technology could have introduced regional imbalances in employment generation in the states. It seems from table 4.11 that on the basis of census data the distribution of agricultural workers among states remained remarkably similar in the two census years, 1971 and 1981, signifying very little change in the regional pattern of agricultural employment generation. It should, of course, be noted that the data are subject to limitations of concept, definition and measurement. These do not reflect the intensity of employment and likely changes in the proportion of the partly employed or the disguidedly unemployed.

Conclusion

It would seem from the evidence presented above that, thanks to the changing technology, agricultural income in India showed a significant rise during the sixties followed by a slower rate of growth during

Table 4.10:
Cross-sectional normalised labour demand functions for districts on farm data
(1973–74)
Dependent variable: Labour days used[a]
Coefficient of multiple regression estimates[b]

| Crop | Model | Intercept | Independent variables[c] | | | | R^2 | F. Ratio | D.W. statistic |
			Output level	Machinery	Fertiliser	Irrigation			
Aman rice (N = 145)	I	2.58	0.50 (0.06	−0.02 (0.0	0.20 (0.04)	0.04 (0.03)	0.78	123.83	1.18
	II[d]	2.17	–	0.16 (0.03)	0.33 (0.04)	0.13 (0.03)	0.67	94–97	1.24
Goro rice (N = 97)	I	2.31	0.66 (0.07)	−0.02 (0.03)	0.25 (0.06)	−0.002 (0.02)	0.87	154.60	1.39
	II[d]	1.04	–	0.11 (0.04)	0.63 (0.06)	–	0.74	134.15	1.78
Wheat (N = 127)	I	2.24	0.66 (0.07)	0.09 (0.04)	0.09 (0.05)	−0.05 (0.03)	0.79	118.18	1.31
	II[d]	0.76	–	0.23 (0.05)	0.38 (0.06)	0.04 (0.03)	0.65	76.18	1.29

[a] Source: Mukhopadhyay, K. Suchin, 1982.
[b] Figures in parentheses denote standard errors.
[c] Normalised by the price of output.
[d] Model II is estimated excluding output level as an independent variable.

Table 4.11
Regional pattern of agricultural workers*

State	Share of state as percentage of total agricultural workers	
	1971	1981
Andhra Pradesh	10.0	10.67
Bihar	11.4	11.18
Gujarat	4.4	4.54
Haryana	1.4	1.50
Karnataka	5.4	6.05
Kerala	2.4	1.90
Madhya Pradesh	9.6	10.43
Maharashtra	9.5	10.19
Orissa	4.2	4.39
Punjab	1.9	1.95
Rajasthan	4.7	4.90
Tamil Nadu	7.2	7.84
Uttar Pradesh	16.8	16.36
West Bengal	5.7	5.86

* Source: Census of India, 1971 and 1981.

the seventies. Agricultural income at the state as well as at the farm level has shown this increasing trend. However, the growth in income has not been uniform across the states. A similar discrepancy exists also in the growth rate of overall state domestic product and state domestic product per capita.

On the employment front, although the new technology seems to have encouraged greater labour absorption in the country, not much statistical information is available to attempt a measurement of the likely increases in such employment. The regional pattern of farm employment does not seem to have undergone any substantial change. The effects of the new agricultural technology in India thus appear to have been relatively more pronounced in the creation of output and income than in employment, although micro level evidence does exist to suggest the employment potential of this technology if properly adopted.

RETURNS TO INVESTMENT IN AGRICULTURAL RESEARCH

In the preceding sections the Indian agricultural research system was reviewed in terms of its organisational structure and evidence was presented on recent changes in agricultural income and employment across the country. These are broad and suggestive approaches to an overall assessment of agricultural research in the country. For more precise and direct evaluation of investment in agricultural research it would be necessary to explore the quantitative magnitude and time path of physical and economic returns to such investment. Specifications of methods and actual estimation of returns are complex tasks, both conceptually and from the point of view of availability of appropriate data. The task is made more difficult by the intractable nature of much of the year to year fluctuations in the levels of production in the agricultural sector in the developing countries. In such settings it would be hard to capture and quantify returns to research as a factor distinguished from the numerous others including weather, that have their impact upon realised production in agriculture.

Reviews of past studies

Nonetheless, studies have been undertaken in many countries including India to estimate returns to investment in agricultural research. These studies in India have used different models and techniques and come out with varying estimates of returns (Saxena 1979). They were all aimed at estimating returns in terms of aggregate agricultural production and not of specific crops. Some of them used time-series data, some used cross-sectional, while some others pooled time-series and cross-sectional data. Evenson and Jha (1973) used a production function model with research and extension as input variables and an index of total factor productivity as the dependent variable. Kahlon et al. (1977) also used a production function model where the dependent variable was agricultural output and the explanatory variables included measured inputs, and proxies for the Green Revolution, and the contribution of research was estimated from the difference in growth rates of output of the pre- and post-Green Revolution periods. Using further modifications in their methodology, based primarily upon the production function approach, Bal and Kahlon (1978) estimated the rates of return to investment in agricultural research in India before and after the Green Revolution. Kahlon (1977) also estimated the contribution of research in the different states of India by the benefit-cost method where the difference in output between two periods of time was related to differences in research investments to calculate the rates of return. The rates of return estimated in these studies are given in tables 4.12 and 4.13.

Three outstanding features of the above estimates of the rates of return to investment in agricultural research are: i) the rates are quite high by any standard of measurement; ii) rates vary widely between time periods; and iii) the productivity of agricultural research shows a high degree of

Table 4.12
Comparative study of returns to investment in Indian agricultural research using different approaches

Study	Period	Returns to 1 rupee research expenditure (Rs.)	Time lag (years)	Internal rate of return
Evenson–Dha	1953–54 to 1970–71	10.88	8	50.00
Kahlon et al.	1960–61 to 1972–73	11.61	5	63.30
Bal and Kahlon	1960–61 to 1964–65	1.91	5	14.00
	1967–68 to 1972–73	14.91	5	71.70

Table 4.13
Statewise returns on investment in agricultural research and development and internal rates of return

State	Time lag (years)	Returns to 1 rupee invested (Rs.)	Internal rate of return (%)
Andhra Pradesh	5	30.29	97.8
Assam	5	8.99	55.2
Bihar	5	17.52	77.3
Gujarat	6	59.99	97.8
Himachal Pradesh	5	4.57	35.5
Dammu and Kashmir	5	12.98	67.0
Karnataka	5	43.12	112.2
Kerala	7	29.43	96.6
Madhya Pradesh	5	40.65	109.8
Maharashtra	5	10.25	59.2
Orissa	5	34.35	102.9
Punjab	5	26.33	92.3
Rajasthan	5	13.32	67.8
Tamil Nadu	5	17.49	77.2
Uttar Pradesh	6	66.44	101.2
West Bengal	5	4.80	36.9

interstate variation. The first characteristic of a high rate of return is almost universally recognised as symptomatic of a chronic under-investment in agricultural research, relatively and absolutely. This fact is often placed before planners for development policy for consideration in the allocation of resources. However, the wide variations in the rates from time to time and from region to region are also an important factor to reckon with in matters of research resource allocation. A degree of caution would, therefore, be necessary in the interpretation of rates of return in view of their sensitivity to the time period and the place of occurrence selected for the study. For example, a simplistic interpretation of the state rates of return would suggest that Punjab has an under-investment in agricultural research relative to West Bengal or Assam, which would be hard to justify in view of the reality. Similarly, the contrast between the pre- and post-Green Revolution periods can hardly be produced as evidence of relative over investment in research in the earlier period. The two periods refer to distinctly different technologies. Returns would depend upon many things including the volume of investment, the crop-wise, factor-wise, region-wise pattern of investment, and the entire environment within which the investment takes place and matures into growth and technological change.

A new estimate

Fully aware of limitations of such exercises, however, it must be recognised that continuous efforts must be made empirically to specify and measure rates of return as accurately as possible. In line with this, a modest attempt has been made here to estimate an agricultural research production function using state-level data. It is assumed here that expenditure on agricultural research and education can be treated as an input that leads to increase in yield per hectare. In a Cobb-Douglas framework, the coefficient of this variable of expenditure on agricultural research and education can be used to get the percentage return in yield to unit (per cent) increase in research expenditure.

The production function has been estimated using average yield per hectare of all cereals (1978-79) as the dependent variable and fertiliser per hectare (1978-79) and proportion of area irrigated to total area under cereals for the same year, apart from expenditure on agricultural research and education (average of 1974-75 to 1978-79) as the independent variables. It may be seen that this specification of research variable implicitly assumes a cumulative effect of such research with each year weighted equally and with a time lag of four years. The data have been used from 14 states of India. The equation estimated by the ordinary least squares (OLS) method follows.

Although the model is simple, the explanatory power of the variables chosen is remarkable (table 4.14). What is also remarkable is the contribution of investment in research which stands out as the single most powerful variable. This estimation may possibly be taken as one more support to the hypothesis of the significant influence that agricultural research could exert upon growth in productivity in India.

ALLOCATION OF RESEARCH: AN ATTEMPTED EXPLANATION

Although the relationship between investment in agricultural research and technological innovation leading to sustained growth is by and large fraught with uncertainties and complexities, attempts have been made to trace this relationship and estimate empirical measures of returns to such investment in a large number of countries including India, as discussed in the preceding section. But what should apparently be an easier and less intractable question has received less attention so far. This relates to the determinants of investment in agricultural research. Recent extensions of the theory of induced innovation from the technological into the institutional field open the scope for considering this important question in formulating an integrated theory of agricultural development. As had been

Table 4.14
Research production for India

Dependent variable: Average yield of cereals (N = 14) per hectare

Independent variables:	Coefficient[a]
Intercept:	-0.90
Expenditure on agricultural research and education	0.22 (0.04)
Fertiliser per hectare	0.19 (0.05)
Proportion of area irrigated to total cropped area under cereals	0.20 (0.06)
R^2	0.94
F-Ratio (3,10)	48.14

[a] Figures in parentheses are standard errors of estimates.

indicated in the second section, the theory of induced institutional innovation seeks to explain the emergence and operation of institutions appropriate to given socio-cultural, economic and physical environment. Decisions to invest in agricultural research may be viewed in this theory as response to the overall physical-cultural institutional environment.

Model and estimates

Hypotheses may be presented to explain variations in allocation of research investment with the help of the milieu of factors – economic, socio-political and demographic. Successful research would be conditioned, to a large extent, by the sensitivity of the resource allocation mechanism to the signals of this milieu.

Having examined the development and structure of the agricultural research organisation in India, and having reviewed the income and employment generation mechanism in agriculture in the different regions of the country in the preceding sections, an attempt is made here to verify how far the allocation of investment in agricultural research in India can be explained with the help of selected institutional factors. Depending on the availability of data and relevance to the problem, 15 determining variables have been selected for the study. These variables can be classified into three sets, broadly defined as (a) economic, (b) socio-politico-institutional, and (c) demographic. Allocation of research fund, the dependent variable, has been measured in five ways depending upon the availability of data. The variables are listed below.

Possible implications and expected signs of coefficients of the independent variables are given in table 4.15.

All the signs suggested in table 4.15 refer to the signs of the partial relationship between the dependent and the respective independent variable. The economic variables except unemployment rate are expected to affect allocation of research investment positively. An economy characterised by high

Table 4.15

Independent variables, implications, and expected signs of coefficients

Variable	Implication	Expected signs of coefficients
x_1	Expression of relative demand-supply position of agricultural products	+
x_2	Lack of economic power of labour force	−
x_3	Economic condition of the State	+
x_4	Level of economic infrastructure	+
x_5	Importance of agricultural sector in the economy	+
x_6	Relative lack of socio-economic power of rural people	−
x_7	Scale of operation	+
x_8	Relative power of perception and articulation of the rural people	+
x_9	Inequality in distribution of farm area	−
x_{10}	Relative labour absorption in agriculture	+
x_{11}	Research awareness of policy makers	+
x_{12}	Population pressure on available resources	−
x_{13}	Lack of family planning, awareness and influence of traditional as against modern values	−
x_{14}	Lack of urbanisation and modernisation	−
x_{15}	Secular trend of population growth	−

unemployment, high rural poverty, and highly skewed distribution of farm size is likely to have relatively low investment allocation for agricultural research. A similar negative relationship is expected between research investment and the demographic variables of population density, family size and secular growth of population – the basic rationale behind this being the lack of enough resources left for investment in a time-consuming uncertain venture like agricultural research.

Estimates have been made of the model separately for the three sets of independent variables, because the number of observations for 14 states do not permit the inclusion of all the independent variables in a single equation. The three sets of equations for the five research variables are presented in tables 4.16–4.18. The results show that all the equations have F-ratios statistically significant at five per cent level, the R^2s showing explanatory power of the variables ranging from 50 to 90 per cent in the economic variables set, from 63 to 97 per cent in the socio-political-institutional set and from 50 to 87 per cent in the demographic set. The majority, though not all, of the individual coefficients conform to the expected signs and are statistically significant.

These estimates with respect to the effects of the critical environmental variables upon investment in research suggest that agricultural research would tend to vary directly with the terms of trade of the agricultural sector against industry, the economic well-being of the state, the economic infrastructure, the importance of agriculture in the economy, the power of perception, articulation and organisation of the rural population, the proportion of the labour force absorbed in agriculture, and the general research awareness of the policy makers. On the other hand, agricultural research would tend to be negatively related to the rate of unemployment and lack of economic power of the labour force, the socio-economic disparity between the rural and urban people, the inequality in income distribution, the pressure of population, the influence of traditional as against modern values, and the lack of

Table 4.16

Explanation of research allocation by economic factors (Coefficient of multiple regression estimates)*

Dependent variable	Intercept	Independent variable (economic)					R^2	F-Ratio (4,9)
		Ratio of price index (food and general)	Unemployment rate	State per capita income (constant prices)	Road length per million population	Agriculture as % of SDP		
Expenditure in agricultural research and education as % of total investment in agriculture (1977–78 to 1980–81)	−2.28	−3.10 (0.47)	−0.09 (0.21)	0.53 (0.32)	0.65 (0.36)	–	0.90	20.51
Expenditure on agricultural research and education per 1,000 population (1977–78 to 1980–81)	−11.07	4.48 (0.73)	−0.25 (0.33)	0.91 (0.50)	1.00 (0.55)	–	0.82	10.46
Expenditure on agricultural research and education per hectare of GCA (1977–78 to 1980–81)	−10.14	1.22 (0.60)	0.55 (0.29)	0.92 (0.41)	0.44 (0.46)	0.30 (0.63)	0.79	5.97
Expenditure on agricultural research and education as % of state agricultural production (1977–78 to 1980–81)	1.68	−0.97 (0.68)	−0.19 (0.33)	−0.14 (0.46)	0.90 (0.53)	−0.85 (0.72)	0.49	1.52
ICAR grant to states (5th & 6th Plan: 1974–75 to 1982–83) per hectare	−9.62	−0.88 (0.43)	0.92 (0.21)	0.16 (0.29)	−0.14 (0.33)	1.34 (0.45)	0.77	5.52

* Figures in parentheses are standard errors of estimates.

Table 4.17
Explanation of research allocation by socio-political factors
(Coefficients of multiple regression estimates)*

Dependent Variable	Intercept	Independent variables						R^2	F-Ratio (6,7)
		Rural-urban poverty ratio	Average farm size	Rural-urban literacy ratio	Gini-co-effi-cient of operated area	Percentage of agricul-tural to total workers	Total research and education expenditure as per cent of tot-al plan expen-diture		
Expenditure on agricul-tural research & educa-tion as per cent of total investment in agriculture (1977–78 to 1980–81)	7.56	-0.73 (0.22)	0.20 (0.12)	0.71 (0.43)	-1.22 (0.56)	-1.29 (0.59)	-0.07 (0.22)	0.96	32.22
Expenditure on agricul-tural research & educa-tion per 1,000 population (1977–78 to 1980–81)	16.38	-1.05 (0.33)	0.29 (0.18)	-0.75 (0.63)	-1.76 (0.84)	-5.08 (0.87)	-0.39 (0.33)	0.94	19.17
Expenditure on agricul-tural research & educa-tion per hectare of GCA (1977–78 to 1980–81)	10.20	-0.69 (0.34)	0.17 (0.19)	0.92 (0.65)	-1.10 (0.86)	-2.83 (0.89)	-0.40 (0.34)	0.88	8.50
Expenditure on agricul-tural research & educa-tion as per cent of State agricultural production (1977–78 to 1980–81)	7.80	-0.40 (0.44)	0.33 (0.24)	0.47 (0.83)	-0.86 (1.11)	-2.06 (1.14)	-0.12 (0.43)	0.63	1.98
ICAR grant to states (5th & 6th Plan: 1974–75 to 1983–83) per ha.	-1.25	-0.81 (0.27)	-0.56 (0.15)	0.43 (0.52)	-0.17 (0.69)	-0.45 (0.71)	-0.34 (0.27)	0.84	6.04

* Figures in parentheses are standard errors of estimates.

Table 4.18
Explanation of research allocation by demographic factors
(Coefficients of multiple regression estimates)*

Dependent variable	Intercept	Independent variable				R^2	F-Ratio (d.f.)
		Density of population	Average size of family	Per cent of rural to total population	Population growth rate		
Expenditure on agricultural research and education as per cent of total investment in agriculture	12.54	0.12 (0.21)	1.38 (0.43)	-2.73 (1.45)	-0.21 (0.49)	0.86	14.24 (4,9)
Expenditure on agricultural research and education per 1,000 of population	11.45	-	-0.94 (0.43)	-2.69 (2.45)	-	0.64	9.71 (2,11)
Expenditure on agriculture research and education per ha. of GCA	4.87	0.76 (0.28)	-0.08 (0.45)	-2.25 (1.91)	-	0.60	4.98 (3,10)
Expenditure on agricultural research and education as per cent of state agricultural production	8.00	-0.11 (0.24)	1.20 (0.50)	-2.80 (1.68)	0.48 (0.57)	0.49	2.18 (4,9)
ICAR grant to states (1974-75 to 1982-83) per ha.	15.20	0.66 (0.21)	-0.35 (0.43)	1.55 (1.45)	0.48 (0.49)	0.58	3.09 (4,9)

* Figures in parentheses are standard errors of estimates.

urbanisation.

The analysis thus supports the hypothesis that the regional allocation of research investment in India can be explained to a large extent with the help of the major economic, institutional and demographic variables selected here.

Principal components model

The above econometric estimates, however, suffer from a major limitation. As is obvious, the number of observations being small, it is not possible to include all the variables in one model although all of them may be quite important. In order to deal with this problem a new set of independent variables has been constructed using the method of Principal Components Analysis (Rao 1965). It uses linear combinations of original independent variables to construct the new set of principal component scores and then proceeds to estimate the regression equation choosing only those of the scores that have the highest degree of explanatory power on the dependent variables in a step-wise analysis. Besides, this model has the additional merit of incorporating properties of all the independent variables in the newly constructed principal components so that none of the variables are really ignored in the ultimate analysis. Using this method and restricting to the choice of two principal components from each of the three sets (economic, institutional and demographic) equations have been estimated for the five research variables.

The results of the Principal Components Model using combination of the three sets of variables, presented in table 4.19 along with the results of the equations using individual variable sets, suggest that a significant percentage of variation in agricultural research allocation can be explained with the help of economic, socio-political and demographic factors. It may be noted that the explanatory power of the model could be increased further had there been a larger number of observations available. The estimates thus seem to provide a useful support for the hypothesis that investment in agricultural

Table 4.19
Results of different models of research determinant functions

Dependent variable	Economic variables		Socio-political insti-tutional variables		Demographic variables		Principal components analysis	
	R^2	F (d.f.)	R^2	F (d.f.)	R^2	F (d.f.)	R^2	F (d.f.)
Expenditure on agricultural research and education as per cent of total investment in agriculture	0.90	20.51 (4,9)	0.96	32.22 (6,7)	0.86	14.26 (4,9)	0.76	3.64 (6,7)
Expenditure on agricultural research and education per 1,000 population	0.82	10.46 (4,9)	0.94	19.17 (6,7)	0.64	9.71 (2,11)	0.81	4.92 (6,7)
Expenditure on agricultural research and education per ha. of GCA	0.79	5.97 (4,9)	0.88	8.50 (6,7)	0.60	4.98 (3,10)	0.86	7.04 (6,7)
Expenditure on agricultural research and education as per cent of state agricul-tural production	0.49	1.52 (4,9)	0.63	1.98 (6,7)	0.49	2.18 (4,9)	0.59	1.72 (6,7)
ICAR grant to states per ha. (1974–75 to 1982–83)	0.77	5.52 (4,9)	0.84	6.03 (6,7)	0.58	3.09 (4,9)	0.90	11.24 (6,7)

research can be viewed as an instrument that might bring about technological innovations, but that would act in response to a host of factors that include meaningful economic, institutional and demographic elements. Successful development policy has to take cognisance of this linkage.

CONCLUSIONS

In an integrated framework of sustained agricultural development, agricultural research may be viewed as providing a link between the socio-political and physical resource endowment on the one hand and technological change on the other. Agricultural research is generated and influenced by the basic cultural-physical environment and perceived needs of the system while it affects the occurrence and course of technological innovation responsible for the dynamic transformation of the economy. Institutions appropriate for given structural conditions emerge and tend to interact with technology in order to bring about meaningful change and growth.

In this chapter an attempt has been made to construct, with the help of some very broad elements, a case study for the Indian agricultural research system, which can hopefully be used as a test for the hypothesis indicated above. It has been suggested that India, which has a dominant position in the developing world in terms of manpower and physical resources, displays rates of technological change and productivity of agriculture that are low both relatively and absolutely. The study has sought to explore this question specifically in the context of the Indian agricultural research organisation and performance. Recounting the findings backwards from the end, it has been shown here that the mechanism of allocation of research resources in India bears the mark of the economic, socio-political and demographic parameters of the system. For instance, this implies that the agriculture-industry terms of trade, the level of unemployment, the level of per capita income and other measures of general economic development, the relative intensity of rural poverty, literacy and

inequality and the relative strength of perception,
mobilisation and articulation of the farm sector, and
also the demographic-historical characteristics of
the rural people would affect the pattern of
investment in agricultural research. Ex post
estimates suggest that investment in agricultural
research has consistently yielded high, though
regionally and temporally divergent rates of return
in terms of agricultural production and yield. These
rates as well as evidence on growth in income and
employment, tend to indicate that it is not only the
volume of investment in research as such that is
important, but the pattern, distribution and
consistency of this investment along with the
efficiency of the organisation of research that are
important determinants of the performance of research
in bringing about growth. There are stresses that
emerge in the process of research administration and
there are complementarities that have to be reckoned
with. In a big country like India such problems
would arise almost inevitably and much of the slack
in research performance might possibly be traced to
these conflicts. However, India has a large
agricultural research organisation with a network of
decentralised units and a vast army of agricultural
scientists. These positive elements have to be
organised in response to the requirements of
technological change and agricultural growth.

The lesson that seems to emerge from this study for
Indian agricultural research policy is that the
persisting sources of conflict should be dealt with
in accordance with the signals of the
cultural-physical system - intrinsic as well as
explicit. Otherwise, even high volumes of investment
in research may not succeed in leading to sustained
growth. The gains in technology, income and
employment are to be translated through the demand
for growth into appropriate policies for the right
volume and pattern of investment in research, which,
in its turn, can lead to generation and diffusion of
technology and to increases in income and
employment. This is a prerequisite for sustained
growth in agriculture.

5 Institutional factors and technological innovations in Bangladesh

WAHIDUDDIN MAHMUD and MUHAMMED MUQTADA

INTRODUCTION

The setting

The economy of Bangladesh provides a 'test case' for development under extreme conditions of poverty, population pressure and landlessness. The country's agrarian base, its reliance on massive food aid, the overwhelming proportion of its rural population, and the fact that so many live near subsistence level – all these demand that development efforts must emphasise growth of agriculture and food production. In the past, food production lagged behind population growth resulting in growing poverty and malnutrition and ever-increasing food deficits. The technological possibilities in food grain production potentially provide some escape from this Malthusian impasse. But the adoption of the new HYV technology in Bangladesh's predominantly subsistence type agriculture requires far-reaching institutional and organisational reforms that would not be easy to bring about.

During the last two decades or so, much has been

written about rural development in general and the amelioration of rural poverty in particular. In reality, however, per capita real income in rural areas has registered a decline and the vast majority of the rural poor have been caught in the grip of increasing impoverishment.[1] Cumulative poverty, landlessness and increasing unemployment are the inevitable outcome of an increasing pressure of population on land along with semi-stagnant productivity being sustained upon a primitive agricultural technology. Agricultural output registered a weak growth rate during the 1950s, although it showed considerable dynamism during the 1960s, when efforts at planned developments began in earnest. Production in agriculture fell drastically during the early part of the 1970s, ostensibly owing to the various dislocations caused by the war of independence. Although there has been a steady increase in agricultural output since the mid-1970s, the annual growth of agriculture over the decade of the 1970s (when estimated from the benchmark year of 1969-70) has lagged far behind the annual growth rate of rural population.[2]

Along with a sluggish rate of growth of agricultural production, there has been a clear trend towards increasing concentration of landownership, and the consequent pauperisation of the peasantry. Between the two Agricultural Census years of 1961 and 1977, the proportion of rural households operating agricultural farms is estimated to have declined from about 70 per cent to about 50 per cent.[3] Most of the households pushed out of their family farms either joined the rank of landless labour households or migrated to urban areas in search of some meagre means of living. According to the Population Census data, the proportion of landless agricultural labourers out of total agricultural labour force increased from 17 per cent in 1961 to 25 per cent in 1974. It is important for our subsequent analysis to recognise that the process of diffusion of the new HYV technology has taken place in this milieu of ever-increasing landlessness and concentration of landholdings.

Although endowed with a generally fertile land,

Bangladesh farming suffers from two important and pervasive constraints that have a bearing on the adoption of agricultural innovations. First is the risk-prone nature of Bangladesh agricultural environment (e.g. unavoidable weather uncertainty connected with droughts and cyclones, and the increasing, and partially avoidable, susceptibility to floods) which gives rise to complex cropping patterns, essentially to hedge against risk. Second is the small size of farm holdings with excessive fragmentation, which renders lumpy investment in agriculture uneconomical for individual farmers.

Although HYVs represent divisible innovations, their adoption is dependent on (or greatly enhanced by) complementary indivisible investment (such as tube-wells). The full utilisation of a deep tube-well or a power pump of one cusec capacity requires about 35 acres of land. The full utilisation of even a shallow tube-well requires 10 acres. According to the 1977 Agricultural Census data, only 5.2 per cent of farm households (controlling about 22 per cent of cultivated land) have holdings larger than 10 acres. A more binding constraint is the excessive fragmentation of landholdings into tiny plots usually scattered over a wide area. According to the same census data, only 3.7 per cent of the farms of 10 acres of larger size were fragmented in less than four plots, while about 80 per cent were fragmented into 10 or more plots. As we shall elaborate later in this study, such fragmentation of landholdings has implications for institutional and organisational reforms directed towards widespread diffusion of agricultural innovations.

Sources of growth in rice output

The entire crop sector of agriculture in Bangladesh is dominated by one single crop, rice, which accounts for nearly 80 per cent of both the gross cropped area and value added in all crops. The growth of rice output, as expected, clearly follows a pattern similar to that of agricultural output as a whole, although annual fluctuations due to climatic conditions make it extremely difficult meaningfully to interpret the long-term trend. During the decade of the 1950s, as in the earlier decades for which data are available, rice area, yield and production in the geographical area now comprising Bangladesh remained virtually stagnant.[4] As a result, the region which at the turn of the present century was dubbed as the 'granary' of India gradually confronted a Malthusian apocalypse. Planned development has had its impact on the growth of rice output (along with the general growth of the economy) only since the early 1960s.

Table 5.1 shows the three-year averages of rice area, production and the implied average yields. Besides the end-periods of the three successive decades, we have also shown the figures for the mid-1960s, since this period can be considered as the vintage point for the introduction of HYV rice in Bangladesh. It appears from the table that the annual rate of growth of rice output was the highest during the period from the end of the fifties to the mid-1960s (about five per cent annually), but dec-lined to about three per cent during the latter half of the 1960s and to only one per cent during the 1970s.

The high rate of growth of output in the earlier period (i.e. early 1960s) was achieved to a large extent through increased yields from traditional varieties only, since there were no HYVs. Falcon and Gotsch (1968), in their analysis of the increase in rice output in the early 1960s, attributed the yield increase to such factors as the higher input of fertilisers, pesticides and water, improved cultural practices, and the use of improved local seed varieties. However, the optimistic forecasts of the growth of rice production, based on such yield increases, were not borne out during the late 1960s, as the increase in the yields of the traditional varieties became increasingly difficult to achieve.

The increase in rice output during the latter half of the 1960s was achieved through an accelerated rate of increase in rice acreage. This was made possible by bringing more land under irrigation which allowed more multiple-cropping (table 5.2). Yet the growth

Table 5.1
Acreage, production and yield of local
and HYV rice in Bangladesh*

	Three-year Averages			
	1957–58 to 1959–60	1962–63 to 1964–65	1967–68 to 1969–70	1977–78 to 1979–80
Area ('000 acres)				
Local rice	20,343	22,183	24,271	21,189
HYV rice	nil	nil	424	3,768
Total	20,343	22,183	24,695	24,958
		(1.7%)	(2.2%)	(0.1%)
Production ('000 long tons)				
Local rice	7,667	9,841	10,745	9,142
HYV rice	nil	nil	616	3,506
Total	7,667	9,841	11,360	12,649
		(5.1%)	(2.9%)	(1.1%)
Yield (lbs per acre)				
Local rice	844	994	992	967
HYV rice	nil	nil	3,254	2,084
All rice	844	994	1,030	1,135
		(3.3%)	(0.6%)	(1.0%)

* Figures in parentheses for total rice area and production and average yield of all rice show the implied annual percentage change over the preceding periods. All estimates are based on official rice production statistics as reported by BBS.

Table 5.2
Trend in consumption of modern inputs
in Bangladesh agriculture*

Period	Chemical fertilisers (nutrient equivalent; lbs per cropped acre)	Irrigated area as per cent of cropped area	HYV rice area as per cent of total rice area
1950–51	nil	1.5	nil
1955–56	0.2	2.1	nil
1960–61	1.8	2.8	nil
1965–66	4.1	4.8	nil
1970–71	10.2	9.1	4.6
1978–79	23.9	11.7	13.6
1979–80	28.0	13.4	19.7

* Source: Bangladesh Bureau of Statistics (1981).

rate of rice output faltered, since the yields of local rice varieties remained stagnant, while the extent of the shift of rice area to HYVs was not large enough to have any significant impact on the overall yield rate.[5]

The low growth rate of rice output between the end of the 1960s and the end of the 1970s (only 1.1 per cent annually) conceals the sharp decline in output in the years immediately following the war of independence. There are also valid reasons to doubt the reliability and the consistency of the official series of foodgrain production statistics.[6] Nevertheless, there is no denying the fact that the diffusion of the HYV technology has not occurred at a sufficiently rapid rate to offset the decline of the rate of growth in area harvested. If we analyse the factors behind the sluggish growth of rice output in the 1970s (compared to the 'benchmark' output of the end-1970s), we come up with the following conclusions: (i) while the gross cropped area under rice continuously expanded during the 1960s, such expansion did not take place in the 1970s. This also reflects the trend in the expansion of total gross area under all crops, since there was no significant shift of acreage between rice and other crops during the entire period under review (see table 5.3); (ii) whatever growth of output took place in the 1970s was due to a shift of rice area to HYVs which accounted for nearly 15 per cent of gross rice area in the late 1970s compared to less than 2 per cent in the late 1960s; (iii) the favourable impact of the shift to HYVs on the average rice yield was to a large extent eroded by the fall in the average yield levels of both the local and the high-yielding varities of rice (see table 5.1).

The analysis draws our attention to certain disconcerting phenomena. While in the 1960s, rapid expansion of irrigated area significantly increased the cropping intensity through multiple-cropping, in the 1970s, there was little increase in the cropping intensity (and therefore in gross cropped area) despite modest expansion of irrigation (see table 5.3). Thus, although irrigation helps in bringing more cropped area under HYVs, the scope for

Table 5.3
Trend in cropped area and rice area
in Bangladesh agriculture*

	Three-year average		
	1957–58 to 1959–60	1967–68 to 1969–70	1977–78 to 1979–80
Net cropped area ('000 acres)	20,253	21,028	20,883
Gross cropped area ('000 acres)	25,883	31,805	31,671
Cropping intensity (%)	128	147	152
Rice area as per cent of gross cropped area	78.6	77.6	78.8
HYV rice area as per cent of gross rice area	nil	1.7	15.1

* Source: Bangladesh Bureau of Statistics (1981).

increasing gross cropped area through irrigation appears to have already been exhausted. The growth of rice output in future has therefore to come almost entirely from an increase in the average rice yield per acre. The increase in the average rice yield would, however, depend not only on the extent of shift of rice area to HYVs, but also on changes in the yield growth rates of different rice crops – both local varieties and HYVs. We have already noted that, while in the early 1960s substantial increases in the yield rates of local varieties could be achieved through more intensive use of inputs, the trend in the yield rates has been in fact downward since the introduction of the HYVs. The stagnation (or decline) in the yield rates of local varieties is to some extent explained by the fact that better quality land has been drawn away from local varieties to HYVs (and that the use of modern agricultural inputs has been mainly on the latter). But, as we shall argue later, government policy is also to be held responsible to the extent that research and extension activities were directed almost entirely to HYVs. However, the more disconcerting phenomenon is the low average yield rate of HYVs realised in the late 1970s (about 38 maunds of paddy per acre) which falls far short of the 'standard' of 50 maunds per acre, set by the experimental stations and extension agencies (see table 5.4).

It may also be noted that, although acreage under HYV rice in the Boro and Aus seasons has been expanding lately, the importance of the different seasonal rice crops in terms of their relative weights in total rice area and production has little changed over the last decade. Still now, nearly 50 per cent of rice acreage and production is accounted for by local Aman rice. This implies that the new varieties have not quite made a dent in the existing crop-mix, but have simply retained and accentuated the existing seasonal structure of crop production. Since the peak season labour scarcity is a familiar phenomenon, such accentuation of the pattern of seasonal cropping activities may not encourage a rapid acceptance of HYVs, especially by large farmers. Lastly, one needs to re-emphasise that less

Table 5.4
Imported and locally adapted HYV rice varieties in Bangladesh

Crop variety	Date of release by BRRI	Crop season	Life cycle (days)	Yield[b] (maund[c]/acre/ paddy)	Characteristics
IR-8	1967	Boro/Aus	160–145	60–70	Dwarf IRRI varieties, susceptible to diseases and pests; can tolerate drought
IR-5	1969	T. Aman[d]	135	50–60	in which case life cycle increases.
Purbachi	1969	Boro	140	55	Imported from China, popular for its early maturity and better rice quality than IR-8; susceptible to diseases and pests.
IR-20	1969	T. Aman[d]	130	50	A better IRRI variety compared to IR-5 with higher disease and pest resistance, but less tolerant to drought.
BR-1 (Chandina)	1970	Boro/Aus T. Aman[d]	160–120 110	50–60	Developed from hybrid materials from IRRI; shorter life cycle, better grain quality and disease resistance compared to IRRI
BR-2 (Mala)	1971	Boro/Aus	140–120	40	varieties. But BR-1 has shorter plant height and poorer drought tolerance. BR-2 has taller seedlings suitable for shallowly flooded Aus field, but unpopular for its genetic instability and higher grain sterility.
BR-3 (Biplob)	1973	Boro Aus/T.Aman[d]	160 140–120	50	Typical IR-8 plant type but with wider disease resistance and much superior grain quality; popular as a T. Aman crop.
BR-4 (Barrisail)	1975	T. Aman[d]	145	60	Out-yields other Aman varieties; tall seedlings, but low photosensitivity and longer life cycle.

[a] Source: Various BRRI reports.
[b] Yield refers to average yield in "controlled" experiments whereas SFTI reports lower yields under "ideal" farm conditions.
[c] 1 maund = 82 lbs; one maund of paddy (unhulled rice) yields approximately 0.67 maunds of rice.
[d] T. Aman = Transplanted Aman. Life cycle is the appropriate number of days from seed to maturity.

than one-fifth of the total cropped area under rice is held by the HYVs. This compares rather unfavourably to the experiences of most countries with the new seeds.

Adaptive research and technology packages

The slow rate of expansion of the HYV rice technology has thrown open a host of issues, mostly institutional, causing concern among planners, economists, agronomists and rice researchers. Since we discuss the global institutional linkages later in this study, here we shall provide a brief assessment of the development of rice research in Bangladesh and discuss some of the views held by the country's rice researchers regarding the sluggish spread of the HYVs.

Ruttan and Hayami (1971) distinguish three different phases of international technology transfer. First, the phase of material transfer characterised by the simple transfer of new materials (e.g. seeds, machines and associated techniques) without any orderly or systematic local adaptation. The process takes place primarily as a result of 'trial and error' by farmers. Second, the phase of design transfer during which exotic plant materials (in the form of certain designs, e.g. formula) are subject to orderly tests and screening. This phase corresponds to an early stage of publicly-supported agricultural research. Third, the final phase of capacity transfer, that is, the transfer of scientific knowledge and capacity to produce locally adaptable technology, following the 'prototype' technology which exists abroad. Although it is possible to trace these different phases in the transfer of HYV rice technology into Bangladesh, it is difficult to say for certain where we stand at the moment. A fair assessment would be that the evolution of the country's research system is nearly past the second stage, although the elements of the first stage (i.e. the 'trial and error' method by farmers) are still found to coexist with those of the later stages (see figure 5.1).

The HYVs were introduced in Bangladesh in the mid-1960s.[7] Before the founding of the Bangladesh

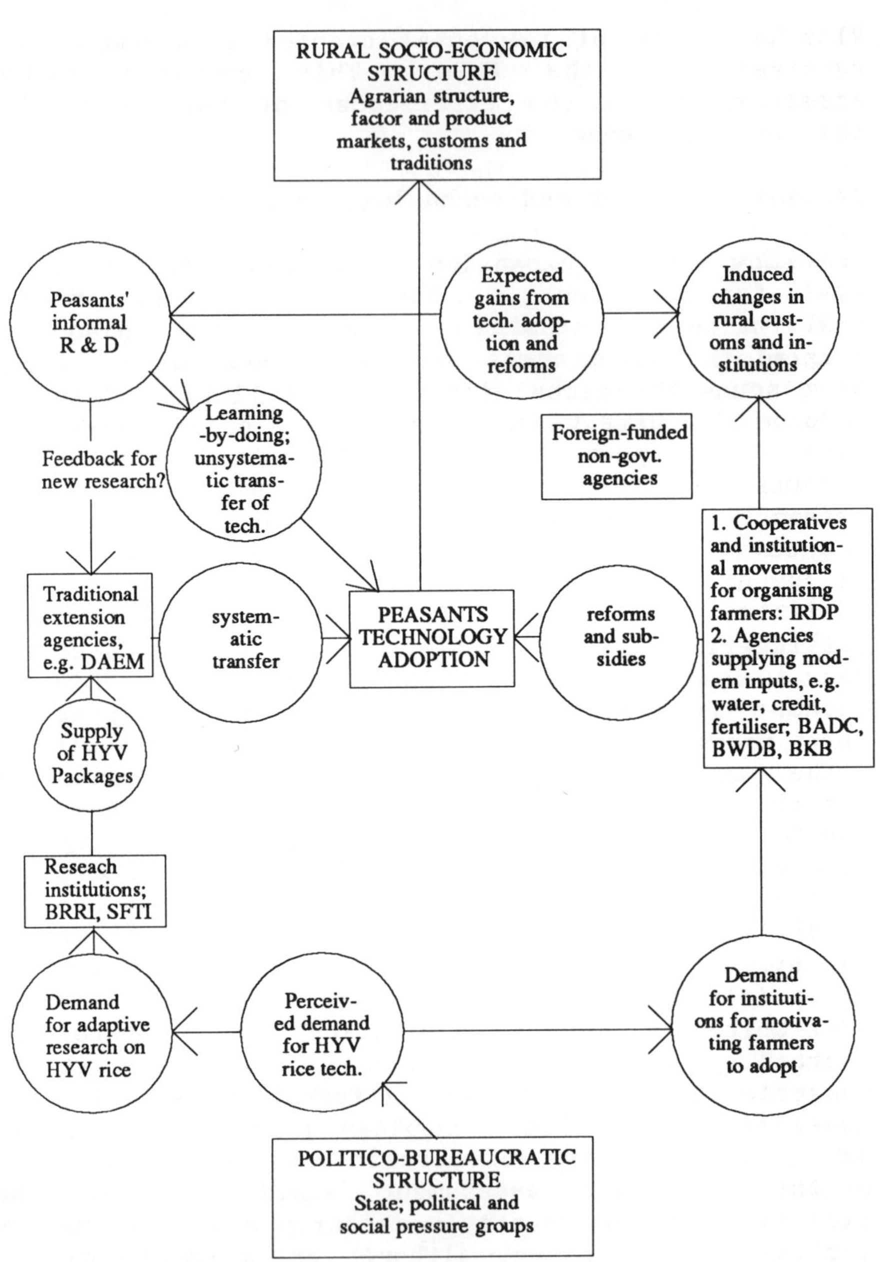

Figure 5.1: Institutional linkages in the generation and diffusion of HYV rice technology in Bangladesh

Rice Research Institute (BRRI) in 1970, the new seeds received 'institutional back-up' through extension agencies, experimental farms and co-operatives (e.g. the Comilla experiment by BARD). Thereafter, BRRI has held singular importance on the question of systematic transfer of the HYV rice technology. The more important of varieties which came into early prevalence were IR-8 (for Boro season), IR-5 and IR-20 (for Aman season). These are the short-strawed IRRI varieties, identified as most suitable for Bangladesh. Other varieties, which through BRRI have been somewhat tailored to suit various climatic and agronomic features of the country, have lately come into currency. Table 5.4 provides some basic information relating to the date of introduction, season(s) in which cultivated, the average yield and the maturity period, along with some characteristics of the plant types. There are also a few other varieties released or being tested by BRRI.[8] A particular variety, called Pajam (a Malaysian cross) was initially considered unsuitable by BRRI, but found a ready acceptance among farmers in different parts of the country.

The major reasons, as cited by the BRRI scientists, for the limited diffusion of the modern varieties are the following: (i) a simple borrowing of genes from a high-yielding plant need not ensure a predictable, constant high yield unless the yield potential is stabilised through a built-in mechanism of resistance to unfavourable climatic factors, pests, diseases, etc.; (ii) most of the varieties that have undergone BRRI tests are unsuitable for deepwater culture. The shorter seedlings of 6 to 8 inches in height become a constraint to transplanting in Aman fields during heavy rainfall in July and August. As a result, the new varieties are unsuitable for nearly 60 per cent of the total Aman transplant area. As for deep-water rice (e.g. Broadcast Aman), there is as yet no HYV variety with the capacity of stem elongation and ability to withstand two to three weeks under complete submergence; (iii) the new varieties, because of their weak photo-sensitivity, cannot perform well when transplanted in late August to September as the days become increasingly shorter followed by a cold spell in mid-November. Such a genetic make-up makes it difficult for the double-cropping of HYVs which is otherwise possible with the local varieties. The BRRI scientists also report other difficulties, such as those related to genetic instability of some plants, higher grain sterility and low drought resistance, under the prevailing agroecological situation. The adaptive research at BRRI has so far met with only limited success, which is one reason why the IRRI varieties (e.g. IR-8) still dominate over the locally adapted varieties (e.g. BR-1, BR-2 etc.)[9]. It is difficult for social scientists to pass judgement on this impasse. It is hoped that solutions to the various problems discussed above are not beyond careful and patient scientific research in the country.

Of yet more concern for our study is the extent of shortfall in the yields of HYVs in the farmer's field compared to those obtained in the experimental stations. The HYV yield figures for the period 1977-78 to 1979-80 as reported in table 5.1 would imply an average yield of 37 maunds of paddy per acre which is nowhere near the 50 maunds 'norm' set for the experimental tests on HYVs (see table 5.4). BRRI concedes that experiments for varietal improvements are conducted under "optimum agro-economic conditions prevailing in farmers' field".[10] In order to avail themselves of the high potential yields of the HYVs, farmers would have to use the 'recommended' combination or 'package' of inputs such as water, fertiliser, etc. The experience so far in Bangladesh would easily allow us to state that the adopter-farmers are not fully versed in the so-called recommended 'package', or if aware are unable for various reasons to apply the appropriate input package.[11] The failure of the 'package' concept (or its limited success) has to be understood in the context of the farmers' field conditions, their resource endowment and other behavioural and socio-economic constraints under which they operate. Without an understanding of these issues, an exercise on scientific enquiries to bring about HYVs within laboratories would defeat the very purpose of adaptive research. To this we shall come back later in the study.

TECHNOLOGICAL CHOICE IN FARMERS' DECISION-MAKING

Apart from the various climatic and agronomic deterrents, there are other important social, economic and institutional forces hindering the rapid adoption and diffusion of the new rice technology. The conventional wisdom is that constraints to the adoption process include such factors as uneconomic size of farm holdings with excessive fragmentation, insufficient human capital, lack of credit, limited access to information, peak season labour shortages preventing timeliness of operations, chaotic supply of complementary inputs such as seed, fertilisers and water, farmers' aversion to risk, especially in areas prone to floods and droughts, and inadequate infrastructural development. Needless to say, the relative importance of these various factors will depend on the particular socio-economic and physical environment in which the farmers operate.

Issues and hypotheses

There now exists a substantial body of literature, both empirical and theoretical, which attempts to explain relative agricultural growth or stagnation. We shall briefly discuss only some of the often cited explanations and hypotheses that have a more direct bearing on our present study. Schultz (1964) for instance, in analysing stagnation in traditional agriculture, forwards his contention as follows: given that there is no prospect of technological advancement in agriculture, the farmers in the traditional agriculture will soon have arrived at a state when opportunities for profitable investment are no longer forthcoming, and hence investment in traditional inputs would no longer pay.[12] The 'profitability' criterion is the key to Schultz's scheme of transforming traditional agriculture. His hypothesis therefore calls for an assessment of relative profitability of the new HYV technology compared to the traditional technology.
While Schultz's hypothesis focuses entirely on the

choice of available technology by the individual farm, the 'induced innovation model' of Hayami and Ruttan (1971) considers the profitability criteria at the institutional level to explain technical change in agriculture (see also Chapter 3 of this volume). Ruttan and Hayami draw upon the historical experience of Japan to suggest that the recent development of HYVs in tropical Asia represents a response by national and international agencies to changes in factor availabilities and commodity prices (especially the fertiliser-rice price ratio). That the technical changes embodied in the new HYVs are biased toward saving the increasingly scarce factor (land) and using the increasingly abundant factor (fertiliser) clearly indicates a 'rational' response of public agencies to economic forces.[13]

In contrast to hypotheses based on 'rational' responses based on profitability, many authors have advanced 'structural' hypotheses to explain lack of technical progress in traditional agriculture. Bhaduri (1973), for example, argues that the tenurial arrangements, under what he calls 'semi-feudal' mode of production, may impede innovations by tenants. Bhaduri develops a model in which the landlord plays a double role, both as a landowner and as a provider of credit to his tenants. Yield-increasing innovations in this situation will reduce the tenants' dependence on the landlord for consumption loans. The landlord will resist innovation if his loss in usury income is less than compensated by the expected increase in his output share. The model has been sharply criticised, both theoretically and on the basis of evidence from several areas.[14] Bhaduri's scheme of share-cropping is far too strict to be real. His ideal landlord-tenant economy, without any other intermediate peasant stratum, is, in reality, hard to come by in Bangladesh, where the predominant status accrues to ownership cultivation.

Mention may also be made of Boserup's (1965) population-based explanation and Raj's (1970) "dualism" hypothesis. Boserup cites population pressure as the most important factor determining technical change in agriculture. This, the author concedes, could only be a 'necessary' condition but

not a 'sufficient' one, since the producers may lack
the resources or knowledge to carry out the technical
change. Raj's "dualism" hypothesis, on the other
hand, lays emphasis on the state of technical
conditions (especially the lack of assured supplies
of water) that may be such as to render the growth of
the HYV technology confined to a few "enclaves".
Although, in practice, the "enclave" phenomenon may
hold true in respect of various country experiences
so far, the hypothesis does not provide any
explanation as to why a widespread diffusion of the
technology is held at bay indefinitely. To this we
may add that it is not simply the availability of the
physical inputs, but the producers' receptivity, and
hence also the institutional framework within which
they operate, that must hold a vital consideration in
a study of the diffusion of the new technology –
whether spatially or among different size groups of
farmers.

Lack of innovations is also sought to be explained
by farmers' aversion to risk – as in Lipton's (1968)
'survival algorithm' hypothesis, or in the
'safety-first' type of models formulated by Roumasset
(1976) and Bell (1972), among others. The adoption
of HYVs may entail a subjective risk (because of the
unfamiliarity with the new technology) and quite
often also, objective risks (due to weather
uncertainty, pest susceptibility and uncertainty
regarding timely availability of crucial inputs). In
the context of the present study, this problem
deserves special attention because of the limited
means of most Bangladeshi farmers, and also because
of the risk prone nature of the agricultural
environment in Bangladesh (e.g. susceptibility of
rainfed HYVs to droughts and floods).

The extent and pattern of HYV adoption

Coming to the field experience with HYVs in
Bangladesh, we find that there is a clear
non-uniformity in adoption rates across regions and
over different size-classes of producers. While the
rice acreage under the modern seed varieties remains
low for the country as a whole, there is a

considerable extent of variation over the different regions of the country. The district level data of the 1977 Agricultural Census show that the percentage of gross rice acreage under HYVs varies from about 40 per cent in Chittagong to only 2 to 3 per cent in Dinajpur. The official series of rice acreage statistics of the BBS shows similar variation over districts, although the adoption rates are shown to be somewhat higher in the BBS series compared to the Agricultural Census data (see table 5.5) The variations in the adoption rate are also reflected in a wide range of disparity in the estimated growth of cereal production for different districts between the mid-1960s and the end of the 1970s.[15]

It is well known that for variations in the socio-economic and physical environment, the HYV seeds have found ready acceptance in certain regions of the country and not in others. The country's climatic, agronomic and hydrological diversity affects the irrigation and cropping options. HYV _aman_, for instance, is double-cropped with HYV _aus_ in varying degrees in different parts of Bogra, Bakerganj, Dhaka, Mymensingh, Comilla, Noakhali, Sylhet and Chittagong. HYV _aus_, on the other hand, is concentrated chiefly in the double-cropped areas of Chittagong, Comilla, Noakhali and Sylhet, while _boro_ HYVs, the clear favourite variety, are grown in most parts of the country where irrigation from low-lift pumps and tube-wells are readily available. The above features need not, however, be interpreted to underscore percentage of land where techno-economic conditions are unavailable for HYV technology to take root and there is still plenty of room for the new technology to expand in tangible proportions in most parts of the country.

Results from several studies are now available regarding the patterns of HYV adoption across different size-groups of farmers in Bangladesh. These studies are based on field surveys conducted in different parts of the country, often covering only one or two villages, so that any generalisation on the basis of these becomes rather tenuous. So far as the adopter ratio is concerned (i.e. the proportion of adopter farmers within each size class), Rahman

Table 5.5
District-level data used in the regression equations[a]

Name of district	HYV	HYV'	HYVB	HYVBS	HYVBL	FLS	IRG	IRGB	IRGBS
	1	2	3	4	5	6	7	8	9
Dinajpur[b]	2.5	6.1	6.7	18.2	21.6	28	3.4	15.4	11.4
Rangpur	3.4	7.8	11.7	16.7	7.9	17	4.8	27.8	35.7
Bogra	7.1	20.3	35.6	36.6	38.6	50	25.5	67.5	62.4
Rajshahi	3.8	6.4	30.3	34.6	29.6	22	13.8	69.4	70.1
Pabna	2.9	5.1	37.9	36.4	31.8	32	4.0	64.3	50.0
Kushtia	12.7	12.5	31.3	33.3	40.0	40	12.4	25.0	33.3
Jessore	3.8	7.5	10.3	5.0	8.8	21	3.9	16.4	10.0
Khulna	4.1	9.7	19.5	6.4	14.6	13	2.2	5.7	4.3
Barisal	8.9	14.2	61.4	53.6	62.7	29	4.7	52.2	42.0
Patuakhali	2.6	8.9	28.8	83.3	13.8	16	2.4	26.7	83.3
Jamalpur	6.5	14.7	31.5	39.5	23.7	21	10.6	54.5	60.5
Tangail	8.8	18.3	58.7	62.0	56.6	32	11.2	67.0	64.9
Mymensingh	13.8	31.7	35.0	38.5	34.6	24	23.8	72.8	70.5
Dhaka	16.0	24.0	48.3	52.6	42.2	62	16.0	56.2	56.9
Faridpur	3.3	3.5	34.2	33.3	35.6	7	12.9	15.1	11.8
Sylhet	11.3	15.4	21.1	16.5	14.9	15	21.9	47.1	44.8
Comilla	21.4	33.2	64.2	68.9	56.4	87	25.4	75.3	77.9
Noakhali	20.4	36.6	70.4	72.1	52.9	46	12.5	77.0	78.7
Chittagong[c]	38.3	52.8	65.4	59.3	81.3	112	44.9	90.0	91.5

(Continued)

Name of district	IRGBL	FSZ	LM	LML[b]	TNC	TNCS	TNCL	LTC
	10	11	12	13	14	15	16	17
Dinajpur[b]	14.9	4.76	1.71	2.97	19.9	34.7	7.2	11.5
Rangpur	23.7	3.52	1.27	2.50	15.2	23.3	6.5	9.8
Bogra	75.4	3.12	1.19	2.61	15.2	21.5	7.3	12.4
Rajshahi	70.0	4.41	1.52	2.85	20.8	26.2	13.4	10.2
Pabna	68.2	4.12	1.31	2.47	16.4	21.8	9.8	8.8
Kushtia	40.0	4.63	1.28	2.36	14.3	19.8	8.4	9.8
Jessore	20.6	4.13	1.36	2.50	14.3	20.1	7.3	20.4
Khulna	10.7	4.31	1.45	3.20	23.4	17.7	20.6	16.4
Barisal	55.0	3.78	1.20	2.84	32.0	30.8	28.3	13.2
Patuakhali	13.8	6.21	1.95	3.68	34.6	33.2	31.7	11.5
Jamalpur	48.4	3.64	1.33	2.39	17.9	26.6	9.7	11.3
Tangail	67.4	3.24	1.05	2.20	19.5	25.0	11.5	11.3
Mymensingh	74.1	3.29	1.19	2.39	13.8	19.0	8.2	9.2
Dhaka	54.6	2.72	0.89	2.22	14.7	15.1	9.9	11.5
Faridpur	18.8	3.32	1.16	2.80	17.1	16.7	20.4	9.9
Sylhet	49.0	4.03	1.35	2.85	15.2	19.8	10.2	7.7
Comilla	72.3	1.96	0.72	2.32	10.0	12.8	4.7	12.8
Noakhali	73.5	2.87	0.93	3.38	30.7	16.6	45.9	12.8
Chittagong[c]	92.2	2.74	0.80	2.49	30.5	35.1	17.7	15.5

[a] Sources: Column (2), Statistical Yearbook of Bangladesh, 1980,BBS. – Column (13), 1974 Population Census Report (National volume). All other columns, Report on the agricultural census of Bangladesh 1977 (National volume).
[b] Variable LML refers to land–man ratio (as defined in the case of LM) for large farmer group, owning more than 7.5 acres of land. TNCS and TNCL represent total rented land as per cent of total cultivated land for small farms (less than 2.5 acres) and large farms (more than 7.5 acres) respectively. All other variables are explained in the text (see "Regression results", below).
[b] For Dinajpur, estimations of HYV adoption and proportion of area irrigated in Boro season include both rice and wheat because of the low Boro rice acreage compared to wheat acreage.
[c] The district of Chittagong Hill Tracts has been excluded because acreage under paddy is relatively very small.

(1981) shows that it is the highest among the large farmers, while no clear size-specific pattern can be discerned from the studies by Muqtada (1975) and Asaduzzaman (1979). Most of the studies, however, seem to have one common finding; that, among the adopter farmers, the intensity of adoption (i.e. the proportion of land devoted to HYVs) is higher for smaller farmers compared to larger farmers. The extent of adoption (i.e. the percentage of land devoted to HYVs by all farmers belonging to each size class) is, of course, the combined effect of both the adopter ratio, and the intensity of adoption by adopter farmers. Here, again, the story is a little anomalous, although the bulk of the findings support higher rates among the small farmers. Khan et al. (1981), Asaduzzaman (1979) and Muqtada (1975) find that the extent of adoption has been at least higher than average for the small farms (farms being classified into small, medium and large), while Rahman (1981) and Ahmed (1981) find no systematic relationship. However, a considerable weight is added to the hypothesis of a negative relationship between farm size and the extent of adoption by the results of the 1977 Agricultural Census as presented in table 5.6. Moreover, some of the studies cited above (Rahman 1981) shows that, particularly for large farmers, it is the intensity of adoption, rather than the initial decision to adopt or not, that is likely to be the more important factor in bringing more rice land under HYV.

Innovative adoption under differential resource endowment

For any technology to prove attractive, a fundamental criterion must be its profitability vis-à-vis the old one. A new technology involves some risk, and farmers ignorant about the outcome may heavily discount the returns from the new technology. Thus, the returns from innovations must be sufficiently high in order to induce tradition-bound farmers to change their age-old techniques. In recent years, there have been numerous studies, based on farm survey data, on the profitability of HYV paddy

Table 5.6
Extent of adoption of HYV rice in Bangladesh[a]
(Percentage)[b]

Farm-size group	Crop season	
	Boro	Aus and Aman
Small (below 2.5 acres)	48.1	9.6
Medium (2.5 to 7.5 acres)	39.2	6.8
Large (above 7.5 acres)	29.5	5.5
All farms	37.9	7.0

[a] Source: BBS, _Report on the Agricultural Census of Bangladesh, 1977_ (National volume).

[b] Extent of adoption is defined as the percentage of total cropped area under rice devoted to HYVs in the respective crop seasons.

cultivation in Bangladesh. A clear consensus arising from these studies is that the yields and net returns per acre are substantially higher for HYVs compared to local varieties. (This observation appears true irrespective of whatever definition one uses for net returns per acre. In most studies, net returns or net income is defined as the gross value of output minus the costs of all material inputs and human and bullock labour, both hired and family-supplied. The cost of family labour is estimated by using the prevailing market wage rate. For consistency of analysis, we shall henceforth use this definition of net returns).[16] The Agro-Economic Research (AER) Section of the Ministry of Agriculture has, for some years now, systematically collected data on costs and returns related to the cultivation of both local and high-yielding varieties of foodgrains. These data (based on farm surveys which do not usually distinguish different size-groups of farms) show that the net increase in yields due to the new seeds ranges from 26 to 90 per cent, while increase in net returns may go up to more than 100 per cent (see table 5.7).

Table 5.7
Costs and returns of local and HYV rice (1878-79)*

	Boro (Mymensingh)		T. Aman (Barisal)	
	Local	HYV	Local	HYV
1. Yield (maunds/acre)	37.1	46.7	30.1	49.5
2. Value of yield (Taka)	2,615	3,336	2,408	3,960
3. Cost of labour (Taka)	1,382	1451	902	1,106
4. Cost of bullock (Taka)	30	32	303	294
5. Cost of material inputs (Taka)				
seed	–	–	33	33
manure	–	–	–.	–
fertiliser	62	104	40	105
pesticides	–	–	3	11
irrigation charges	78	93	–	1
6. Total cost (Taka) (3+4+5)	1,552	1,680	1,281	1,549
7. Net returns (Taka/acre) (2-6)	<u>1,063</u>	<u>1,656</u>	<u>1,127</u>	<u>2,411</u>

* Value of yield has been arrived at by taking the ruling market paddy prices of the respective crops for the year concerned. Labour costs include those incurred for hiring labour as well as for using family labour (imputed at the ruling market wage rate; see text for rationale). The figures are extracted from <u>costs and returns</u> surveys conducted by AER, Ministry of Agriculture, for individual crops.

209

Thus the relative profitability of HYVs appears beyond doubt, at least for all farm size-groups considered as a whole. We are thus left with the apparent paradox of high profitability of HYVs along with low rates of adoption which calls for a reinterpretation (or a rejection) of the Schultzian 'profitability' hypothesis.

It hardly needs to be emphasised that farmers' choice of technology depends on their resource endowment and on their access to markets for various agricultural inputs. Due to the imperfect nature of the rural input markets (e.g. markets for labour, draft power, credit, etc.), the effective prices of these inputs cannot be the same for farms of different sizes and tenures. Hence one can expect variation in input use and land productivity among farm-size groups, depending on the technology used in production (which determines the possibility of substitution among different inputs). Most authors on Bangladesh agree that, in the case of traditional technology, farm-size is negatively related to land productivity, cropping intensity and the use of labour per acre.[17] This result is hardly surprising, since in traditional cultivation techniques, labour is the dominant input and hence it results in higher output per acre on family-based small farms which have a favourable endowment of family labour (and possibly of draught power) relative to land. There is less agreement, however, (and there exists far less empirical evidence) regarding how the differential resource endowment affects the nature of input use and land productivity under the new HYV technology with considerable scope for intensive use of modern inputs such as fertiliser and water. The AER studies mentioned earlier do not provide information according to farm-size groups. On the basis of the few studies that are available on the size-group-specific costs and returns for both local rice varieties and HYVs, we can make the following tentative observations[18]: (i) the yields and net returns per acre of HYVs are substantially higher compared to local varieties for all groups of farmers. This appears to be the case for share-croppers as well, even without any cost-sharing arrangements; (ii) all groups of farmers use more of both labour and material inputs (e.g. fertiliser, pesticides, etc.) on HYVs compared to local varieties. Also the proportion of total cost (especially the cost of material inputs) to gross value of output is higher for HYVs compared to local varieties. Thus, although the new technology is scale-neutral, it is not resource-neutral; (iii) small farmers apply more labour and less material inputs per acre, compared to large farmers, both in the case of local and high-yielding varieties. Moreover, the percentage increase in per acre labour input on HYVs over local varieties is much higher for small farms compared to large farms. The converse is possibly true in the case of increase in the cost of material inputs; (iv) it is difficult to generalise, on the basis of available evidence, regarding the percentage increase in yields and in net returns of HYVs over local varieties for different size-groups of farms. We feel inclined, however, to conclude that the increase is higher for larger farms. When it comes to the percentage increase in yields (or net returns) per labour hour, this is clearly higher for larger farms compared to smaller farms.

The above observations are clearly in conformity with the resource endowment of different size-groups of farmers. The higher cost of material inputs for HYVs across all farm size groups is, of course, explained by the well-known fact that the profitable adoption of HYVs requires the application of a minimum dose of complementary inputs such as fertiliser and water. However, smaller farmers, because of the financial constraints under which they operate, tend to economise on the use of material inputs and, instead, adopt a more labour-intensive method of growing the HYVs (compared to larger farmers). As a result, the negative relation between farm size and labour input per acre becomes even stronger in the case of HYVs compared to local varieties.[19] This would suggest that the HYV technology, as adopted by small farmers, offers greater scope (compared to local crop varieties) for increasing productivity by adopting such labour-intensive methods of cultivation as more intensive land preparation, water management, weeding and other intercultural practices. (Some recent studies show that in the case of traditional technology, higher labour-intensity on small farms is mainly due to more multiple-cropping and the choice of labour-intensive crops on these farms; the negative effect of farm size on labour intensity at the individual crop level is either weak or absent.)[20]

Landownership and innovativeness: The subsistence pressure hypothesis

It is often alleged that, for various socio-economic reasons, landownership and entrepreneurship do not always go together in Bangladesh. This is indeed evidenced by the relatively lower extent of adoption of HYVs by larger farmers, which we discussed earlier. There is no doubt that many non-economic factors such as social customs, attitudes and traditions may coexist with (or dominate over) the pure profit motive in farmers' decision-making. Nevertheless, we have seen in our previous discussion that the input combinations chosen by farmers of different size groups (while adopting HYVs) can be explained to a large extent by such 'economic' and institutional factors as farmers' resource endowment and their access to markets for agricultural inputs. This same set of factors could also underlie the decision-making by different farmer groups (including the decision to adopt innovations) in achieving their respective set of objectives.

As for large farmers, they can generate internal surplus, and they have also easy access to institutional credit at low rates of interest. But, since the adoption of HYVs requires hiring more labour, they face the problem of so-called 'supervision cost'. They have, on the other hand, the option of investing their surplus funds either in agricultural innovations or in other activities (e.g. usury capital, speculative investment, purchase of land etc.) and they have also the option of renting out the whole or part of their land to other farmers. There are very few in-depth empirical studies on savings and investment behaviour of different groups of cultivators. From the few studies that are available, large landowners are found to spend only a small fraction of their savings on agricultural investment.[21] Instead, they spend most of their surplus funds for purchase of land, construction of houses, investment in speculative business and social ceremonies. A significant proportion of their income also comes from the share of output on rented-out land, which they find 'uneconomical' to cultivate themselves. Because of the excessive fragmentation of landholdings, even the very large farmers in Bangladesh do not find it economical to invest on their own in modern irrigation equipment such as tube-wells. The full utilisation of such irrigation equipment requires formation of user groups which is often difficult under the prevailing structure of the rural society, characterised by factionalism and patron-client relationship. (Bangladeshi large farmers are thus different, in many respects, from their counterparts, say, in the Indian Punjab, where large landowners have played the crucial role in bringing about the so-called 'Green Revolution' and in promoting private capitalism in agriculture).[22]

It may be noted that large farmers are usually found to play more of a 'leadership role' in a few areas where special support programmes are undertaken by the Government for the diffusion of the new technology. This has been the case, for example, in the Comilla Kotwali thana (the birthplace of the so-called Comilla co-operatives) and in certain areas covered by the IRDP programme.[23] In these areas, massive institutional effort and support have been given towards the introduction of the new technology in the form of infrastructural facilities and the subsidised supply of inputs. Since larger farmers have far more access to such subsidised inputs, it apparently alters their 'profitability' calculations, and they adopt the new technology to take advantage of these support programmes. The above phenomena may explain the seemingly conflicting results regarding HYV adoption across farm-size groups as reported by different studies discussed earlier. Some of these studies, which are based on farm-surveys conducted in areas covered by the Government's special support programmes, show a higher rate of adoption by larger farmers, while most other studies report contrary results. Farm size is after all a surrogate for many potentially important factors which affect farmers of different size-groups differently. Since the influence of these factors varies in different areas and over time, so does the relationship between farm size and adoption behaviour.

What may appear to be rather intriguing is that the smaller farmers, who operate under a severe constraint (and have less access to government-supplied inputs), should register a relatively higher rate of adoption. We have sought a rationale for this in what we call the 'subsistence pressure' hypothesis. The only resource the small farmers have in surplus is family labour; and we have already seen that the HYV technology, as adopted by small farmers, offers them the scope for stretching the limit of gainful employment on their own farm. (They can, of course, seek employment on larger farms, but working as hired labour is socially looked down upon, and it is difficult to get employment during slack seasons, when small farmers have higher excess labour.) Seeking the possibilities in the new technology thus appears almost like the last straw in the small farmers' bid to survive as a family farm rather than sell their land (and other agricultural assets) and join the rank of landless labourers. This thesis may be traced, albeit partially, to Boserup's (1965) "population-based" explanation which we discussed earlier, and to Hayami's (1981) contention that "the Green Revolution can be considered an induced innovation in response to growing population pressure on land". This is also in contrast to Lipton's (1968) "survival algorithm" hypothesis according to which smaller producers may prefer a small output with certainty than a larger output entailing high risk. Our explanation would imply that a 'smaller output with certainty' may still be a risky proposition in terms of a global perspective of survival. At this point, small farmers may be inclined to take a 'limited risk', by adopting HYVs to a limited extent and by operating on a minimum package of inputs, as permitted by their resource constraint. Clearly, the 'survival algorithm' can work either way: just as taking too much risk may jeopardise small farmers' chance of survival, taking no risk may equally do so.

One must be cautious in interpreting the 'subsistence pressure' hypothesis as discussed above. The hypothesis does not imply that poverty conditions must prevail throughout for the new technology to have a universal appeal. Nor does it imply that a breakthrough in agricultural output can be achieved through a process of adoption under subsistence pressure alone. On the contrary, we have advanced the hypothesis to explain a 'perverse' pattern of the diffusion of HYV technology which is accompanied by increasing poverty and population pressure on land (as has been the experience with HYVs in Bangladesh). Increasing poverty is, however, a result not of innovations, but of insufficient extent of innovations. Since small farmers would be willing to take only 'limited risk', it is only up to a certain range that the appeal of the new technology could be met through subsistence pressure. Khan et al. (1981) described their finding about the relatively higher rate of adoption by smaller farmers in Bangladesh as 'reassuringly surprising'. In the light of our above discussion, it would appear that the phenomenon is, in fact, neither surprising nor very reassuring.

TECHNOLOGY ADOPTION: AN ECONOMETRIC EXERCISE

Although the rice acreage under the modern seed varieties remains low for the country as a whole, we have already noted a considerable extent of variation in the HYV adoption rate over different regions of the country. The cross-sectional district level data are therefore suitable for studying econometrically the relationship between agricultural innovations and such other factors as farm size, tenancy, literacy, etc. In this section we shall report some regression results obtained on the basis of such data. The estimated variables used in the regressions, along with the sources of data, are reported in table 5.5.

In running any econometric test on the determinants of agricultural innovations, it is important to have clear _a priori_ hypotheses about which types of innovations may be specified and considered as endogenous variables from the point of view of farmers' decision-making. The decision to adopt one particular innovation (e.g. the use of more fertiliser) and the decision to adopt another (e.g. to bring more acreage under HYVs) are generally simultaneous decisions and thus, probably subject to the same random disturbances. Therefore, in a regression analysis, the use of one to 'explain' the other would render the regression coefficients biased and inconsistent. Such econometric tests cannot also bring out the underlying factors which work behind both the simultaneous variables, and the high value of R^2 in such cases merely suggest the complementarity of innovations.[24]

In the context of our analysis, fertiliser-use and HYV adoption can be treated as representative endogenous decision variables. But we have rejected the idea of treating irrigation as a dependent variable because it is well-known that, given the present level of technology, the provision of modern irrigation in Bangladesh is almost wholly external to individual farmers' decision-making.[25] Thus, while HYV adoption and fertiliser use are used as endogenous decision variables in separate regressions, it is quite valid to use irrigation as an exogenous explanatory variable in each case.

Regression results

The first set of regressions uses the percentage of gross rice acreage (total of Aman, Aus and Boro rice) under the HYVs as the dependent variable. The adoption rates, estimated both from the BBS annual series of data (for 1978-79) as well as from the 1977 Agricultural Census data, are used separately. Among the explanatory variables are irrigation, farm size or land-man ratio, tenancy and literacy. All the variables are estimated from the 1977 Agricultural Census data, with the exception of the literacy rate, which is estimated from the 1974 Population Census data. The following results are all based on simple linear regressions.[26] (One asterisk sign denotes the significance of the coefficient at less than 1 per cent level, two at 5 per cent level, and three at 10 per cent level; t-statistics are shown within parentheses).

HYV acreage (Agricultural Census data)

(1) HYV = 17.15 + 0.49*IRG − 11.01**LM

$\qquad\qquad\qquad$ (3.77) $\qquad$ (2.31)

$\quad R^2 = .759$

(2) HYV = 12.54 + 0.47*IRG − 13.82*LM + 0.42*TNC

$\qquad\qquad\qquad$ (4.65) $\qquad$ (−3.65) $\qquad$ (3.42)

$\quad R^2 = .865$

(3) HYV = 4.44 + 0.49*IRG − 9.47***LM + 0.94*LTC

$\quad R^2 = .807$

HYV acreage (BBS data)

(1) HYV' = 26.41 + 0.71*IRG − 15.03**LM

$\qquad\qquad\qquad$ (3.57) $\qquad$ (−2.17)

$\quad R^2 = .731$

(2) HYV' = 19.65 + 0.68*IRG − 19.15*LM + 0.62*TNC

$\qquad\qquad\qquad$ (4.28) $\qquad$ (−3.23) $\qquad$ (3.21)

$\quad R^2 = .841$

(3) HYV' = 5.52 + 0.71*IRG − 12.15***LM + 1.55**LTC

$\qquad\qquad\qquad$ (3.95) $\qquad$ (−1.87) $\qquad\qquad$ (2.13)

$\quad R^2 = .791$

Where HYV = Gross HYV acreage as percentage of gross rice acreage based on 1977 Agricultural Census data

HYV' = Percentage of HYV acreage based on BBS data

IRG = Gross area irrigated as percentage of net cultivated area

LM = Net cultivated area divided by number of male population of farm households (excluding landless labour households), 10 years of age and above

TNC = Total rented land as percentage of total cultivated land

LTC = Percentage of rural male population over 25 years of age having more than five years of schooling, based on 1974 Population Census data.

Judged by the coefficient of determination (R^2), the regression results are highly satisfactory. Its quality is not much affected by whether the BBS or the Agricultural Census data are used. It can be seen from equations (1) and (1') that irrigation and land–man ratio alone account for nearly 75 per cent of the variation in adoption rate. Adding tenancy and literacy as explanatory variables, one at a time, improves the value of R^2, and the coefficients of both variables are significant. However, compared to tenancy, literacy appears to have a less significant effect, as judged by the value of R^2 and the level of significance of the respective coefficients.

As expected, irrigation is shown to have a strong positive effect on HYV acreage. However, the coefficient of land–man ratio is consistently negative. When the land–man ratio is replaced by the average farm size in the regressions (the variable FSZ in table 5.4), the coefficient is always negative and in most cases, highly significant. (This is what is to be expected since the two variables are highly correlated, and can be considered almost a proxy variable for each other.) However, since a better regression fit is always obtained by using the land–man ratio instead of the farm size, we have reported here the results in the case of land–man ratio only. Contrary to popular hypotheses, tenancy is found to have a favourable effect on HYV adoption.[27] On this point, we shall have more to say later. The coefficient of the literacy variable has the expected positive sign.

In the next set of regressions, we use as the dependent variable the proportion of rice acreage under HYVs in the Boro season only. It may be worthwhile to analyse the adoption of Boro HYVs in particular, since these have been grown for a relatively long period of time (compared to the HYVs introduced more recently in the Aus and Aman seasons) so that the farmers' adoption behaviour is now more well established.[28] Also, the extent of adoption is much higher in the Boro season and shows more regional variations. The adoption of Boro HYV rice is further analysed by running regressions for small and large farmers separately, the two groups being defined as those cultivating less than 2.5 acres of land and more than 7.5 acres respectively. The 1977 Agricultural Census data permits the estimation of the relevant variables separately for each of these farm-size groups, except for the literacy variable, for which we use the same data as in the equations (1) to (3). The regression results are presented below:

<u>HYV Boro acreage</u>
<u>All farmers</u>

(4) HYVB = 53.52 + 0.35**IRGB − 26.61**LM

        (2.60)   (−2.30)

   R^2 = .660

(5) HYVB = 50.60 + 0.24**IRGB − 39.32*LM + 1.23*TNC

        (2.48)   (−4.70)   (4.43)

   R^2 = .853

(6) HYVB = 20.97 + 0.38*IRGB − 21.43***LM + 2.14 LTC

        (2.98)   (−1.90)   (1.73)

<u>Small farmers (less than 2.5 acres)</u>

(7) HYVBS = 6.59 + 0.67*IRGBS

$$(5.55)$$

$$R^2 = .644$$

<u>Large farmers (7.5 acres and above)</u>

(8) HYVBL = −30.58 + 0.57*IRGBL + 3.27**LTC

$$(4.98) \qquad (2.43)$$

$$R^2 = .662$$

Where HYVB =	Boro HYV acreage as percentage of total Boro rice acreage	
IRGB =	Percentage of Boro rice area irrigated	
HYVBS and HYVBL	are the same as HYVB, but defined for farms of the size of below 2.5 acres and above 7.5 acres respectively.	
IRGBS and IRGBL	are the same as IRGB, but defined for farms of the size of below 2.5 acres and above 7.5 acres respectively	
TNC and LTC	are defined as before.	

The results for the adoption of HYV Boro rice as shown by equations (4) to (6) are more or less similar to those obtained earlier in the case of HYV adoption for all rice seasons combined. The coefficients of irrigation, land−man ratio and tenancy have the same sign as before and are highly significant. However, the coefficient of the literacy variable, although of the right sign, is not significant at the 10 per cent level (equation 6).

Coming to the separate regressions for small and large farms (equations 7 and 8), it can be seen that we are no longer able to obtain as high a value of R^2 as in the earlier regressions. The reason for this is not difficult to guess. By considering farms of particular size groups, we take away one of the major sources of variation in the adoption rate, namely, the land−man ratio or the farm size. The land-man ratio for the small farm group shows too little interdistrict variation to be used as an explanatory variable. For this group, irrigation alone accounts for about 65 per cent of the variation in adoption, while neither tenancy nor literacy is found to have any significant effect. For large farmers, besides irrigation, only literacy is found to have a significant effect. Although the land-man ratio for large farms shows some variation (the farm size being open-ended), its coefficient is not found significant. Neither is the coefficient of tenancy found significant at the 10 per cent level.

Implications for HYV adoption

We are now in a position to analyse the significance of the regression results obtained so far. Considering that we have not taken into account many socio-economic as well as agronomic factors (e.g. availability of credit, suitability of land, etc.), these results are remarkable for the explanatory power of the variables considered. The effect of the land-man ratio on the adoption rate deserves particular attention in view of some of the hypotheses we have advanced earlier regarding the set of forces that work behind HYV adoption for different farmer groups. We cited evidence that in areas with strong institutional support, big farmers tend to lead small farmers in the rate of adoption, at least for the initial years. Since the availability of irrigation facilities is also an indicator of the institutional support provided by public agencies, the coefficient of irrigation in our regressions probably picks up some favourable effect of such institutional support on HYV adoption by large farmers. The coefficient of land-man ratio, on the other hand, may largely show the effect of what we have called 'the subsistence pressure' in the case of small family farms.

It must be admitted that the negative coefficient of the land-man ratio (or of farm size) does not, by itself, provide any conclusive evidence in support of the 'subsistence pressure' hypothesis discussed earlier. We have argued that a different set of factors lies behind adoption decisions by different size-groups of farmers. A negative relationship between farm size and the adoption rate may as much reflect the socioeconomic factors that hinder adoption by large farmers as the 'subsistence pressure' that induces small farmers to adopt. In particular, the negative coefficient of the land-man ratio can pick up the effect of peak season labour shortage faced by large farmers as well as the effect of 'subsistence pressure' (i.e. the need to find productive employment on family farms) in the case of small farmers. Several authors have indeed stressed the importance of 'peak season' labour shortage as an impediment to HYV adoption.[29] That the land-man ratio is not found to be a significant explanatory variable in our separate regressions for large farms suggests that the coefficient of this variable in the overall regressions does not probably reflect the effect of 'labour shortage'. There is also ample indirect evidence to suggest that the peak-season labour scarcity is not a serious bottleneck to HYV adoption in Bangladesh. Such scarcity of labour would have induced large farmers to mechanise cultivation, which has not happened to any significant extent in Bangladesh. Although the wage rate for hired labour in agriculture rises to some extent in the peak seasons, the overall trend in the real wage rates of agricultural labourers has been downward since the late 1960s.[30] We have also cited evidence that, at the prevailing wage rates, HYV adoption is in fact highly profitable for large farmers who depend mostly on hired labour.

The results about the positive effect of tenancy on HYV adoption may appear rather unexpected and therefore need further scrutiny. Lipton (1978) observes that "the limited 'available' empirical work indicates no consistent relationship between tenancy and HYV adoption". Various studies on the pattern of HYV adoption in Bangladesh, discussed in the previous chapter, do not throw much light on the subject, since the effect of tenancy is not systematically studied by controlling the effects of other variables (e.g. farm size, access to inputs, etc.). In the context of our regression analysis, one may however think of some plausible explanations for a positive effect of tenancy on HYV adoption. It may be noted that the coefficient of the tenancy variable is not found significant in the separate regressions for either small or large farms. While, for any particular size group, HYV adoption may not be significantly related to the incidence of tenancy, it is quite plausible that tenancy may have an overall favourable effect by transferring land from less innovative larger farmers to smaller and more innovative farmers. In any case, the coefficient of tenancy variable in our overall regressions reflects, to a large extent, the effect of tenancy in the case of middle-sized farms (i.e. 2.5 to 7.5 acres) which account for nearly 60 per cent of all rented-in land (cf. Agricultural Census, 1977). It is likely that these middle farmers, who rent in land mostly from large farmers, are less handicapped by the so-called 'supervision cost' (compared to large farmers) and suffer from a less severe resource constraint (compared to small farmers).

It must be emphasised that the above results regarding the positive effect of tenancy on HYV adoption do not necessarily challenge the so-called 'inefficiency' of the institution of share tenancy. What may be true is that, within the present agrarian set-up, tenancy perhaps enables more intensive (as distinct from efficient) use of certain resources (particularly excess family labour) in order to be able to adopt the new technology. In this respect, it may be again an indirect manifestation of the phenomenon of adoption under 'survival pressure' which, as we have argued earlier, cannot be relied upon for a technological breakthrough on any large scale.

Lastly, the coefficient of literacy in our regressions shows that formal education has some favourable effect on agricultural innovations. Surveying a number of studies, Schultz (1975) found that education plays a strong role in determining rates of adoption of new technology. The regression results presented here show that while the effect of literacy is quite pronounced in the case of large farmers, there is no significant effect of this

variable on the adoption behaviour of small farmers.
This is what is to be expected, since the literacy
variable, as defined, does not much concern small
farmers, few of whom would have in any way completed
five years of schooling.

Results on fertiliser use

To explain the factors behind farmers' decision to
use fertiliser, we have again used cross-section
district-level data. The estimated average
fertiliser use per gross cropped acre, which shows
considerable variation over regions, is taken as the
dependent variable. The best regression fit is
obtained in the case of the following equation:

$$(9) \quad FLS = 10.48 + 1.24*IRG - 28.15***LM + 3.78**LTC$$
$$(3.00) \qquad (-1.84) \qquad (2.28)$$
$$R^2 = .741$$

where FLS is the amount of fertiliser used per acre
of gross cropped area. Availability of irrigation,
which increases the effectiveness of fertiliser use
(both on local and high-yielding varieties of crops),
is found to be the most important explanatory
variable. As in the case of HYV adoption by big
farmers (equation (8)), the literacy variable is
found to have an equally significant positive
coefficient. The land-man ratio has a negative
coefficient, although significant at only 90 per cent
level. In fact, when the land-man ratio is replaced
by farm-size in equation (9), its coefficient is not
found significant at all. Thus the negative effect
of land-man ratio (or farm size) is not found as
strong (or to exist at all) in the case of intensity
of fertiliser use as in the case of HYV adoption.
The 'subsistence pressure' may have some positive
effect on fertiliser use in an indirect way, that is
by inducing smaller farmers to adopt more HYVs which
require more fertiliser compared to local varieties.
But the effect is weakened (if not eliminated) by the
commonly observed phenomenon that larger farmers use
more fertiliser per acre for any single crop variety
compared to smaller farmers. It may also be noted

that the 'subsistence pressure' favourably affects the intensity of fertiliser use (per net acre of cultivated land) in another indirect way, that is through higher cropping intensity on smaller farms. But this effect is not captured in equation (9), which considers fertiliser use per acre of gross cropped area.

INSTITUTIONS, RESEARCH AND TECHNOLOGY ADOPTION – THE INTERRELATIONSHIPS

In the previous section, we dealt at some length with an analysis of the important factors affecting farmers' decisions to adopt the new technology. This has been done with a view to unfolding the broad constraints on the rate of adoption, which although seen to be relatively higher among small farms, is not very encouraging as a whole. It is common knowledge that every technology, or for that matter, state of art, is nourished and sustained within a particular institutional framework (which in turn may change its character upon the impact of the technology). Hence, whenever the former fails to flourish, it calls for a close scrutiny of the institutional arrangements – whether it is the paucity of institutions, ineffectiveness of their role, or lopsidedness of their development.[31]
While there is no doubt that there exists a very close interrelationship between technology and institutions, the issue often appears whether a breakthrough should occur in the former or the latter. Ishikawa (1980), for instance, brings out the following typologies: (i) technical change without any institutional change; (ii) technical change with minor adjustments in institutions; (iii) technical change only after a thorough (radical) institutional change; (iv) technical change itself unleashing forces that would affect institutions. It is our belief that such categories need not be mutually exclusive; rather these may co-exist or may be sequential. In the analysis of agricultural transformation, both technology and institutions should be treated as variables and should be allowed

to interact. As Ruttan (1980, p.27) puts it:

> "Dissent over the priority between technical and institutional change is unproductive. Technical change and institutional change are highly interdependent and must be analysed within the context of interdependence."

But let us now try to elaborate this point further in the context of Bangladesh's experience.

The institutional linkages – A schematic representation

In figure 5.1, an attempt was made to represent schematically the entire gamut of institutional linkages in the generation and diffusion of the HYV rice technology in Bangladesh. This represents a modification of Ruttan's (1980, pp.30–40) 'induced innovation model' (see Chapter 3 of this volume) as applied to Bangladesh's context. Ruttan's model is designed to capture the interaction of technical and institutional innovations with the socioeconomic and political environment in which the research system operates. However, Ruttan's conceptual framework is different from ours in certain important respects.[32]

While Ruttan deals exclusively with the induced demand for, and the outcome of, scientific research, we have defined institutions in a much broader sense, so as to include not only a few prestigious scientific institutions, but also the institutions that are indispensable in motivating and helping farmers to adopt innovations. The latter would include the various agencies involved in supplying modern inputs to farmers, as well as the grass-roots organisations where the "new technology is made known to peasants and tested, adapted and popularised with the participation of peasant users" (Chapter 7 of this volume).

In Ruttan's scheme, the generation of technological innovations is conceptualised as a process of 'circular and cumulative causation'.[33] In his model, the demand for innovations is conditioned by the 'pay-off matrix' that identifies the gains and

losses of various interest groups from alternative
technological choice. The actual supply of
innovations would then be determined by the
performance of the innovation-producing institutions
(e.g. scientific research institutions). In
contrast, the circular chain of causation is much
weaker (if not entirely absent) in our scheme,
particularly in respect of the articulation of the
perceived gains of the interest groups into demand
for appropriate scientific research. This mainly
reflects upon the underdeveloped stage of indigenous
scientific research in Bangladesh, and the
predominance of imported technology (which so far has
undergone only limited local adaptation).

The linkages are, however, stronger in the
evolution of the institutions that motivate farmers
by bringing about structural reforms and by supplying
subsidised inputs. (For example, the issue of input
subsidies has been much more of a subject of
political controversy than the question of quality
and direction of research conducted at the
agricultural research institutions.) As we shall
elaborate further, there is in general a dialectic
interaction or an 'interface' (instead of any
'circularity of causation'), involved in the
so-called process of induced institutional innovation
which ultimately determines the performance of the
institutions and the extent of diffusion of the new
technology. The circularity of causation, as
conceptualised by Ruttan, is mostly absent in
Bangladesh, again due to a lack of systematic
articulation of the latent demand of the rural
interest groups into state-level policies and
objectives. This in turn is the result of an
urban-bureaucratic elitist bias of the successive
regimes in Bangladesh and erstwhile Pakistan.
Moreover, political instability since Bangladesh's
independence has made it difficult clearly to
identify the interest groups as represented by the
particular political regimes.

In what will follow, we shall look more closely at
the mode of operations of the multiple constellation
of institutions related to the diffusion of the new
technology, as identified in figure 5.1. An

elaborate description of the entire gamut of institutions involved in agricultural development will be of course unwieldy, and is also outside the scope of the present study.

Objectives of the Government

The search for food self-sufficiency has been a top priority issue for all successive regimes in power since the country's independence. The food problem is perhaps the most critical aspect of Bangladesh's struggle to achieve economic development. It bears upon the rate and structure of growth, the rate of inflation, poverty and nutrition, the trade balance and the Government's fiscal position. In the past, foodgrain production lagged behind population growth resulting in growing hunger and ever-increasing food deficits. The fact that Bangladesh is one of the world's largest recipients of food aid, although foodgrain production is the single predominant occupation of her population, is symptomatic of the extreme nature of Bangladesh's situation. In point of fact, the severity of the food problem is well known and is such that the fundamental stability of the state and the administration depends on the continuing impetus for its solution. Moreover, achieving self-sufficiency in foodgrains is a desirable objective which also fits well with the Government's other expressed objectives of planned development, such as improving the balance of payments and reducing rural poverty and unemployment.[34]

Given the prospects that the new HYV technology would increase yields, would absorb more labour (at least in the initial stages of its adoption) and, because of its divisibility, would be acceptable across the entire range of small and large farmers, planners and policy makers were quick to respond to the idea of a transfer of technology. Accordingly, the setting up of an institutional framework for the systematic adaptation and diffusion of the HYV technology was emphasised in the Government's agricultural development strategy. The Government, therefore, embarked on a programme of technology adoption through the following: (i) mobilisation of resources toward research and experimentation; (ii) development of the necessary infrastructural framework and institutions that would be responsible for supplying modern inputs (e.g. seed, fertiliser, water, etc.); and (iii) organisation of motivational and institutional movement to encourage producers' receptivity. Despite its zeal, however, the administration finds itself in precarious balance because of the rather slow impact of the new technology on the desired objectives.

Research, extension and input supply

The lower left-hand side of figure 5.1 shows the channel through which the imported technology undergoes adaptive research and is systematically transferred to the farmer's field. Although several agencies are engaged in scientific research on the suitability and breeding conditions of various crops in Bangladesh, BRRI perhaps holds the single most important authority on the new rice breeds. Once new seed varieties are developed by BRRI, experiments conducted by the Soil Fertility Testing Institute (SFTI) under farmers' field conditions form the basis of the recommended 'package' of inputs (especially the recommended fertiliser use).

Although, to some extent, the research institutions are also responsible for the popularisation of their varietal tests, the major onus of disseminating the results of research to farmers lies with the extension agents, mostly state-sponsored. Traditionally, this has been the responsibility of the Directorate of Agricultural Extension and Management (DAEM) under the Ministry of Agriculture. The main objectives of DAEM include the following: (i) disseminating the results of research and demonstrating scientific farming techniques to farmers; (ii) motivating and aiding the farmers to adopt the new agricultural innovations in order to enhance productivity; (iii) providing information and communication service from various departments of the Agricultural Ministry to the farmers; (iv) maintaining liaison between farmers and various government and non-government agencies, etc. Although DAEM is making a notable effort to carry the extension and communication channel on a nationwide basis, the crucial training part is yet to be established for want of infrastructural facilities and other logistics. As conceded by DAEM, the system has also certain inherent weaknesses which render the traditional methods of technology transfer rather ineffective. These include: (i) lack of effective linkage between research and extension; (ii) inability of an extension agent to cover the jurisdiction allotted to him (coverage extends to about 2,000–3,000 families per agent); (iii) expansion and co-existence of a large number of single crop extension-oriented agencies providing, through their respective network, quite contradictory information and motivation. As a result, the impact of 'traditional' extension work on the diffusion of the new technology has been minimal.

To help and motivate farmers to adopt the new technology, there are a number of state-sponsored agencies which supply modern inputs to farmers, at subsidised prices. Several agencies are simultaneously involved in supplying agricultural credit (e.g. commercial banks, BKB, KSS, etc.) and irrigation equipment (BADC and BWDB). Besides supplying credit, some of these agencies also conduct field experiments and provide a non-traditional channel for extension work. The performance of the BADC with respect to fertiliser distribution has been quite impressive (if such performance is to be gauged by the increase in the amount of fertilisers distributed annually). The increase in irrigated area has been rather modest (see table 5.2) and it would take quite some time before the country can come up with a comprehensive solution to the country-wide irrigation and water management problem.

Institutional reforms and peasants' socio-economic environment

While, on the one hand, efforts are aimed at adaptive research and the supply of modern inputs, there is also the need for such institutional and structural

230

reforms that would cater to the demands of the new technology. The co-operative movement represents the most important of such institutional reforms. Although the co-operative movement has a long tradition, the impact has never been very encouraging. As such, co-operatives would require a fresh impetus to rise to a changing set of agrarian forces. It is to this end that the present co-operative movement, under the banner of the IRDP, has set a country-wide network in an effort to replicate the results of the so-called 'Comilla experiment'.[35] Because of the importance attached to it, both by national and international sponsors, we take a close view of the performance of the IRDP co-operatives in the following section.

Further motivation towards technology adoption is provided by other development agencies or programmes outside the formal public institutional framework. Most of these programmes are funded by foreign sponsors although the operational responsibility lies with local organisations. The purpose of these programmes is diverse and multifarious, most of them having an expressed willingness to support the improved standard of living of the poor and attainment of higher levels of agricultural production through a successful transfer of technology. The organisations, whatever their merits, have a distinct impact on the social, economic and cultural forces in the countryside, although the geographical coverage of these, considered individually, may not be high. Most carry out "ideas" in an experimental way through pilot schemes covering one or few villages and/or thanas. Some of the resourceful ones attempt replication over wider areas in successive phases.[36] Because of the huge number of these organisations and particularly because of the fragmented nature of their operational process, it is extremely difficult to assess comprehensively the impact these have had on the issue of technology transfer.

The effectiveness of the Government's efforts at motivating farmers to innovate would depend on the rural socioeconomic structure, which in turn would determine the expected gains of different farmer

groups from the attempted reforms (see the upper
right-hand side of figure 5.1). While government-
sponsored agencies can attempt to resolve the supply
side constraint (by supplying inputs at subsidised
rates), it is much more difficult to cope with
another set of constraints, imposed by the prevailing
rural socioeconomic institutions. This complex
scenario of rural institutions, which has evolved
over generations through the interaction of social,
economic and demographic forces, is characterised by
such features as the patron-client relationship,
personalised relationships coexisting with imperfect
commodity and factor markets, the institution of
share-cropping, and the concentration and
fragmentation of landholdings. We have already
discussed in the earlier sections how these factors
work differently on different groups of farmers in
their response to the appeal of the new technology.

The social and economic institutions, mentioned
above, have a significant role to play in
conditioning peasants' choice in production
decisions. While external efforts are geared toward
a systematic transfer of technology through formal R
& D and extension and supply of material resources,
the peasants themselves may still not find it easy to
adapt to changing circumstances. They may also
evolve their own strategies by sticking to the
traditional and well-practised technology, shifting
to the new technology, or "creating" a hybrid between
the two. This is tantamount to 'learning-by-doing',
what one would call peasants' informal R & D. (This
is shown in the upper left-hand side of figure 5.1)
The nature of informal R & D, one must note, is
initially conditioned by the peasant's own resource
position, and the socioeconomic framework within
which he operates. Such a process of innovation,
however, has its own obvious limitations, especially
with regard to efficiency considerations. The proper
role of informal R & D is to provide feedback to
research institutions so that scientific research can
be directed to suit the farmers' socioeconomic
environment.

INSTITUTIONAL BIASES AND DISTORTIONS

In the previous sections, we have attempted to provide a schematic view of the institutional linkages in the process of generation and diffusion of the new technology. Here, we shall argue that, besides the various limitations of the system already discussed, there are strong biases and various distortions in the thrust on institutional development in this country. These biases, by and large, tend to work in favour of the larger farmers who, we have seen, have not shown any encouraging signs of adoption, and against the smaller farmers, who despite formidable resource constraints, have registered better adoption rates.

Broad-based agricultural strategy: The role of co-operatives

We have seen that the co-operatives constitute the single most important nation-wide institutional network at the grass-roots level, directed toward peasants' adoption of the new technology. The nature of the co-operative based agricultural growth strategy, implemented through the IRDP, is essentially structured on the widely-known "Comilla model", first started in Comilla Kotwali thana in 1961. The initial 'success' of the model encouraged the Government to extend it to all the administrative thanas of Comilla, and later, in the early 1970s, over entire rural Bangladesh. We do not have adequate data to provide a comprehensive account of the IRDP's performance, although an overall assessment may be made from the long experience of the Comilla experiment which, after all, has provided the basic structural guidelines to the IRDP.[38] The 'integrated approach' as envisaged in the Comilla model consists of four different components: (i) a two-tier co-operative structure – village-based primary co-operatives, known as Krishi Samabaya Samiti (KSS) and their federation at the thana level, known as Thana Centre Cooperative Association (TCCA); (ii) the Thana Training and Development Centre (TTDC), responsible for providing advanced knowledge on better farming techniques to 'higher' officials and 'model' farmers of KSS; (iii) the Thana Irrigation Programme (TIP) for providing subsidised irrigation equipment (especially tube-wells and low-lift pumps); (iv) a rural works programme (RWP), to mobilise idle manpower for infrastructural investment.

The fundamental objective of the Comilla model was to protect the small, potentially viable farmers and to make available to them cheap credit and subsidised inputs to enable them to participate in the new technology strategy. In many ways, the Comilla model may be seen as having produced a sharp impact on the farmers of the Comilla Kotwali Thana. The actual breakthrough came after 1968 when the HYVs were introduced, and the co-operative members, relative to the rest of the cultivators, were in principle better equipped to adopt the new technology. The percentage of total acreage under irrigation rose from 2.2 per cent in 1964/65 to about 32 per cent in 1973/74, thus boosting up the yield rates of boro paddy in particular. Over the decade, total production of rice doubled. This is indeed a rather impressive performance which, however, must be weighed against the costs of the programme: (i) the Comilla experiment received a strong booster, not only in the pioneering efforts of Akhter Hamid Khan, but also from the Government's massive allocation of cheap credit and subsidised inputs. Besides, the investment funds for TIP and RWP were also largely provided by the Government. Thus the success of the programme must be seen in the backdrop of abundant and heavily subsidised supply of inputs and other public investments; (ii) despite attainments in higher productivity and income, the dependence on cheap credit grew, which enabled the members to relax efforts at pooling private resources for productive investment. There was also an increasing number of defaulters, mostly among the larger farmers, which seriously affected the credit scheme. Khan (1979), commenting on the defaulters, maintained that "it was not so much the inability to repay as the probability of getting away with it that determined the degree of default"; (iii) with respect to the public investments in the area (e.g. tube-wells and power pumps) a high degree of unutilised capacity was reported; (iv) the proportion of members to non-members was seen to increase with farm size; the 'managing committees' of the co-operatives also contained a higher proportion of members from the large landholding groups. Given this situation, it is not unlikely that the benefits offered by the co-operative would largely accrue to the higher farmers, hence defeating a major policy objective of the system. Further, it has been observed, both in the case of Comilla Kotwali Thana and other IRDP project areas, that the members of the co-operatives have not performed any better than the non-members with respect to the adoption of HYVs. While farmers outside the co-operatives (usually smaller farmers) tended to lag behind the member farmers in the early years of the programme, these lags typically disappeared within a few years. This would perhaps indicate that, given proper infrastructural development and availability of complementary inputs, some growth and technology adoption can be achieved anyway, independent of such institutional movement and without incurring such enormous costs in subsidies. One may, however, concede that such organisational structures may provide a basis for spreading knowledge about the new technology and improved techniques of production (mostly among the so-called 'progressive leader' farmers) such that the non-members, too, can emulate. Once the new technology is rooted among the smaller non-member farmer, they are seen to perform no worse (if not better) than the larger farmers. Such performance, however, falls short of the potential that the new technology carries, primarily because of its resource intensiveness which the small farmers often fail to meet.

A reversal of the institutional bias in favour of the small farms would inevitably require changes in the agrarian structure which would indeed prove formidable and depend on how far it is feasible and politically expedient. This would entail some kind of land reforms, which may range from marginal to radical measures. One must note, however, that a policy of land reforms by itself would not be effective, unless such reforms are accompanied by other complementary institutional support. We have contended that, although small farms' adoption rate is relatively high, this may only be true up to a certain extent. The bulk of the small farmers may still not be capable of accepting the challenge of the resource-intensive technology. As such, egalitarian policies need not always be efficient. One may, however, still make a strong case for 'distributivist' land reform by associating it with the current agrarian policy of the IRDP-led co-operativisation. We argued that one of the reasons why the co-operatives were inefficient was that these were mostly managed by larger farmers, who, in spite of receiving the greater share of the benefits, did not engage in higher investment in the new technology. By breaking the domination of the rice (through distributive policies) and by public investment (through IRDP) in infrastructural development along with the assured supply of inputs (which need not be subsidised at present rates), one may seek the possibility of the HYVs spreading among the smaller farms quite rapidly.

The above strategy would obviously call for a Draconian step on the part of the administration – a step perhaps beyond its will and political ability. Until that happens, future scientific research must contend with the problems posed by the existing rural socioeconomic structure.

Biases in scientific research

Within the context of the prevailing socioeconomic structure, the direction of scientific research in the country suffers from many biases, most of which again work to the disadvantage of the small farmers. However, as we have already argued, these biases arise from certain exogenous factors and institutional weaknesses, rather than from any 'circularity of causation' linking the perceived gains of the political groups with the direction of research, as hypothesised by Ruttan (1980). The base of scientific research in the country is still too

weak to respond to any induced demand for technology
changes generated within the country. Even if the
research capability is there, due to the
institutional weaknesses, there is little feedback
from the farmers' field to the research stations (see
figure 5.1). The attitude of the scientific
community is also partly to be blamed. The standard
of evaluating achievement in scientific research is
set in the light of the training they receive abroad
- getting maximum yield under ideal input use, rather
than maximising yield subject to farmers' resource
constraint.

The lack of a 'feedback' process is demonstrated by
the fact that many locally-improved rice varieties,
which evolved through farmers' own informal R & D,
did not attract the attention of BRRI breeders until
very recently. Yet, some of these varieties (Pajam
being most notable among these) are extremely popular
in certain parts of the country, particularly when
compared to the 'official' HYVs. The spread of these
varieties has taken place without any government
patronage. It is only very recently that the rice
breeders in the country have started working with the
aim of incorporating the yield potential of the IRRI
varieties in some of these locally-improved varieties.

One of the reasons, often cited by the extension
agents, for the popularity of the many
locally-improved varieties (compared to the official
HYVs) is the relatively low input costs of these
varieties. The yield from these varieties can be
improved by demonstrating to farmers the value of
using balanced fertiliser doses and of better
cultural practices (e.g. line-sowing, etc.). There
is clearly a need for more research to evaluate the
performance of the locally-improved varieties with
improved husbandry methods, keeping in view the need
for developing low-cost packages suitable to small
farmers.

As regards the officially released HYVs, the
so-called 'input package' mainly comes in the form of
recommended doses of fertilisers along with the use
of water. There has been little research in the
field experiment stations to evaluate the
quantitative impact of labour-intensive farm

management techniques on HYV yields, or to evaluate the yield response to fertiliser at the lower end of the application range. This has rendered the so-called 'input package' quite irrelevant for the smaller farmers who use more labour-intensive techniques and less material inputs. Another drawback of the present strategy of HYV research and extension is that it refers to single-crop input packages and not to a full agricultural-year cycle. This again works to the disadvantage of smaller farmers who cultivate their land more intensively through multiple cropping, in order to utilise their excess family labour.

In the light of our above discussion, we can refer back to Hayami and Ruttan's (1971) contention that the recent development of HYVs in tropical Asia represents a 'rational response' by public agencies to factor availabilities and commodity prices (especially the fertiliser-rice price ratio). Given the international fertiliser-rice price ratio, HYVs are indeed immensely profitable from a social cost-benefit point of view. However, rural institutions and infrastructure may not be sufficiently developed to reflect such profitability in the decision-making of farmers of different size-groups with different resource endowments and having differential access to commodity and input markets. Financial constraint renders profitability almost an irrelevant criteria for small farmers. Induced transfer of technology in this context should then imply either developing institutions to make resources available to the farmers who matter, or to conduct research such as to find varieties requiring less of scarce material inputs (here, e.g. fertiliser) and more of the abundant factor (labour) - a situation that reflects the production conditions for the majority of Bangladeshi farmers.

SUMMARY AND CONCLUSIONS

There is a growing volume of literature on the role of institutional factors in the generation and diffusion of technological innovations in agriculture. This study represents an effort to achieve a better understanding of these institutional factors in Bangladesh's setting, characterised by extreme conditions of poverty, landlessness and population pressure. Methodologically, the study attempts an application of Professor Vernon W. Ruttan's induced innovation model which seeks to explain the direction of technological and institutional changes in terms of the national resource and cultural endowments. The study, however, subscribes to a broad concept of institution. On the one hand, there are the various organisations carrying out scientific research and extension work, as well as those engaged in distributing modern inputs and motivating farmers at the grass-roots level. On the other hand, there are those rural institutions of tenancy, markets, etc. which operate within the prevailing socioeconomic structure. For an understanding of the institutional parameters of technology adoption, it is essential to analyse the linkages and interaction between these two broad categories of institution.

Bangladesh farming suffers from two important and pervasive constraints that have a bearing on the adoption of the new seed-fertiliser technology: first is the risk-prone nature of the Bangladesh agricultural environment; second is the small size of farm holdings, fragmented into tiny plots. The nature and extent of technological innovations in Bangladesh agriculture must also be seen against the backdrop of an increasing pressure of population, and accompanied by a clear trend towards increasing concentration of landownership and consequent pauperisation of the peasantry.

The entire crop sector in Bangladesh is dominated by one single crop, rice. An analysis of the factors behind the growth of rice production over the last two decades reveals certain disconcerting phenomena. Although the HYV rice technology was introduced in Bangladesh as early as in the mid-1960s, its expansion has been extremely sluggish; less than one-fifth of the total cropped area under rice was held by the HYVs by the end of the 1970s. The diffusion of the HYV technology, instead of helping in achieving a breakthrough in foodgrain output, has been in fact accompanied by a lower rate of output growth compared to the period immediately before the introduction of the HYVs. While the process of expansion of total cropped area has virtually come to an end, the yield rates of local rice varieties have registered a declining trend, and there remains a large gap between the potential and actual yield rates of the HYVs.

The process of diffusion of HYV rice technology in Bangladesh agriculture exhibits simultaneously many features which are commonly associated with different 'phases' of international technology transfer. Thus, while government efforts are geared towards a systematic transfer of technology through formal R & D, extension and the input supply network, farmers are found to evolve their own strategies, through 'learning-by-doing' or what one could call farmers' informal R & D. Although the imported HYVs have very high yield potential, these are not well-suited to the country's agro-ecological environment. Indigenous scientific research aimed at adapting the HYVs to local conditions has so far met with only limited success, which is one reason why directly imported HYVs still dominate over locally adapted varieties. The so-called 'technology packages', in the form of recommended input combinations, do not also have much appeal to farmers adopting the HYVs.

Evidence from numerous studies, based on farm survey data, suggests that smaller farmers devote a higher proportion of their land to HYVs compared to larger farmers. This is also true of the 1977 Agricultural Census data. Another important finding from these studies is that it is the intensity of adoption, rather than the initial decision to adopt or not, that is likely to be the more important factor in bringing more rice land under HYVs. That this is found to be more true in the case of larger farmers deserves further research.

A clear consensus arising from the various studies is that the yields and net returns per acre are substantially higher for HYVs compared to local varieties. This appears to be the case for all groups of farmers, including share-croppers. The percentage increase in per acre labour input on HYVs over local varieties is much higher for small farms compared to large farms. (The reverse is possibly true for the increase in the cost of material inputs.) This would suggest that the HYV technology, as adopted by small farmers, offers greater scope (compared to local varieties) for increasing production by adopting more labour-intensive methods of cultivation. This is an area where more scientific research should be done to evaluate the quantitative impact of labour intensive farm management techniques on HYV yields, keeping in view the need for developing low-cost packages suitable to small farmers.

The pure profit motive is conditioned by social customs, attitudes and traditions in farmers' decision-making regarding the adoption of agricultural innovations. Owing to various socio-economic factors, landownership and entrepreneurship do not always go together in Bangladesh. In contrast to the experience of the so-called Green Revolution in many developing countries, big farmers in Bangladesh, by and large, have failed to play the 'leadership role', except in a few areas where massive institutional support programmes have been undertaken. As for the smaller farmers, they are found to be more innovative, despite formidable resource constraints under which they operate. A rationale for this is sought in what may be called the "subsistence pressure" hypothesis. Amidst the scenario of ever-increasing landlessness and concentration of landholdings, seeking the possibilities in the technology appears almost like the last straw in the small farmers' bid to survive as family farms (rather than sell their land and join the ranks of landless labourers). However, since small farmers would be willing to take only limited risk, no breakthrough in agricultural output could be achieved through a process of adoption under subsistence pressure alone. On the contrary, the hypothesis is advanced to explain a 'perverse' pattern of the diffusion of HYV technology which is accompanied by increasing impoverishment of the peasantry.

Cross-sectional district-level data are used for running econometric tests on the relationship between agricultural innovations and such factors as farm size, tenancy, literacy, etc. Of the two most important variables, the provision of irrigation is found to have a strong positive effect on the extent of HYV adoption, while farm size (or, alternatively, the ratio of land to family labour) is found to have a negative effect. It is argued that the negative effect of land-labour ratio is more a manifestation of 'subsistence pressure' than of peak-season labour shortage. Contrary to popular hypotheses, tenancy is found to have a favourable effect on HYV adoption. While this result does not resolve the question of so-called 'inefficiency' of share tenancy, it does suggest that, within the present agrarian set-up, tenancy perhaps enables a more intensive use of family labour for adopting innovations – again an indirect manifestation of the 'survival pressure'. The regression results also show that formal education has some favourable effect on the adoption of innovations. As for the intensity of fertiliser use, the provision of irrigation and literacy are found to be the most important explanatory variables. However, the negative effect of land-labour ratio (or of farm size) is not found as strong (or to exist at all) in the case of fertiliser use as in the case of HYV adoption.

The classical model of how research, extension service and the farmer should relate to each other (often represented by a flow chart) is not found to work well in Bangladesh's case because of many institutional weaknesses, biases and distortions. The analysis presented in this study can be considered a partial validation of Ruttan's 'induced innovation' model, in which the supply and demand mechanism for technological innovations is conceptualised as a process of 'circular and cumulative causation'. Compared to Ruttan's scheme, however, the circular chain of causation is found much weaker in Bangladesh's context, particularly in respect of the articulation of the perceived gains of various interest groups into demand for appropriate scientific research. This mainly reflects upon the underdeveloped stage of indigenous scientific research and the predominance of imported technology. The linkages are, however, found much stronger in the evolution of the institutions that are engaged in supply inputs and motivating farmers through various reforms. The development and performance of these institutions is seen to be conditioned by the interactions between the socio-economic environment on the one hand and the politico-bureaucratic structure on the other.

The country's thrust on institutional development suffers from many biases and distortions which, by and large, tend to favour the larger farmers. The co-operatives, essentially patterned after the widely-known "Comilla model", constitute the single most important institutional network at the grass-roots level, directed towards peasants' adoption of the new technology. While the Comilla model has been successful in augmenting income and production, it has made a heavy claim on public funds and administrative resources. The co-operatives have a poor record in terms of mobilisation of savings, recovery of public loans, and utilisation of installed irrigation capacity. Besides, the benefits offered by the co-operatives accrue mainly to larger farmers, thus defeating a major policy objective of the programme, namely to protect the small and potentially viable farms. A reversal of the biases in favour of small farmers would inevitably require changes in the prevailing agrarian structure. The combination of a far-reaching land reform with public support programmes directed at increasing agricultural productivity seems to hold out more chances of success than either of the programmes pursued separately. But such a strategy would obviously draw strong opposition from powerful rural interests.

Within the context of the prevailing socioeconomic structure, the direction of scientific research in the country suffers from many biases, most of which again work to the disadvantage of small farmers. Not only is the research capability lacking, but there is also little feedback from the farmers' field to the research stations. Scientific research is mainly directed at getting maximum yield under an 'ideal' input package, rather than maximising yield subject to farmers' resource constraints. There is clearly a need for more research directed at seeking improved varieties for use by small farmers under sub-optimal input use. Moreover, since the socioeconomic, climatic and agronomic conditions vary over different parts of the country and across farm-size groups, a decentralised research system is likely to be more responsive to the needs of various regions and different farmer groups.

Finally, any great disparity in farmers' resource endowments makes the task of adaptive research enormously difficult, since the type of technology demanded by different farmer groups would be quite different in respect of labour and capital intensities. Such a situation also calls for a closer co-ordination between scientific research activities and the institutional efforts directed at removing the socioeconomic constraints under which farmers of different size-groups operate.

NOTES

[1] See Khan (1977) and Mahmud (1983).
[2] See Mahmud (1983).
[3] See Mahmud (1983).
[4] See Hossain (1981).
[5] For analysing the growth of rice output in Bangladesh, Mahmud (1983) follows a decomposition scheme that distinguishes the contribution of different elements such as changes in area, yield and the crop-mix.
[6] For a detailed discussion on this, see Pray (1980).
[7] The introduction of superior rice varieties, such as those from Japan and Egypt, dates back to a much earlier period, but these varieties received only limited acceptance by farmers.
[8] For a detailed discussion, see BRRI (1977).
[9] See BRRI (1977).
[10] Ibid.

[11]   The farm survey data of the AER, as reported by Khan (1981) shows substantial shortfall in the use of fertilisers from the recommended dose.

[12]   See Schultz (1964).

[13]   See Hayami and Ruttan (1971).

[14]   See, among other studies, Griffin (1974).

[15]   See Hossain (1980).

[16]   While yields per acre can be estimated unambiguously in physical terms, the estimates of net income differ as regards how to impute cost of family labour and whether to use full cost or the cost of cash-purchased inputs only.

[17]   See Hossain (1981).

[18]   Rahman's (1981) study of Comilla farmers is a typical example which conforms to all these generalisations.

[19]   See Ahmed (1980).

[20]   See Hossain (1981).

[21]   See Rahman (1979) and Hossain (1981).

[22]   See Mahmud (1983).

[23]   See, for example, Rahman (1981) and Abdullah (1976).

[24]   Feder et al. (1981) quotes many econometric studies on agricultural innovations which neglect this problem of simultaneous decision variables.

[25]   This is because of the small size and excessive fragmentation of land-holdings.

[26]   When the percentage of HYV acreage is to be explained, ordinary linear regression entails some specification bias (since the value of the dependent variable is bounded by 0 and 100), and can occasionally produce nonsensical predictions outside the interval 0 to 100. The problem of mis-specification is probably less acute in our case, since the _observed range_ of variation in the adoption rate is much smaller in the district-level data that we have used, compared to data on individual farms which are commonly used in empirical studies. In fact, in some of our regressions, we have tried the alternative specification

$$Y = \frac{1}{1 + e^{-a_0 - a_i X_i}}$$

which satisfied the boundedness condition for Y (in the limit 0 to 100) and can be estimated by the OLS method by logarithmic transformation. Compared to the ordinary linear regressions that we have reported, there was no systematic variation in the value of R^2 and the significance of coefficients was not affected in most cases.

[27] One may suspect here some multicollinearity problem to exist, since the incidence of tenancy may be positively related with land–man ratio and the concentration of land–holdings; see Hoosain (1980). In our case, however, the simple coefficient of correlation between TNC and LM is only .23, so that there is no serious multicollinearity problem.

[28] Many studies have found that the pattern of adoption may depend on the length of time since the introduction of HYVs; see, for example, Schulter (1971).

[29] See, for example, Harriss (1972).

[30] See Khan (1977) and Mahmud (1983).

[31] See Ruttan (1980).

[32] Ibid.

[33] Ibid.

[34] See Mahmud and Osmani (1980); also see Mahmud (1980).

[35] For a discussion on this, see Abdullah (1976).

[36] The DANIDA scheme in Noakhali, for instance, is being considered for replication by the DANIDA/SIDA-sponsored 100-thana Intensive Rural Works Programme.

[37] See Biggs and Clay (1981).

[38] Khan (1979) critically discusses the experience of the "Comilla experiment" while Abdullah (1976) evaluates the performance of the IRDP co-operatives.

6 National research systems and the generation and diffusion of innovations: Horticulture in Kenya

M. J. DORLING

INTRODUCTION

For a proper understanding of the role and performance of institutions in the research system in Kenya, particularly the component responsible for generating and diffusing innovations in the horticultural industry, the scope of the chapter includes extension, government administration and policy-making linkages.

Horticulture in Kenya is served mostly by a separate and well defined institutional research structure. It therefore constitutes a convenient sub-sector for study. But it is also important from food production and nutritional standpoints, showing different crops and production techniques from those in the rest of agriculture. Consequently, findings from the study contribute towards knowing what affects a country's ability to develop and adopt agricultural technology, a key aspect in improving productivity and increasing employment. In doing this the study specifically focuses on the most intensive crops grown, namely: vegetables, fruit and ornamentals. This particular emphasis and the

analytical approach used are likely to have wide
interest for those concerned with horticultural
innovation in developing countries.

Objectives

The general objective of the study is to understand
the institutional structure, conduct and performance
of the national research system, in its function of
furthering the generation and diffusion of technical
innovations in the horticultural industry in Kenya.
  More particularly the study helps to:
(a) show whether Kenya has used successfully its
    capacity in national research to generate,
    develop and innovate appropriate horticultural
    technology in order to facilitate growth in
    productive employment;
(b) show what types of technology have been stressed,
    e.g. yield-increasing technology, and whether
    these types are the most advantageous under
    existing conditions;
(c) show the part played by economic and political
    factors in the success or failure of
    institutional development in research to provide
    suitable technology;
(d) show whether the national research structure is
    adequate in terms of complementing and
    supplementing introduced technology.

Definitions

Horticulture in Kenya is defined by the fruit,
vegetable, flower and foliage crops that are
officially recognised and classified as horticultural
produce. The constituent horticultural crops are
easily identified by reference to statistical reviews.
The concept of an institution will also include that
of an organisation. Following Ruttan (1979), the
concept of an institution refers to:
(a) the behaviour of a particular institution;
(b) the relationship between an organisation and its
    environment;
(c) the rules that govern behaviour and relationships
    in an organisational environment.

Within this framework induced institutional change and innovation can take place. While induced institutional innovations are of interest in a horticultural research system, they are not to be confused with technical innovation in horticulture.

Hypotheses

The three main hypotheses stated by Ruttan (1980) are used and tested as guides for the analysis. These arise from the classification of institutional factors, determining the productivity of national agricultural research, into those concerning (a) the structure and organisation of the research or extension system itself, and (b) the economic and political organisation of the economy for which the research and extension is being developed. The consequent hypotheses are:

(1) The development and the performance of agricultural research systems are conditioned by the national resource and cultural endowments. This general hypothesis permits the formulation of specific hypotheses regarding (a) the path of technical change, (b) the economic and political demand for technical change, and (c) the political system influencing the latent demand for technical change.

(2) An agricultural research system which is decentralised with respect to location, management and funding, tends to be more responsive to the resource and cultural endowments, and hence more productive in supplying new technology to farmers, than a highly centralised system.

(3) An agricultural economy characterised by a great disparity in size (or income) of farms will be less effective in generating a sustained demand for a productive national agricultural research institution than an agricultural system characterised by reasonable equity in farm size distribution.

Methodology

Qualitative and quantitative analysis is used to
establish essential facts, perform tests of
hypothesis and enable relevant conclusions to be
drawn. Interest centres mainly on contemporary
experience, but the approach allows an historical
perspective to be given to the development of the
horticultural research system and its influence on
innovations. The effects of agrarian structure, and
economic and political factors in this process are
given consideration.

Initially a comprehensive literature review is
attempted with the purpose of extracting essential
information from available sources. An interview
schedule was also used to record answers to a set of
core questions, put to senior personnel in (a)
horticultural research stations, and (b) important
government, parastatal, educational and industry
bodies having direct links with the research stations.

THE HORTICULTURAL INDUSTRY

Horticultural crops

A traditional method of identifying horticultural
crops is simply to categorise produce into fruits,
nuts, vegetables, cut flowers, foliage and pot
plants, and nursery material. In general this
approach serves very well and is a reminder that much
of commercial horticulture is the intensive growing
of garden-type crops on a field scale for
profit-making purposes.

Table 6.1 draws on quoted sources listing the most
important commercially produced horticultural crops
in Kenya today. The items include potatoes, sweet
potatoes and coconuts, which are important crops in
Kenyan agriculture, yet do not fit easily into a
purely horticultural framework. In fact the same can
be said about vegetables, such as peas and cabbages,
which can be grown with farm crops and become
somewhat separate from the horticultural context.
The list does not include coffee and tea, perennial

crops which are vital to Kenya's export earnings from
agriculture, and which under some classifications are
considered branches of horticultural activity. But
these crops have qualified from the early days of
their introduction in Kenya for specialised and
separate research station facilities, and therefore
can be seen as field crops in the same way that other
crops, such as cotton, pyrethrum, rice and tobacco,
are classified. Cassava and cocoyams have also been
excluded on the grounds that both are usually
classified as field crops, the former being a staple
food crop in many parts of Kenya and the latter
confined to stream banks and swampy areas.
Recognition of less important horticultural crops is
given in the footnote to table 6.1.

Annual horticultural crops, which include many of
the vegetables and cut flowers, afford a high degree
of flexibility and diversification. Intermediate in
this sense are asparagus, pineapples, bananas,
passion fruit and strawberries. The most fixed and
inflexible crops are citrus, tree fruits and nuts.

The wide variety of vegetables, fruits, nuts and
other horticultural crops which are grown includes
tropical and temperate types. In short, Kenya, which
straddles the equator at varying elevations, offers a
wide range of climates and soils, and these, in
conjunction with irrigation, allow many horticultural
crops to be grown successfully on commercial and
smallholder scales, and under very intensive
conditions. The reason for the importance of
temperate region vegetables in commercial
horticulture is one of coincidence of favourable
growing conditions for producing high yielding and
nutritious crops. These are generally to the taste
of urban and rural dwellers alike, and frequently
offer the best opportunity for improving dietary and
food consumption levels.

Economic importance

While first-hand observation of holdings and country
and city markets affirms that horticultural activity
is an important part of the agricultural economy in
Kenya, there is a lack of economic data permitting a

Table 6.1
Commercially produced horticultural crops in Kenya, 1981[a],[b]

Vegetables	Fruits	Cut flowers, foliage and nuts	Nursery pot plants	Nursery material
Asparagus	Avocados	Cashews species and varieties	Numerous	Cuttings
Artichokes: Globe Tuberous	Bananas	Coconuts		Plants
Beans: Broad Green	Citrus fruits: Grapefruit Limes Oranges Tangerines	Macadamias		Scions
Beetroot	Mangoes			Seeds
Brassicas: Broccoli Brussels sprouts Cabbage Cauliflower Kale	Melons Sweet Water			Stocks
Brinjals	Pawpaws			
Capsicums (Sweet peppers)	Pineapples			
Carrots	Temperate fruits Apples Grapes Peaches Pears Plums Strawberries			
Celery Chillies (Hot peppers)				

Vegetables	Fruits	Cut flowers, foliage and nuts	Nursery pot plants	material
Cucurbits:				
Courgettes				
Squash				
Cucumber				
Marrows				
Leeks				
Lettuce				
Mushrooms				
Okra				
Onions:				
Bulbs				
Green				
Peas				
Potatoes				
Rhubarb				
Radish				
Spinach				
Sweet corn				
Sweet potatoes				
Tomatoes				
Turnips				

[a] Sources: Horticultural Crops Development Authority (1974); Agricultural Information Centre (1979); Horticultural Crops Development Authority (1979).

[b] Many other horticultural crops occur in addition to those mentioned under the headings, although in general they are of less importance and often less well known. For example: vegetables (including Asian types): dudhi, Chinese cabbage, karelas, loofah, mooli, cow peas (green), fennel, horseradish, kohlrabi, parsley, parsnips, pumpkins, watercress; fruits: apricots, blackberries, cape gooseberries, custard apples, guavas, kumquats, litchis, loquats, mulberries, nectarines, plantains, quince, tree tomatoes.

rigorous quantitative assessment of its components and overall position. But some hard quantitative evidence is available and selected aspects of it are discussed below.

Value of production

Table 6.2 shows the values of different crop categories in Kenya. While a concise summary of horticultural contribution is not given, it can be seen that potatoes, fruit and vegetables, bananas and plantains, pineapples and cashew nuts account for around 12 per cent and 9 per cent in 1976 of actual total crop and total agriculture production values, respectively. Estimates for 1978 and target figures for 1983 suggest that these contributions would not change much. A large share of horticultural produce, particularly fruit and vegetables, is directly consumed on smallholdings by the rural population, and, although included in the assessment, it may well be underestimated. Another source of underestimation can arise from the many decentralised transactions which characterise rural marketing arrangements.

Production by farm size

Breakdowns of agricultural statistics in Kenya are affected by the polarised size-structure of farms, namely large farms and small farms. There were 3,350 large-farm households in 1976, compared to 1,417,926 small-farm households and a further 4,750 small households in irrigation schemes (Crawford and Thorbecke 1978). In the large-farm areas the average size of farm is over 700 ha, although some relatively small units do occur. Small farms are generally between 0.2 ha and 12 ha in size, but if larger they do not exceed 20 ha.

Table 6.3 gives a breakdown of land utilisation for large farms in 1977. Root-crops and vegetables amount to only 3,781 ha, with most of the area being accounted for by the Rift Valley, Central and Eastern Provinces. Coconuts and cashew nuts, with 846 ha and 1,300 ha respectively, occur solely in the Coast

Table 6.2
Total value of production of agricultural commodities in Kenya, 1976, 1978, 1983*

Items	1976 Actual	1978 Estimate	1983 Target
	(K£'000 at 1976 prices)		
Food crops:			
including – Potatoes	20,400	22,200	27,400
Fruits and vegetables	8,346	9,399	14,469
Bananas and plantains	11,600	12,650	16,550
Total	197,792	213,225	262,317
Industrial crops			
Total	16,582	17,606	33,820
Export crops:			
including – Pineapples	1,314	1,823	3,562
Cashew nuts	1,159	1,546	2,318
Total	142,746	175,260	214,779
Livestock products:			
Total	129,436	137,319	171,703
Total agriculture	486,556	543,410	682,619

* Source: Republic of Kenya (1979).

Table 6.3

Large farm land utilisation in Kenya, 1977*

Crops	Tl ha
Cereals:	
Wheat	80,853
Maize	85,442
Barley	10,883
Oats	3,941
Others	800
Total	181,919
Temporary industrial crops:	
Sugarcane	32,198
Pyrethrum	2,313
Sunflower	3,652
Others	464
Total	38,627
Root crops and vegetables	3,781
Temporary fodder crops	7,989
Other temporary crops	126,149
Temporary meadows and fallow land	145,708
Permanent industrial crops:	
Sisal	67,507
Tea	24,679
Coffee	29,982
Wattle	11,694
Coconut	846
Others	2,132
Total	136,840
Fruit:	
Cashew nuts	1,300
Pineapples	5,400
Others	998
Total	7,698
Uncultivated meadows and pastures	1,873,600
All land	2,522,311

* Source: Central Bureau of Statistics (1979).

Province. Pineapples occupy 5,400 ha and are concentrated mostly in Central Province. Even though some horticultural use of land on large farms cannot be detected in the breakdown, it is clear that less than 1 per cent of the total land area in this category of farms is concerned with horticultural production. In fact the large farms engaging in horticulture tend to be few and highly specialised. Pineapple and apple growing are examples of large-scale operations, with one company in each case accounting for practically the whole commercial output. Furthermore, large-scale production of vegetables, mainly for local factory drying and dehydration, and of carnations takes place on the shore-line of Lake Naivasha. Also it is evident that some cashew nut and coconut production at the Coast is of a large-scale nature, although, as with vegetables and cut flowers, the greater share is accounted for by small producers.

Turning to the small-farm sector, table 6.4 gives land utilisation for crops in the 1974/75 year. The data show clearly that the major part of horticultural output comes from small farms. English potatoes were produced on 261,200 ha, mostly in the Central and Eastern Provinces. Sweet potatoes took up 32,600 ha and were concentrated in Central and Nyanza Provinces. Bananas are almost exclusively grown on smallholdings in Kenya and accounted for 130,400 ha. Other fruits used 13,500 ha and vegetables 56,000 ha. Lastly, coconuts and cashew nuts claimed 51,300 ha and 53,500 ha respectively. These are large areas by any reckoning, and, as might be expected, by far the largest shares of all the crop areas are associated with mixed cropping as opposed to pure stands.

The other category of small farms, which is of considerable importance in contributing to horticultural production, is that concerned with irrigation schemes. Of the five major schemes only Perkerra near Lake Baringo is of interest, because it specialises in producing onions and chillies along with small quantities of melons and other horticultural crops. The tonnage of onions and chillies from the Perkerra scheme in 1978/79

Table 6.4
Small farm land utilisation in Kenya, 1974/75[a]
(Integrated rural survey)[b]

	Total '000 ha	
	Pure	Mixed
Cereals:		
Local maize	224.6	970.0
Hybrid maize[c]	258.2	242.6
Finger millet	30.5	47.4
Sorghum	16.8	189.6
Other cereals	18.5	93.4
Pulses and nuts:[d]		
Beans	49.9	713.6
Cow peas	11.7	259.5
Pigeon peas	0.1	115.2
Field peas	4.1	12.3
Groundnuts	3.5	14.3
Other	1.1	36.3
Root crops:		
English potatoes	48.9	212.3
Sweet potatoes	10.9	21.7
Cassava	41.2	28.7
Other	17.7	24.4
Fruit, vegetables and oils:		
Bananas	19.6	110.8
Other fruits	1.2	12.3
Vegetables	4.0	52.0
Oilseeds	13.0	11.5
Temporary industrial crops:		
Sugarcane	55.0	8.7
Pyrethrum	22.4	4.7
Cotton	25.0	45.1
Other	2.6	3.4
Permanent crops:		
Coffee	92.0	19.3
Tea	59.0	5.8
Coconuts	2.0	49.3
Cashew nuts	5.5	48.0
Other	23.1	10.5

[a]   Source: Central Bureau of Statistics (1979).
[b]   Includes areas planted in both long and short rain seasons and excludes pastoral and large farm areas.
[c]   Includes yellow and synthetic maize.
[d]   Excluding cashew nuts.

258

represents a major source of supply in the country
for these popular crops. In addition to Perkerra
several small-scale irrigation schemes are either
producing horticultural produce or are planned to
start shortly.

Consumption, marketing and processing

With so large a share of the population in Kenya
engaged in or dependent on smallholder agriculture,
it is to be expected that intensive horticultural
crops of all types form an important part of the diet
in rural areas. Table 6.5 shows the average value
per smallholding of household food consumption in the
1974/75 period to be 2,594 Kenya shillings, composed
of own produced and purchased items. Of this total
around 10 per cent was accounted for by English
potatoes and other crops (including vegetables) from
the holding, while about 3 per cent went on purchased
fruits and vegetables. Nutritionally, this
horticultural produce forms an essential source of
vitamins in the food of many families which are
heavily dependent on traditional staple crops.

In the larger towns and the three large cities,
Nairobi, Mombasa and Kisumu, the high, medium and low
income groups are all heavy consumers of fruits and
vegetables. Each of these income groups tends to
have its own specialised demands for different types
of fruit and vegetables. Also in the urban centres,
as well as in remote areas sometimes, the hotels and
restaurants, catering to an increasing domestic
clientele and tourist trade, create a constant demand
for fresh and processed horticultural produce.

The market structure for fresh horticultural
produce is really a subject in itself, and it has
come in for sporadic comment and formal study as the
inadequacy of facilities, in the face of expanding
supply and demand, has become a necessary target for
action.[1] The market system can be divided broadly
into (a) the three municipal wholesale markets in
Nairobi, Mombasa and Kisumu, and (b) the numerous
retail markets run under the control of local
authorities in the urban and rural centres of
population. Market structure and efficiency, both at

Table 6.5
Percentage distribution of household food
consumption by type of food, 1974/75*
(Integrated rural survey of small farms in Kenya)

	Per cent
Own-produced items:	
Maize	14.85
Finger millet	0.66
Sorghum	1.66
Beans	6.32
English potatoes	4.43
Other crops	5.86
Beef	0.96
Other meat and poultry	3.66
Milk	11.57
Total consumption of own produce	50.00
Purchased items:	
Dairy products and eggs	1.77
Grains, flours and root crops	19.20
Meat and fish	9.10
Fats and oils	3.20
Sugar and sweets	6.63
Fruits and vegetables	3.39
Drinks and beverages	5.40
Salt and other flavourings	1.35
Total food purchases	50.00
Total food consumption	100.00
Total value of consumption (KSH) per household	2,594

* Source: Central Bureau of Statistics (1979).

the wholesale and retail levels, play a crucial part
in the economic success and expansion of the
horticultural industry. But Kenya is fortunate in
the sense that while market facilities are often
found inadequate in many aspects of quality
maintenance, regulation, modern facilities and
administration, there is nevertheless the continuing
stimulus provided by a rapidly expanding population

and a modern transportation system, which, under the favourable profit margins, leads to impressive review and planning studies, market flexibility and overall perceptible development.

Without processing certain horticultural crops would find only limited home and export markets in fresh sales. A good example of this kind of reliance is that shown by passion fruit, which as fresh fruit is demanded only in small quantities. However, given the existence of facilities for juice extraction, a very attractive product can be made. Noticeably increased ranges of products from processing occur for tomatoes, pineapples and mangoes, where canning and juicing concerns all three, and jam can be made from the latter fruits. In the case of vegetables, not only canning increases the product range. Drying and dehydration processes permit this type of produce to be concentrated in forms which are suitable for packet soup manufacture and other food preparations.

At present the main processing plants for horticultural produce in Kenya are few in number. They are located near the main areas of supply, close to large centres of population. The list runs as follows:
(a) Kenya Cashew Nuts Ltd. - Kilifi (Coast Province);
(b) Kenya Orchards Ltd. - Machakos (Eastern Province) (canning, jams, juices);
(c) Kenya Canners Ltd. - Thika (Central Province) (pineapple - canning, jams, juices);
(d) Kenya Fruit Processors Ltd. - Thika (Central Province), Kisii (Nyanza Province) (passion fruit - juice);
(e) Pan African Vegetable Products Ltd. - Naivasha (Rift Valley Province) (vegetable - drying and dehydrating);
(f) Kabazi Canners Ltd. - Nakuru (Rift Valley Province) (tomatoes - canning, juice).

Imports and exports

Kenya is quite self-sufficient in all types of fruit, vegetables and decorative plant material, and is far more concerned with the rewarding exports for this produce than with the low-level imports. Exports of horticultural products have been constantly on the increase in recent years, thereby proving to be a valuable earner of foreign exchange and adding to the list of long-standing agricultural export crops for which Kenya is well-known, e.g. coffee, tea, sisal, pyrethrum. Horticultural exports can be divided into fresh and processed produce, the latter becoming increasingly important. Treating these categories together, the export values of fruit and vegetables shown by the Census Bureau of Statistics (1979) were around K£20 and K£18 millions in 1977 and 1978.[2] These figures are up from about K£4 millions in 1970, illustrating the impressive growth that has taken place.

With this short introduction of the horticultural industry in Kenya, in terms of production, marketing, processing and export operations, the background has been established, against which, development, institutions, and research in particular can be discussed. The next section will attempt to summarise important aspects of development policy and planning, which will then lead on to the study of the national research system serving horticulture.

INSTITUTIONS AND DEVELOPMENT PLANNING

Horticultural Crops Development Authority and
Horticultural Cooperative Union Limited

The Horticultural Crops Development Authority (HCDA) was set up under statutory Order in 1967. This organisation is responsible to the Ministry of Agriculture, and is seen as an instrument for formulating policies for the development of the industry. The Authority has been granted wide regulatory and controlling powers and is intended to have a central position of influence in the horticultural industry. Understandably, though, such a body must take time to evolve into an effective organisation, and so far it has faced serious constraints in the form of inadequate finance and qualified staff.

Zelenka (1975) in a wide-ranging study states the

following significant contributions made by the HCDA:
(a) development of procedures in exports (air space allocation, standardisation of containers, influencing air freight rates, etc.);
(b) establishment of packing stations for smallholders in different areas;
(c) assistance to specific crop production and marketing, e.g. passion fruit, macadamias, onions, new crops (trials), etc.;
(d) development of market information system;
(e) assisting the Ministry of Agriculture in defining the extension and research needs of the industry.

The areas of main commitment for the HCDA in the future are seen by Zelenka as: operational support of the industry, policy formulation and recommendations, extending credit facilities, and initiating horticultural projects lacking private enterprise support.

The Horticultural Co-operative Union Limited (HCU) has an earlier history than the HCDA, and operates directly on behalf of horticultural growers. It was first registered in 1952 and serves 56 primary societies and two district cooperative unions. The aim of the HCU is to serve horticultural producers (both primary co-operative societies and large-scale farmers) by (a) advising on what produce to grow and when to grow it, in collaboration with the Ministry of Agriculture, and (b) marketing the produce at a reasonable price and paying promptly. Its marketing operations are carried out locally and abroad, and advice is given on the grading, packing and quality that is needed, particularly in the world market. While wishing to become a larger and more effective organisation in terms of better facilities and more personnel, it also sees its role as supporting the HCDA, under its advice and guidance, in disseminating information to growers and traders on market trends.

HORTICULTURAL DEVELOPMENT GUIDELINES AND PLAN

In 1973 the Ministry of Agriculture set up a committee with the idea of drawing up horticultural development guidelines. The report by that name, Ministry of Agriculture, HCDA, FAO (1974), had as its main objective the presentation of details of crop policy for a selected list of vegetables, fruits and flowers. The discussion on each crop looked at special features and problems, stressing needed research and bottlenecks in production and marketing.

These guidelines formed a most useful foundation for the Development Plan for Kenya's Horticultural Industry, Zelenka (1975). This plan in turn influenced subsequent development policy for horticulture contained in the Kenya Development Plan 1979–83 (Republic of Kenya 1979). Some noteworthy views on horticultural research appearing in the 1975 plan are listed below:

(1) Recognition was given to the main problems of horticultural research in Kenya identified in the report of the Working Party on Horticulture, set up by the Government in 1969. These ran accordingly: (a) limited funds and personnel, (b) research not being adapted to needs, (c) lack of basic research, and (d) lack of dissemination of results. Among the Working Party's recommendations for combating weaknesses were the needs to create posts in virology, nematology, and plant physiology within Kenya's agricultural research services; and concentrate research in horticulture at the Thika National Horticultural Research Station and at Naivasha, where multi- and bi-lateral assistance could be used to secure the services of skilled workers.

(2) Staffing at the main National Horticultural Research Station, Thika, and at the Molo and Mtwapa sub-stations, was thought to be inadequate. Strengthening of specialist staff at the Thika station should involve the appointment of additional researchers in fruits and vegetables, plant pathology and entomology. Areas of responsibility for the station should extend to nursery stock inspection, plant propagation, plant introduction and testing, and help with the demonstration plots in the horticultural development areas.

(3) Research should be done on practical problems

confronting the horticultural industry and be based on the adaptation to Kenyan conditions of plant material and growing techniques successfully developed in other countries. In the past, however, Kenya has placed excessive reliance on using practices developed abroad, and too often the local problems in horticultural technology have been ignored.

Kenya Development Plans

The Development Plan for Kenya's Horticultural Industry was obviously influenced by the Government's overall planning strategy at that time contained in the Kenya Development Plan 1974-78 (Republic of Kenya 1974). The latter asserted that there were excellent prospects for the rapid increase in output of many horticultural crops. For the period 1974-79 the increase in pineapples was expected to be at an annual rate of 18 per cent, while that for fruits and vegetables was put at 10 per cent per annum. Hand in hand with this growth in horticulture there would be development in local markets, processing and exports. Horticulture would be assisted by the Government through intensification of horticultural research, increased emphasis on extension and training, consolidating the HCDA's role in coordination and development of the industry, and establishing Horticultural Production Centres (HPC's) around the country with a view to making them export oriented.

Continuing the process of national stocktaking and policy re-formulation for the economy, the Kenya Development Plan 1979-83 (Republic of Kenya 1979), singled out small farmers as one of the five target groups for alleviation of poverty. It went on to say that the technology constraint was expected to be more severe during the period of this fourth development plan than during the previous 20 years.

Research emphasis

The Kenya Development Plan 1979-83 discussed a strategy for overcoming the constraints on agricultural development. One of the elements in the strategy is of particular interest in the present study, because it stresses acceleration of the development of appropriate technologies. The following stated objectives appear to be specially relevant to horticultural production:

(a) to develop new technology for achieving more intensive land-use, involving both yield increasing practices and higher yielding varieties of already established crops;

(b) to undertake research which concentrates on the development of new varieties, techniques and farming systems for the high potential areas;

(c) to devise effective research programmes by the study of the socioeconomic setting of farm activities and related economic analysis at headquarters and field research stations;

(d) to attend more to labour-intensive technology and less to capital-intensive technology.

Employment generation

Table 6.6 shows the estimates cited in the Kenyan Development Plan 1979-83 for increased employment in 1983 compared to 1976 figures, resulting from expanded acreage and intensification of horticultural crops. These estimates indicate that additional employment in horticulture could represent around 7 per cent of the total increase for agriculture. In 1983 the contribution of horticultural crops to total employment in agriculture is put at just over 3 per cent. No indication is given in the table of the relative contributions of area and intensity effects to increased employment in the important fruits and vegetables category. Since such crops are mostly labour intensive anyway, it is understandable if the area effect is the larger contributor to increased employment arising from expanded production.

Nutrition levels

With regard to food and nutritional targets, the Kenya Development Plan 1979-83 was optimistic that the population can enjoy an improved food intake and achieve a satisfactory nutritional balance. Emphasis on smallholder oriented programmes will ensure a wider range of crops, particularly fruits and vegetables, which will afford a more varied diet. It is observed that horticultural development in the past has followed conventional lines of income generation, which have not raised nutritional standards perceptibly. New measures therefore will aim to correct this situation.

Overall effects of horticultural industry development

The forward linkages of the horticultural industry with processing, packing and marketing (domestic and export) operations have already been shown up as having significant direct income and employment effects in the economy. Furthermore, it is clear that these direct effects are augmented by horticulture's backward linkages with input supply and service firms. The development plans for increased horticultural production in Kenya therefore highlight the important multiplier effect (combining direct and indirect contributions to income and employment) of increased investment in this form of activity. With the data at hand this cannot be quantified in the study, but for reasons of production intensity and widespread linkages it is likely to compare most favourably with that caused by investment in the rest of agriculture.

In a simpler industry context, it is instructive to compare horticulture's estimated contributions to total agricultural production value and total agricultural employment. As stated previously, these are around 9 per cent (1978) and 3 per cent (1983) respectively. On this evidence, the relative contribution to income seems greater than that to employment in recent years. However, further horticultural development could close the gap, as is indicated by its 7 per cent share of estimated additional agricultural employment between the two years 1976 and 1983, shown in table 6.6.

Table 6.6
Estimated growth of agricultural employment in Kenya, 1976 and 1983[a]

Activity	1976 total hours (millions)	1983 total hours (millions)	Additional hours (millions)	Per cent	Area effect (per cent)[b]	Intensity effect (per cent)[c]
Potatoes	57	71	14	1	86	14
Fruits and vegetables	54	102	48	4	No estimate	No estimate
Bananas and plantains	41	59	18	1	100	0
Pineapples	8	20	12	1	100	0
Cashews	1	3	2	–	50	50
Total crops[d]	3,117	3,956	839	60		
Total livestock[d]	2,773	3,088	315	23		
General framework	609	840	231	17		
Total activities	6,499	7,884	1,385	100		

[a] Source: Republic of Kenya (1979).
[b] Based on assumptions as to hours of work due to additional hectarage (area effect).
[c] Based on assumptions as to hours of work due to higher yields (intensity effect).
[d] Increase in work hours for crops and livestock is mainly related to on-farm investments.

TECHNOLOGICAL DEVELOPMENT AND RESEARCH, EXTENSION AND EDUCATION

Theories have been advanced to explain technological development in countries and regions of the world. Consequently, hypotheses exist on what influences resource allocation to different types of agricultural research. For instance, the theory of induced innovation developed by Hayami and Ruttan (1974), involving the effects of changes in relative prices of resource inputs, has drawn a lot of attention in recent years. Using the historical perspective over a long period of time, these researchers have identified (as elaborated in Chapter 3 of this volume) an emphasis on the development of

(a) biological technology (yield increasing) in Japan when land has been scarce relative to labour, and (b) a mechanical technology in the United States when labour has been scarce relative to land.

Thus in the case of biological technology, which includes the breeding of higher yielding varieties of crops, improvement of growing practices and the introduction of particular inputs, e.g. fertiliser and pesticides, the effect is to overcome the land constraint by permitting more intensive production. One might see this development as an indirect substitution of land by increased application of other inputs. In contrast, under a mechanical technology a substitution of labour by machines takes place, thereby increasing the average productivity of the labour which remains employed.

These observations therefore have led Hayami and Ruttan to explain the development of technology in terms of a response to changes in relative factor prices. Hence, it can be argued that farmers will always take notice of changing resource prices and will seek to adjust the use of inputs over time in a way that maximises a particular objective function (say profit) not only under a given technology, but also under alternative technologies. This being so, it is held that pressure will be exerted directly by farmers on the research system, or indirectly by government on their behalf, for it to innovate in ways that recognise the economic logic of factor-price relationships. This induced innovation theory has been used in the present study to formulate an hypothesis which can be tested in relation to horticultural research in Kenya.[3]

A review of historical perspective of agricultural research in Kenya reveals that long past Independence there continued to be a bias in the allocation of research resources towards cash crops such as coffee, tea, wheat, maize, and to research stations serving former European settler areas. The Kenya Development Plan 1974-79 did recognise the need to correct this imbalance in the allocation of resources to agricultural research. Therefore while the funding of research affecting peasant farmers has been somewhat gradual, important changes have nonetheless occurred, with research emphasis under the Kenya Development Plan 1979-83 being placed on food crops and marginal areas, involving particularly the outlying research sub-stations. This realignment of research priorities in agriculture in Kenya has resulted from a complexity of forces, largely arising internally but also influenced by foreign aid.

Adaptive research programmes are considered to be the all important means by which co-operation between farm economist and biologist can lead to meaningful recommendations, standing a good chance of adoption by farmers. Adaptive research is based on possible solutions which the biologist can agree to as being suitable to the natural conditions of farm operation, and the economist can say are economically feasible. However, it must be made clear that adaptive research is a complement to applied research and certainly does not preclude it (see figure 6.1).

One example of an adaptive research project using a joint agronomic and economics approach is that reported on by the National Horticultural Research Station (1979) and by Gathee (1980), an agricultural economist. With respect to maize and bean growing under semi-arid conditions where planting time is crucial, Gathee states that the introduction of a simple and efficient planting device might well be able to remove the labour constraint affecting the adoption of highly intensive labour systems. In short the research pin-pointed a serious resource constraint on farms and turned attention to devising a planting technique for overcoming it. Muchiri in a recent work, Ahmed and Kinsey (1984), does in fact offer specific recommendations for solving this problem. An equipment innovation package is suggested which is thought to be superior to the traditional system of planting and cultivation.

QUANTITATIVE ANALYSIS

Farm size and income inequality

Lack of data for identifying farms engaged in horticultural activity, either for the home or

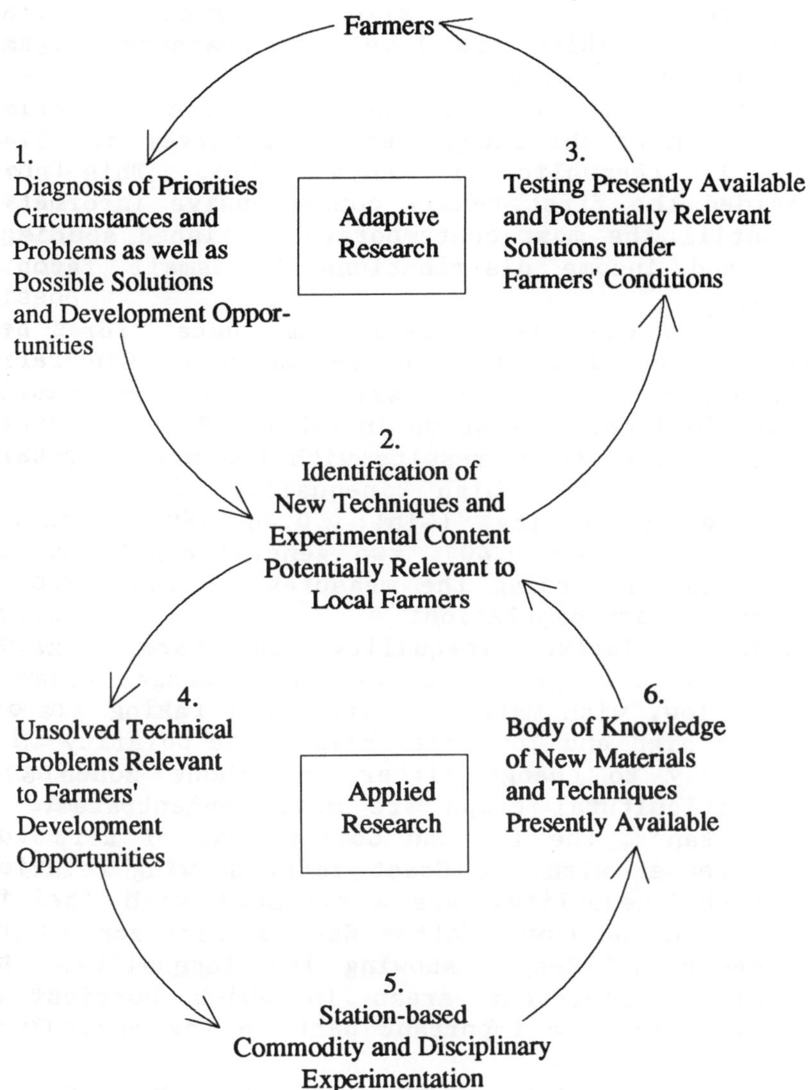

Figure 6.1: Interactions between applied and adaptive
research*

* Source: Collinson 1980.

market, makes it necessary to take data across all types of small farms, in order to say anything at all about the disparities in size (hectarage) and income (household), which are likely to characterise farms growing horticultural crops.

Of interest in this respect is a compilation of data given by the Integrated Rural Survey of 1974-75 (IRS 1) (Republic of Kenya 1977). This survey provided the first really comprehensive information, and still the most contemporary published source, on size and income distributions for small farms in Kenya.[4]

Using the IRS small-farm data for eight agro-ecological zones, the estimates of Gini ratios as measures of relative variation for farm size and household income are shown in table 6.7.

Looking at these results with the use of Todaro's (1977) criteria: high inequality (Gini: 0.50), moderate inequality (Gini: 0.40-0.50), and low inequality (Gini: 0.40), two general conclusions can be reached regarding the measures in table 6.7 for the small-farm population:

(1) The relative inequality in farm size in individual agro-ecological zones ranges from high to low, with marked polarisation taking place in the high and low categories. The polarity is not likely to change either for those zones where horticultural crops are most concentrated. For instance, the Tea and Coffee West of Rift zones together with the Coast zone, showing relatively high inequality, are contrasted with the Tea, Coffee and Lower Cotton East of Rift zones in the centre of Kenya, showing low inequality. Both groups represent areas in which horticultural crops play an important part in the agricultural economy.

(2) The Gini ratio estimates support the conclusion that the two groups of zones, important horticulturally and contrasting with regard to farm size inequality, again differ in levels of household income inequality. But now the West and Coast zones, while still showing high inequality in two cases and moderate inequality in another, are more closely matched by the three

Table 6.7
Gini coefficients for farm size and household income distributions for Kenyan
smallholders in agro-ecological zones 1974-75

Type and source of estimates	Tea West of Rift	Coffee West of Rift	Upper Cotton West of Rift	Tea East of Rift	Coffee East of Rift	Lower Cotton East of Rift	Coast	High Altitude Grassland	Total Zones
Farm size[a]	0.57	0.54	0.48	0.39	0.39	0.39	0.54	.	0.48[c]
Household income[b]	0.50	0.55	0.49	0.49	0.48	0.49	0.55	0.22	0.53[d]

a   Estimated from IRS data tabulations summarised by Dorling (1979).

b   Estimated from IRS data tabulations summarised by Lijoodi and Ruthenberg (1978). Households with negative
    incomes included in total, but as many lower income groups merged as necessary to show a non-negative income
    contribution for the newly formed first group in each distribution.

c   Excluding High Altitude Grassland farms.

d   Including High Altitude Grassland farms.

273

central zones, all with moderate household income inequality.

The study by Dorling (1979) also suggests a third general conclusion: namely that if farm operating surplus (difference between total value of production and total farm costs) was used instead of household income for the income distributions, then greater levels of inequality would be shown generally for the zones but with their relative positions remaining much the same.

From the standpoint of trying to assess how the demand for a productive national horticultural research organisation may be affected by disparities in farm structure, the above findings for the population of small farmers are thought likely to represent similar conditions for the constituent small horticultural producers, who are responsible for the major part of horticultural output in Kenya.

Agricultural input prices

Since the theory of induced innovation encompasses the effects of relative price structures of resources, it is useful to observe the movement of prices of important categories of agricultural resources over time in Kenya. At the same time a fuller understanding of some of these price movements can be obtained by referring to corresponding levels of resources used in agriculture.

Published government statistics for agricultural input prices and levels of use in Kenya convey a disjointed and incomplete picture over time. Comprehensive data of this nature has to be brought together from several different sources, involving both internal (mostly government) and external (UN bodies: FAO and Economic and Social Council) compilations. Weber (1981) has published a set of relevant data from which summaries are made in tables 6.8 and 6.9 for input levels and prices.

Table 6.9 shows agricultural wages and land prices (Kiambu District, Central Province, used as proxy for representing land in the High Potential Areas) in terms of constant 1970 prices, and the agricultural labour force (persons) per hectare of

agricultural land over the past 20 years.
Semi-logarithmic graphs constructed from these data
(Dorling 1982) show that for the period 1962-72 the
trends in agricultural wage increases and land prices
were somewhat similar in percentage terms.
Immediately before this period wages had risen
steeply with land prices at first rising and then
falling steeply. But of real importance are the
opposite trends in land prices and wages between 1972
and 1979, when land prices more than doubled and
wages fell back gradually. In current price terms
reflecting inflation, Weber gives the price of land
in Kiambu District as 3,993 KSh/ha in 1972 and 27,219
KSh/ha in 1979. In contrast the agricultural wage
rate for Kenya rose from 181 KSh/month to 272
KSh/month.[5]

The increasing scarcity of land relative to
labour is borne out by the steep increases in land
prices and agricultural labour (active persons) per
hectare of agricultural land (Dorling 1982).
Agricultural labour rose from around 0.45 person per
hectare in 1960 to over 0.80 person per hectare in
1979. Weber (1981) shows the total agricultural
hectarage of land (arable, crops and pasture) in
Kenya in 1978 as 6.04 million hectares compared to an
active population of 4.68 million in agriculture.
This situation is even more emphasised in the upward
trend of agricultural labour per hectare of arable
land and crops, a somewhat more realistic measure
than that obtained per hectare of total agricultural
land (Dorling 1982).

The wage rates shown, influenced as they are by
the government's minimum wage rate legislation, take
no account of farm family labour remaining at home
and therefore not competing in the hired labour
market. This category of non-wage labour is by far
the larger part of the total labour force in
agriculture and is expanding rapidly because of the
high birth rate. Indeed many of the peasant farmers
and their families have, if not zero, very low
off-farm opportunity costs, a situation which
contrasts greatly with the high opportunity costs of
land in agricultural areas.

The conclusion (Dorling 1982) from the above

Table 6.8

Agricultural input levels, Kenya 1960–78*

	1960	1970	1975	1977	1978
Active agricultural population (1,000 persons)	2,868	3,753	4,154	4,409	4,682
Arable land and crops (1,000 ha)	1,670	1,745	1,765	2,270	2,270
Total agricultural land (including pasture) (1,000 ha)	5,614	5,560	5,545	6,040	6,040
Nitrogen (N) (tonnes)	1,800	22,000	19,400	22,477	25,284
Phosphate (P_2O_5) (tonnes)	3,500	24,200	29,400	27,262	21,196
Potash (K_2O) (tonnes)	200	3,100	4,000	4,217	5,042
Pesticides (tonnes)	.	4,337	2,873	8,555	7,976

* Source: Weber (1981).

analysis is that agricultural land in Kenya is becoming increasingly scarce relative to agricultural labour, and that this situation is reflected in a decreasing ratio of wage rate to land price over recent years. Hence, the increasing labour/land ratio is favourable to labour-intensive forms of agriculture, e.g. horticultural enterprises. At the same time the analysis by Dorling (1982) indicates that nitrogen inputs are also being applied more intensively under a decreasing ratio of nitrogen to land prices. A similar situation is recognisable in the use of phosphate, potash and pesticides. Therefore the use of agricultural chemicals, as a category of capital inputs, reflects an increasing agro-chemical/land ratio, which is not inconsistent with an increasing labour/land ratio. Greater intensity in the use of both labour and

276

Table 6.9

Development of prices for purchased agricultural inputs at constant (1970 prices,
Kenya 1960–79*

	1960	1970	1975	1978	1979
Agricultural labour (KSh/month)	37	129	132	132	134
Agricultural land (Kiambu District) (KSh/ha)	2,593	3,086	5,085	7,730	9,673
Nitrogen (N) (KSh/100 kg of plant nutrient)	237	185	354	272	283
Phosphate (P_2O_5) (KSh/100 kg of plant nutrient)	201	115	290	232	n.a
Potash (K_2O) (KSh/100 kg of plant nutrient)	147	115	197	162	n.a
Pesticides (KSh/tonne)	n.a	9,304	11,381	8,748	n.a
Petrol (Nairobi retail) (KSh/litre)	0.95	1.03	1.47	1.11	1.24
Tractors (KSh/tractor) (av. CIF import values)	20,029	31,655	26,666	32,667	n.a

* Source: Weber (1981).

277

agro-chemicals is a complementary condition in obtaining increasing production per unit of land.

So far the data has concerned the whole of Kenyan agriculture. But by far the larger part in area and population terms concerns peasant agriculture. An even higher labour/land ratio can be expected for the small farm sector alone, and the same may well be the case for the agro-chemical/land ratio. This is because it is now commonly recognised that, due to market imperfection in resource acquisition in the rural sector, the relative price structures for inputs facing small and large farmers are significantly different. This leads to different patterns of resource utilisation and in the case of small farms to the higher ratios described.

In light of the theory of induced innovation, the relative price structure now recorded takes on special importance. Certainly it would argue for labour-intensive methods of production, but this can also go hand in hand with greater agro-chemical intensity. The need clearly is to increase agricultural output per unit of land and horticultural enterprises are inherently suited to this objective.

The question therefore arises as to how the economic logic of factor-price relationships in Kenyan agriculture affects the national research system, particularly in horticulture, and eventually influences forthcoming innovations. This aspect, however, is best left until after discussion of the information and communication channels between farmers, extension personnel and research establishments has taken place.

Planned expenditure in horticulture

Some indication of the emphasis placed on horticultural expenditure in the two planning periods is gained by comparing planned development expenditures on horticultural projects as given under (a) Research Division and Crop Production Programme headings for the 1974-78 Plan, and (b) the crops development heading for the 1979-83 Plan. The research, extension and other projects for

horticulture (table 6.10) account for 3.66 per cent
of planned crop development expenditure (table 6.11)
in the 1974–78 period and 4.18 per cent in the
succeeding 1979–83 period.

Research expenditure

Since no data was obtained on the funding of
individual research stations engaging in
horticultural activity, use is made of some of the
figures estimated on the funding of publicly
supported research stations for work in horticulture
(Dorling 1982). For instance, assuming that the 4.18
per cent share of crop development expenditure going
on horticultural projects also applies to total
expenditure (including recurrent expenditure), then
for the period 1979–83, K£7.72 millions (see
qualification in note 6) goes to horticultural
projects. Since this figure contains K£4.55 million
for the Horticultural Production and Marketing
Project, and K£0.18 million for horticultural
extension and potato storage and marketing,
activities which would not involve the research
stations to any large extent, it would seem that the
remaining K£3 million (approx.) over the five–year
period could represent a reasonable estimate of the
funding figure for horticultural work on research
stations in Kenya.
The Joint Research Services contribution to total
expenditure in 1979–83 is K£10.617 million, the
nature of which has been explained previously, and is
not of course covered above by the crops development
category. Since no direct information was obtained
for allocating the Joint Research Services total
expenditure to its constituent activities, the best
estimate that can be made of the horticultural
component (including related plant material
quarantine and East African Herbarium expenditures),
would be to assume in the first place that the
breakdown between livestock development and crops
development work was in the ratio exhibited by these
two categories alone, i.e. approximately 33 per cent
and 67 per cent respectively. Applying the latter
share to the Joint Research Services total

Table 6.10
Comparison of planned horticultural project development expenditure*

	Development plans	
Horticultural projects	1973/74 – 1977/78 (1973/74 prices) K£ '000	1978/79 – 1982/83 (1978/79 prices) K£ '000
Research		
Horticultural Research Project	131	367
Tropical tree crop research	68	–
Potato Research Project	64	53
Silk worm development	26.7	221
Passion Fruit Research Project	–	188
Extension		
Horticultural extension	194.5	59
Other		
Coast crop research and development	–	560
Potato storage and marketing	–	122
Horticultural Production and Marketing Project	–	4,550
Cashew nuts processing plant	200	–
Mtwapa Central Tree Crop Nursery	9.6	–
Bukura Horticultural Nursery	7	–
Thika Horticultural Nursery	7	–

* Source: Republic of Kenya (1974, 1979).

Table 6.11

Planned crop development and recurrent expenditures, Kenya Development Plans 1974–78 and 1979–83*

1974–78 (1973/74 prices) K£ '000		1979–83 (1978/79 prices) K£ '000	
Development			
Research Division Programme	1,699	Crops Development Programme	146,354
Crop Production Programme	17,619		
Subtotal	19,318	Subtotal	146,354
Recurrent			
Research Division Programme	3,874	Crops Development Programme	38,407
Crop Production Programme	10,027		
Subtotal	13,901	Subtotal	38,407
TOTAL EXPENDITURE	33,219	TOTAL EXPENDITURE	184,761

* Source: Republic of Kenya (1974, 1979).

expenditure gives a figure of K£7.11 million for crops development. If then it is assumed that 4.18 per cent of this total goes to horticultural work, on the same basis as the assessment described earlier, then another K£0.297 million is added to the previously estimated K£3 million, to obtain approximately K£3.3 million in total planned allocation over the 1979–83 period to research stations for horticultural work.[6]

This estimate of K£3.3 million represents about 6.7 per cent of the five-year research institution funding total of K£49.2 million from the 1979–83 plan (Dorling 1982). Such a share, and it is quite possibly an underestimate, denotes an important commitment of funds to the horticultural sector. It would go mainly to the National Horticultural Research Station at Thika and the Coast and Molo Research Stations, where horticultural research is concentrated.

Parity and cost–benefit aspects

To gain extra insight it is helpful to apply parity and cost–benefit concepts to expenditure in horticulture. One form of parity ratio is obtained by dividing the percentage allocation to horticulture of total crop research and development expenditure by the percentage contribution of horticulture to total crop output. In the 1974–78 period the planned allocation to horticulture of total crop development expenditure (including research, extension and other projects) was previously estimated as 3.66 per cent. In 1976 the contribution of horticultural crops to total crop output value was found to be around 12 per cent. This share was not expected to change in 1978, and for that matter not expected to change by very much in 1983. Applying these percentages gives a parity ratio of 0.3, and if 4.2 per cent is substituted for 3.66 per cent the ratio for the 1979–83 period is little changed at 0.35. If however 4.2 per cent for the period 1979–83 is an underestimate and the allocation is 7.28 per cent, then the parity ratio rises to 0.6.

Another form of parity ratio can be calculated for

the 6.7 per cent allocation of research institution funding to horticultural activity in the 1979–83 period. This ratio is 0.56 which is similar to the previous value. In summary it would seem that research and development work in Kenyan horticulture is being funded less than proportionally to its contribution to agricultural output. This is something that planners should take note of, although it may well adhere to the present planning priorities.

With the restricted data available in the study, a cost–benefit assessment of expenditure in horticulture will not be attempted. What can be said is that the parity ratios are no indication of the relative net benefit which arises from expenditure in horticulture. But one of the advantages of these measures is that they focus the need to justify their existence in the allocation of scarce resources. The justification seems conspicuously absent in official references.

In fact there is good reason to suppose that investment in research and development in horticulture can show at least as high, and probably even higher, cost–benefit ratios as those in the rest of agriculture. For instance, the lower base of horticultural activity on most small farms would suggest that the potential which can be realised in horticulture from existing technology is very favourable. But the exploitation of new technology in Kenyan horticulture also offers many profitable opportunities in an industry which is striving for greater self–sufficiency in inputs, and rapid expansion in output and marketing facilities. The nutritional benefits from expanded consumption of fruits and vegetables are also immense in a country where since the 1970s there has been falling per capita output from agriculture and a very high annual population growth rate.

The important conclusion arising from the parity ratios calculated is that there has been substantial underfunding of horticultural research and development, relative to the industry's contribution to total agricultural output. This is in sharp contrast to the situation in the United States, a country with a highly developed agriculture, where

research expenditure on horticultural commodities has
a high share of funding in relation to their output
value contribution (Ruttan 1983). Probably, the main
reason for the underfunding in Kenya is the poor
organisation of horticultural producers, which is
disclosed in a later section, leading to inadequate
direct representational pressure on the government's
research funding system.

INSTITUTIONAL FRAMEWORK FOR HORTICULTURAL RESEARCH

Since the organisation of horticultural research and
the related institutional structure involve complex
relationships and procedures, only those features are
discussed which are thought to have the most
important bearing on the role and performance of this
particular type of research. The information given
derives from details obtained from interviews and
literature references.

Sources of horticultural research

Figure 6.2 indicates that there are three main
sources of horticultural research effort in
Kenya.[8] The first source is that of government
research stations, with the Research Division of the
Ministry of Agriculture acting as the main centre of
administrative control. Other divisions of the
Ministry of Agriculture can undertake limited
research but their involvement in horticulture is
small. KARI also comes under this source. One of
the distinguishing characteristics about much of the
government-sponsored research in horticulture is that
it appears mostly aimed at the smallholder rather
than the large-scale grower. And since the greater
part of Kenya's horticultural output derives from
smallholdings, this strategy is not only equitable
but it also represents an important contribution to
achieving increased rural employment.

The second source comprises the centres of higher
instruction in horticulture. This mostly involves
the Faculty of Agriculture at Kabete, University of
Nairobi, with Egerton College near Nakuru in the Rift

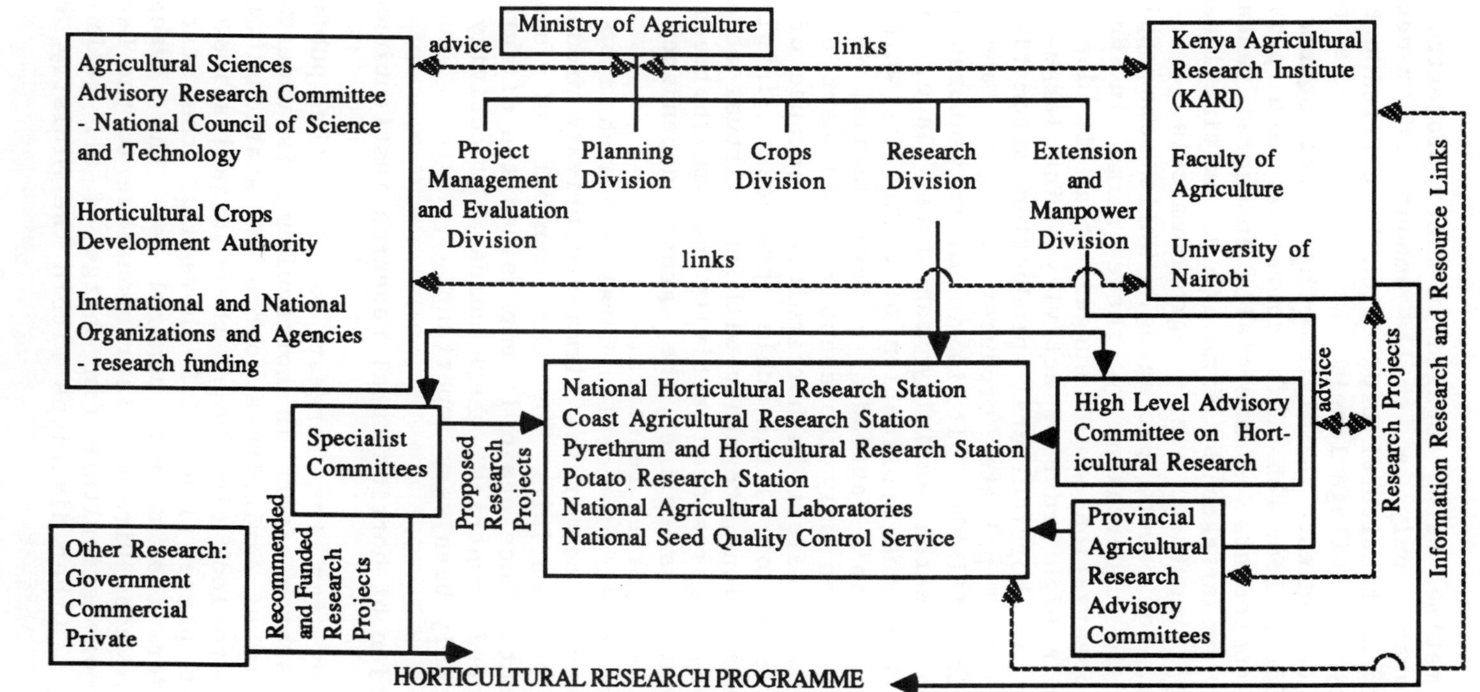

Figure 6.2: Organisation determining horticultural research programme, Kenya 1981.

Valley sometimes being engaged in minor projects. In all, though, only a small amount of research benefiting horticultural production is carried out at university and college level.

It is convenient to divide the third source of horticultural research into commercial and private producer components. Commercially sponsored research is mainly undertaken by firms (e.g. multinationals) supplying inputs, i.e. agro-chemicals, seeds, machinery, equipment, packaging. While some applied research is undertaken by these firms in Kenya, general observation would suggest that it does not amount to a high order of activity. The large-scale companies in horticultural production automatically fit into the private producer category and are readily identifiable. It is in these ventures that important research activity can take place, quite separate from the application of technology which is imported from abroad. Nevertheless, this contribution can only represent a small part of the total research effort in horticulture, because of the small number of large-scale producers involved. These observations, together with the virtual absence of independent research activity by smallholders in horticulture, underline the great dependency on government-supported research.

The low level of horticultural research in Kenya contributed by commercial firms and private producers is hardly surprising in the context of under-development. Once again it emphasises the difference in developed countries, where comparable activity has traditionally been very important.

Organisation of horticultural research institutions

Because by far the largest part of the publicly funded research effort in horticulture is undertaken by Ministry of Agriculture controlled stations, it is important to look closely at their organisation. In the first place figure 6.2 shows that the stations come under the direct control of the Research Division, whose Director is responsible to the Deputy Director of Agriculture (in charge of all technical services). Other divisions within the Ministry, e.g.

Crops Division, Horticultural Branch, also have horticultural industry interests of a technical and economic nature. They have means, via committees within and outside the Ministry, of communicating with research station staff.

There is only one National Horticultural Research Station, with headquarters and field stations at Thika and a small sub-station at Gituamba. It was established at Thika in 1948, having been transferred from Molo in the Central Rift Valley, where it was conveniently situated for dealing with fruit and vegetable crop problems facing the settler farmer. The Thika research station has a coordinating function under its Director for research in horticulture in the whole country. This particular task is made easier by the specialisation which occurs at the different stations by virtue of geographical location and particular line of inquiry. Examples of this specialisation are:

(a) Potato Research Station at Limuru, established in 1969;

(b) National Agricultural Laboratories in Nairobi, specialising in soils and agricultural chemistry, Kenya Soil Survey, entomology, plant pathology, irrigation and drainage;

(c) National Horticultural Research Station, Thika, with its programme in exotic vegetables, vegetable seed production, fruit and crop protection;

(d) National Pyrethrum and Horticultural Research Station, Molo, set up in 1942 and today, apart from its concern with pyrethrum, concentrating on temperate fruits and related nursery stock, and supervising vegetable trials at the Perkerra Irrigation Scheme sub-station, near Lake Baringo;

(e) Coast Agricultural Research Station, where research activities at the main Mtwapa headquarters go back at least to 1958, and horticultural attention centres on tree crops – tropical and citrus, vegetables and entomology;

(f) National Seed Quality Control Service, Nakuru, with its horticultural crops unit within the Seed Inspectorate (formerly the Kenya Inspection Service for Seeds) becoming operational in 1975.

To complete the main organisational picture for horticultural research in Kenya, it should be noted that both KARI (Muguga) and the Faculty of Agriculture (Kabete), have their own self-contained field-station research facilities. At the same time they can also arrange for the use of facilities on other research stations when the need arises.

The Board of Management of KARI is responsible to the Ministers of Agriculture, Livestock Development and Natural Resources. Through its Director it draws up its research programme, which because of established areas of responsibility, e.g. plant quarantine and the herbarium, and committee linkages with the Ministry of Agriculture, is made complementary with the rest of the horticultural research programme in Kenya. The same is largely true for the Faculty of Agriculture, which although existing separately from the other publicly-funded research stations, does nevertheless have close association with them through informal liaison and committees.

Determining the horticultural research programme

At the level of research programme determination the institutional structure and relationships become more complex. Firstly, three influences play an important part in drawing up the horticultural research programmes of publicly-funded institutions. These are discussed below:

(1) The National Council of Science and Technology comes under the Office of the President. It is responsible for formulating scientific policy for the country which eventually gets expressed in the current Kenya Development Plan. This body with its Agricultural Sciences Advisory Research Committee undertakes a national coordinating role for all agricultural research. Research clearance and permission are obtained through this channel too.

(2) The Horticultural Crops Development Authority represents an important institutional innovation. Set up under statutory order in 1967 and made directly responsible to the Ministry of

Agriculture, it was conceived as an advisory, regulatory and policy formulating body with wide powers to aid in the rapid development of the horticultural industry.

(3) Similarly, international and national organisations granting foreign aid, e.g. FAO, World Bank, ODA, USAID, through their links with the Ministry of Agriculture and the research centres, advise and in turn are consulted on research project funding.

Within the Ministry of Agriculture three types of committee play a crucial part in helping to determine the horticultural research programme. The committees also influence research at the institute and university level. They are as follows:

(a) High Level Advisory Committee on Horticultural Research (HILAC);

(b) Provincial Agricultural Research Advisory Committees (PARACs);[9]

(c) Specialist Technical Committees (STCs).

The HILAC is appointed under the Research Division of the Ministry of Agriculture with members representing high level research interests in horticulture, i.e. Directors of research stations and KARI, Faculty of Agriculture representative, and when necessary specialists from foreign aid organisations. It advises particularly the Director of the National Horticultural Research Station, Thika. The committee periodically reviews the horticultural research programme, projects and results, and brings to notice research and technological developments at home and abroad, thus keeping the horticultural research establishments abreast of events.

The PARAC in each province in Kenya is organised under the Extension and Manpower Division, Ministry of Agriculture. It has as its chairman the Provincial Director of Agriculture and for its secretary, the Officer in Charge of the main local Ministry of Agriculture research station. Other members include Heads in the Provincial Director of Agriculture's Office, District Agricultural Officers, and Research Officers at local research stations.

The PARACs usually meet twice a year (just before

the planting seasons) with the objective of evaluating and developing the overall research efforts in each province. At the first meeting in the year the proposed research programme is presented by the institutions involved and discussion is especially called for so that research-extension-farmer interests and linkages can be taken into account. The second meeting of the PARAC reviews the past year's research programme and results. It also calls for suggestions for the coming year's programme.

Thus a mechanism exists for a continuous interchange of views between the research establishment and the technical field services. One main criticism of the procedure is that there is insufficient participation from all sides of the technical field services. It will be recalled that Warui and van Eijnatten (1979) have drawn attention to the failure of PARAC meetings in the Coast Province (and no doubt more widely) to engage fully the participation of invited front-line extension workers. Consequently, the grassroots opinion on farmers' needs in the PARACs is probably inadequately represented and the received wisdom correspondingly deficient in the full range of information. It should be emphasised that the influence of the PARACs extends beyond just the Ministry of Agriculture research structure. Participation at committee meetings by KARI and university personnel can and does take place, although just how active the communication is, it is difficult to assess.

Lastly there are the STCs, formed at the invitation of the Research Division and composed of research scientists. These committees in various fields of research are asked to consider project proposals for their suitability to funding. In this way Directors of research stations move forward their proposed annual research programmes and budgets for acceptance.

Research projects recommended by the STCs are sent forward for final selection for funding under the Ministry of Agriculture's research station budget appropriation. In this way and along with research projects funded under KARI and the Faculty of Agriculture, the total horticultural research

programme comes into being. It is a continuing
process which ideally should be flexible enough to
meet the needs of the rapidly expanding horticultural
industry.

Political and civil service environment

It would appear that the institutional structure
underlying research organisation and determination of
the research programme in horticulture shows
flexibility and the capacity to change and innovate.
The political and civil service systems show
themselves willing to recognise the importance of
providing institutional arrangements and scientific
manpower for furthering horticultural research and
development in line with priorities laid down in the
development plans. Just how successful these efforts
have been in serving the technological requirements
of the industry raises questions to which answers
must now be attempted in the following sections.
Horticulture is a good example of an industry,
including research, which has gained from the
knowledge of overseas specialists on the scientific,
economic, technical and administrative fronts. There
is every reason to believe that this situation will
continue within the present political and
bureaucratic structures. These, it will be
remembered, have shown great stability since
Independence. Naturally Kenya will aim to replace
expatriate skilled workers with her own people when
these are available. However, the signs are there
that Kenya will continue to engage the knowledge and
experience of officials from abroad in a way which
benefits the rapid development of the country.

Institutional structure and induced innovation

With the foregoing commentary it is now possible to
test in some measure the postulate, based on the
theory of induced innovation, that pressure is
exerted on the research system, directly by farmers
or indirectly by government on their behalf, to
generate new technology which recognises over time
the economic logic of factor-price relationships.

It would appear that small horticultural producers in Kenya are not organised strongly enough at present to bring much direct pressure to bear on the research establishment, regarding desirable technology and innovation under given economic conditions. This is so for a number of reasons listed below:

(1) Primary co-operative societies specialising in horticultural products are few in number and small in membership. Even co-operative societies dealing with horticultural produce as part of their operations are few in number. Hence, unlike the sustained pressure from influential co-operatives serving coffee growers, which represent forcefully the opinions and problems of growers and organise direct financial support from producers to the Coffee Research Foundation, concerted pressure by growers in horticulture on research programmes is impossible. Even in the Coast Province where some fruit and vegetable co-operatives exist, no evidence is provided by Warui and van Eijnatten (1979) of any direct influence by growers on research programmes.

(2) Most small producers in horticulture in Kenya have insufficient education and time for taking part in organised activity of the kind which is necessary to forge representative opinion and action on research needs.

(3) Even if small horticultural producers had the ability and wish to express views on research needs, it is likely that other pressing matters in their life would gain precedence. Any move to counteract this tendency would need considerable support and organisation from government sources.

(4) It has been noted already that the small number of large growers in horticulture, who would find it easy to express views on research requirements, are often in a position to provide or buy-in improved technology and thus operate independently of the national research system. This may not always be the case but so far they do not appear to have brought much direct pressure to bear on research programmes.

In connection with the reasons given above, it is clear that government influence to redress the underorganisation of horticultural producers has either had little effect or been lacking. In particular this suggests that management of horticultural research stations in Kenya may well be failing in science "entrepreneurship". For it might be expected that directors of these stations would seek to organise the producers they serve (a) to make the extension of new technology more effective, and (b) as an interest group for bringing pressure on the political system to support horticultural research.

Any pressure that is brought to bear on the research system for responding to economic conditions facing horticultural growers seems to arise almost totally and indirectly from government employed officials (including extension and research staff) through institutional committee arrangements, with all their attendant shortcomings. However, although the evidence does not refute the ability of government officials to take notice adequately of factor-price relationships in determining research programmes, it is doubtful whether this has occurred sufficiently in the past.

The relatively poor organisation of horticultural producers in Kenya is noticeably in contrast with what has been achieved in developed countries. There, horticultural producers are among the best organised, and over the years have often exerted crucial influence in the setting up of research stations and provision for research interests. It would seem that public sector research funding in any branch of agriculture is unlikely to receive its fair share unless there is an active and well articulated demand for it from producers themselves. Passive reliance on government officials to represent the best interests of producers is seldom enough. It is for this reason that the suggestion was made earlier, that the apparent underfunding of horticultural research in Kenya was likely to be the result of poor producer organisation and the consequent lack of direct representational pressure.

Once an assessment has been made of research projects, technology and innovations, something will be said about how innovation in Kenyan horticulture has actually responded to the economic logic of resource-use over time.

HORTICULTURAL RESEARCH INPUTS AND PROJECTS

In addition to institutional information, details were also recorded from interviews about horticultural research inputs and projects. It is appropriate at this point to say something about the nature of interview response before summarising the remaining findings.

Interview responses

It should be made clear that several factors worked against the interviewing procedure used. In the first place research station, government and other officials in Kenya are extremely busy people, who are at the same time very conscious of possible political and administrative repercussions in answering questions put to them by an independent academic researcher. Hence, the easiest information to obtain was that which called for immediate opinions or easily recalled facts. When it came to financial and other input data and more intricate information, which cannot be referred to in annual reports, then the result was less successful. In addition the setting up of the Kenya Agricultural Research Institute has led to considerable debate within the Research Division, Ministry of Agriculture, as to mutual roles and relationships between KARI and the Ministry of Agriculture administered research stations. Questions concerning or touching on such matters sometimes received rather guarded replies. Furthermore, it must be recorded that arranging appointments where exceedingly long distances are involved in travelling is never easy at the best of times, with the inevitable delays, cancellations and re-scheduling of interviews that sometimes take place.
    In conclusion, although the interviews definitely gained worthwhile information, some of the difficulties encountered would have been overcome if the inquiry (a) had involved a group of people conveying more official importance, and (b) had not

Table 6.12
Research stations engaged in horticultural research – hectarage, staff, research project emphasis[a]

Research Station	Total hectarage[b]	Senior and technical staff[b]	Horticultural research project emphasis	
National Horticultural Research	1981 – 364 ha	1976 – 33 (22) 1975 – 23 (9)	1974–76:	Variety trials, cultural practice trials. Vegetable seed production and evaluation. Plant protection
Potato Research Station, Tigoni	1981 – 1 ha (1) (2 ha planned)	1981 – 1 (1)	1978 and earlier:	(as above), also plant breeding, food technology, virology, bacteriology
Coastal Agricultural Research Stn., Mtwapa, Matuga, Msabaha	1981 – Mtwapa 243 ha Matuga 81 ha Msabaha 40 ha	1981 – 39 (3) 1981 – 2 1981 – 12	1979:	Variety trials, cultural practice trials, plant protection
National Pyrethrum and Horticultural Research Stn., Molo	1981 – 170 ha (21)	1981 – 21 (4)	1973–75:	(as above), also vegetable seed production and evaluation
Perkerra Irrigation Horticultural Sub-unit	1981 – 12 ha (12)	1981 – 3 (3)	1973–75:	Variety trials, cultural practice trials. Vegetable seed production and evaluation
Kenya Agricultural Research Institute (KARI)	1981 – 600 ha	1981 – 159 (1)	1975:	Plant quarantine, seed testing, pathology, propogation, compost

Faculty of Agricul- ture, University of Kenya	1981 – 250 ha (10)	1981 – 36 (4)	1979–81:	Seed production, plant breeding, tissue culture – propogation, mixed cropping
National Seed Quality Control Service, Nakuru			1978–79:	Potato variety description and per formance. Vegetable seed inspection. Seed testing
National Agricultural Laboratories, Nairobi			1978 and earlier:	Soils, entomology plant pathology
Western Agricultural Res. Stn., Kakamega			1978 and earlier:	Vegetable agronomy trials
Nyanza Agricultural Res. Stn., Kisii			1978 and earlier:	Vegetable agronomy trials
Embu Agricultural Research Station			1978 and earlier:	Nematode control, rotation trials

[a] Source: Interviews and Annual Reports.
[b] Horticulture in brackets when specified.

come close on the heels of other recent studies
dealing with aspects of resource allocation and
productivity in agricultural research in Kenya.
However, it must be remembered that some offsetting
advantages exist for the one-man inquiry, such as
flexibility and directness, which it is hoped were
put to good effect in the interviews carried out.

Research station inputs and horticultural research

<u>Research station hectarage and staff size</u> The main
observations arising from the information given in
table 6.12 are as follows:
The headquarters and main centre for horticultural
research in Kenya is the National Horticultural
Research Station, Thika. It contributes by far the
largest hectarage and complement of research
scientists and technicians engaged on horticultural
projects. However, not all of the 364 ha (1981) is
devoted to purely horticultural activity, since other
agricultural research projects are carried out at
this station, e.g. concerning grain legumes. Of the
other research stations carrying out horticultural
research, only the Potato Research Station at Tigoni
(near Limuru on the outskirts of Nairobi) and the
Perkerra Irrigation Horticultural Sub-unit (near Lake
Baringo), involving very small hectarages are devoted
entirely to horticultural research.
The Potato Research Station at Tigoni was started
in 1969, with potatoes treated as a special crop
under the NHRS-Thika. In 1972 the station was
transferred as a special unit to separate
administration under the Director of Research,
Ministry of Agriculture. The station has some
temperate fruit trials and in 1979 incorporated an
ornamental (flowers and shrubs) section, scheduled to
expand to 5 ha eventually.
The Coastal Agricultural Research Station, composed
of three locational units, is concerned with tropical
fruits and vegetables, but this research work only
requires a small part of the total hectarage devoted
to research. Again the very large hectarage at the
Muguga headquarters of KARI is largely devoted to
non-horticultural crops. The same can also be said

for the Faculty of Agriculture's field station, University of Nairobi, where only 10 ha out of 250 ha is concerned with horticultural crops, excluding coffee of which there is 50 ha. The other research stations shown in table 6.12 as carrying out horticultural research are either laboratory intensive, as in the case of the National Agricultural Laboratories, Nairobi, or mainly agricultural research stations, which give over a small hectarage to horticultural research, as the need arises; usually at the request of the NHRS-Thika for the purpose of local-based trials.

This relative allocation of research station land to horticultural research is closely paralleled by the disposition of senior and technical staff. A glance at table 6.12 makes it clear that by far the highest concentration of qualified research scientists and technicians working in horticulture is based at the NHRS-Thika.

A closer inspection of the figures for senior and technical staff working in horticultural research highlights the following conditions:

(a) while the small numbers of qualified research staff and technicians in horticultural research are increasing slowly year by year, there is a serious shortage of them relative to development policy and the goals set for the industry;

(b) the numbers of horticultural research staff shown in table 6.12 include a small fraction of expatriate staff, who on average have more experience and better qualifications than their Kenyan colleagues. But the expatriate fraction is not only decreasing over time, it also represents a smaller number each year;

(c) given the small numbers of research personnel in horticulture, the constant need for some of them each year to seek further qualifications under staff development programmes means that research stations often find themselves facing exceedingly tight staffing situations;

(d) research personnel sent for further training, especially for higher degrees, sometimes find it advantageous at the end of their training period, or shortly afterwards, to leave their posts and take up employment in parastatal organisations, other government divisions and ministries, commerce and industry. This type of job mobility constitutes a serious loss in research expertise as well as the denial of opportunity to someone previously excluded; and

(e) comparison of the staff numbers shown in table 6.12 for the several research institutions engaged in horticultural research shows clearly the large differences between the well-endowed KARI station and most of the rest. The well-endowed nature of KARI is to a considerable extent due to pre- and post-Independence foreign aid policy. Thus aid preference has often been given to international research stations as opposed to national ones.

<u>Research project emphasis</u> In seeking information concerning research project emphasis it was necessary to make a detailed study of research station annual reports. But no annual reports were available later than 1979, and the latest annual report available for NHRS-Thika was that for 1976.

In general, therefore, a problem existed for the research stations involved in horticultural activity because of their inability to bring out annual reports on time and thus avoid serious delays. These reports are a most useful digest of research results and progress, without which dissemination of valuable information and data is adversely affected. It is necessary for the directors and officers in charge of research stations to make sure that adequate reporting of research, both written and verbal, is prompt and of the highest quality, since it is only in this way that important research objectives — the timely creation and imparting of new knowledge — can be met.

Obviously the short periods over which research project emphasis is listed in table 6.12 cannot in any way do justice to the changing and longer-term emphasis over decades. The point to remember though is that the rapid development of commercial horticulture and allied research in Kenya is very much a post Second World War feature, and most of it is post-Independence (1963) and related to successive development plans. Consequently, it is thought useful to study the research project emphasis in recent periods in the 1970s. As a result of this exercise the following main points can be made:

(1) Horticultural research projects in the periods stated in table 6.12 noticeably emphasise:
  (a)   variety trials;
  (b)   cultural practice trials;
  (c)   vegetable seed production, evaluation and testing;
  (d)   plant protection – fungicides and insecticides.

(2) Much of the research project emphasis on horticulture for fruits, nuts and vegetables appears to follow closely that recommended by the Horticultural Guidelines (1974). Furthermore, research development in floriculture, although still small-scale in nature, is also following along the lines suggested in this document.

(3) Most of the research in horticulture is of a strictly applied nature. Very little basic research is attempted.

(4) Much of the research in horticulture is yield-increasing in nature and implies greater intensity of resource-use, including labour.

(5) There appears to be little research designed to establish profitable cultural practices under labour-intensive techniques.

(6) There is a growing concern in horticultural research circles with creating domestic sources of vegetable seed production.

(7) Research projects show little if any inquiry into different and more intensive forms of crop mechanisation.

<u>Research emphasis and employment</u>   Interviewed researchers expressed repeatedly the view that greater intensity of production and higher yields in horticulture would account for considerable increases in labour inputs under existing technology. Poor performance in horticultural crop production on many smallholdings across the provinces was cited as leaving much room for improvement and higher levels

of labour-use. No clear ideas were presented as to how this would affect seasonal peak demands of labour in various types of horticultural production. In all probability it would increase the peaks but also decrease the troughs.

It is interesting to note that rather than explore the bounds of profitable labour-intensive techniques for different horticultural crops, the predominating interest of the research stations interviewed appeared to be in how to overcome difficulties caused by seasonal labour shortages on individual farms.

Given the existence of labour constraints on farms and the fact that more intensive horticultural production calls for higher labour inputs anyway, it would appear that the research priorities are established in favour of studying technologies which fit into the existing widespread patterns of resource availability and relative costs, and for which resource combinations adhere to economic logic on the farm. If this is the case it explains why little attention in horticultural research is given to substitution of agro-chemicals by maximum intensive labour techniques.

<u>Specific horticultural research contributions</u>
Following the general discussion of research emphasis it will help with deeper understanding to give a brief account of specific examples of contributions of research to horticultural production on farms.

Over the whole range of vegetables (including potatoes) a lot of work has been done on testing the performance of varieties under different production conditions. These varieties have usually been imported. Consequently, this work on selection and growing recommendations is crucial in bringing forward the best and newest varieties through commercial channels. Variety testing followed by variety breeding of vegetables in the future are expected to nurture eventually a thriving commercial seed industry in Kenya.

In fruit and nut research many examples exist where results benefit producers. Of particular note is the testing and selection of citrus, avocado, macadamia, and temperate fruit rootstock. Along with this has gone the development of suitable grafting techniques. These types of fruit and nut with the addition of cashews have also been given local recommendations, based on continuing trials, for fertiliser, pest-control and irrigation treatments. Special attention has been paid to passion fruit variety selection and growing methods with the result that juice extraction now takes place on a large scale. Research work on grapes has given promising results for a few cultivators, although there has been little introduction of this fruit to farms so far. From 1957 to 1975 pineapple research at NHRS-Thika, dealing with variety selection, growing methods and nematode control, laid the foundations for today's large-scale production.

Invaluable research has been carried out on disease and pest control for horticultural crops. It should also be mentioned that KARI is responsible for all plant quarantine operations in Kenya and has continued to screen horticultural plant material with the latest techniques and highest standards.

This is by no means a complete specification of research contribution in horticulture. For instance the efforts to promote a vigorous commercial interest in floriculture have met with great initial success. Also research station efforts in maintaining nurseries for plant propagation material and in making it available to producers, especially in the case of fruit and nut stock, are proving an essential service in conditions where plant material is scarce and commercial interests are slow to be attracted.

INNOVATIVE TECHNICAL CHANGE AND PUBLIC SECTOR RESEARCH

Contribution of research to innovation

Addressing the question as to what has been the contribution of public sector research to innovation in Kenyan horticulture, the foregoing discussion and presentation of information from interviews and research station annual reports leads to several main observations listed below:

(1) Conceptually it is useful to think of innovations as being divided into two categories, (a) those which have been imported, and (b) those which have been discovered or developed through improvement in Kenya.

(2) With this distinction in mind, it is clear that some innovations in Kenyan horticulture fall in the category of direct imports.

(3) From the evidence assembled in this report it would appear that the main contribution of public sector research to innovational techniques in horticulture has been one of improving on imported technology and innovation, mainly by way of adapting them to specific growing conditions, and to resource quality and availability in Kenya.

(4) In improving on earlier techniques and recommending specific innovations, horticultural research has performed a valuable service to the industry.

(5) Particular categories of innovative contributions are:[10]

  (a)   fertiliser recommendations for crops, and under specific soil and climatic conditions;

  (b)   plant protection and weed control recommendations for crops under specific soil and climatic conditions;

  (c)   crop planting, growing and cultivation methods under specific soil and climatic conditions;

  (d)   propagative methods and plant material;

  (e)   testing and selection of new crops;

  (f)   tillage, soil erosion control, rotation, intercropping and mixed cropping system methods;

  (g)   harvesting and storage techniques for particular crops.

Further aspects of research and innovation

Horticultural research is likely to become more concerned with problems hitherto not emphasised and requiring sometimes more basic scientific inquiry. Examples of this general broadening and greater depth of inquiry are the fields of plant breeding; plant protection agents and methods; crop mechanisation; harvesting, storage and processing techniques.

And while research priorities will continue to centre on the needs of small farmers far more than those of large-scale producers, there will nonetheless be problem areas where the interests of both are served. It can also be expected that as horticultural research efforts grow, more attention will be paid to large-scale production needs. The latter will almost certainly increase in number if for no other reason than the fact that large-scale enterprises are increasing in number. These will involve a larger measure of Kenyan business interests and management than at present. Under these conditions it is not difficult to see how powerful horticultural business interests under Kenyan ownership can argue for public research services. Their contribution to the agricultural economy will become too large to be ignored.

The information offered on research publication and contacts with extension personnel and farmers tended to give the impression that there was sometimes serious slackness in these activities, caused by poor institutional management and administrative procedures. On occasion too much can be left to personal initiative, discretion and academic freedom. The difficulties which characterise so many of the government research institutions in developing countries, such as financial stringency, budget changes, staff movement and shortages, shortages of materials and transport, lack of plant and transport maintenance, postponement and interruption of research projects, can create an atmosphere of inertia. In this situation the will to formulate objectives, keep them in mind, and attain them according to schedule, becomes quickly eroded.

Therefore, the interviews conducted certainly do nothing to contradict the previously summarised views of Warui and van Eijnatten (1979) on matters of research publication and research contacts with extension personnel and farmers. This raises serious doubts as to the efficiency with which diffusion of innovations in horticulture takes place.

Findings and induced innovation theory

It has been shown that horticultural research in Kenya has stressed yield-increasing technology, implying more intensive use of labour and agro-chemicals. This emphasis has tried recently to take note of serious resource constraints on farms and to find ways of overcoming them.

In general it is possible to conclude that horticultural research and the resulting innovations, over Kenya's rather short history of development, have followed the logic of factor-cost relationships, reflecting the overriding scarcity of high potential land relative to labour and other inputs. In this matter observation is consistent with the theory of induced technical innovation.

What is noticeable though, as mentioned earlier, is the lack of direct pressure being exerted on the horticultural research system by growers for the generation and development of the most profitable technology. The presence of such pressure can be an important ingredient of the theory as expounded for developed country agriculture. So too can the indirect pressure from government officials and organisations on growers' behalf fit into the theory. It is this last type of pressure which shows up in the case of Kenyan horticulture, although it is difficult to say just how intense it really is.

SUMMARY AND TESTING OF HYPOTHESES

The study has been organised to meet the overall objectives set for the volume and to test the three guideline hypotheses covering aspects of horticultural research and technical innovation. Since these tests have not yet been performed it is necessary that they should now receive attention, so that the outcomes can be summarised along with the rest of the main conclusions of the study. In attempting this it is useful to begin by stating briefly some of the dominant features of the Kenyan agricultural economy and horticultural industry which have come to light in the course of the general inquiry.

Role of horticulture

In 1976, approximately 80 per cent of the population
in Kenya were located in the agricultural sector and
around 62 per cent of the population were from small
farm or rural landless households. The 1979-83 Kenya
Development Plan points out that 76 per cent of total
agricultural employment in 1976 was provided by small
farms. Judging by the figures for poor small farm
and landless households, almost a third of the
population face serious poverty in the agricultural
sector. This is borne out by the high and moderate
inequalities in farm size and household income, which
existed for smallholders in 1974/75 in the majority
of agro-ecological zones in Kenya (table 6.7).

The wide range of climates and soils in Kenya,
aided by irrigation, allow many horticultural crops
to be grown profitably on commercial and
home-consumption scales. In contributing around 12
per cent of total crop production value in 1976
(1979-83 Kenya Development Plan), the horticultural
industry in Kenya is affected by the dominance of
small farms. Tables 6.3 and 6.4 clearly show that
the major part of horticultural output comes from
this category of farms.

Expansion of horticultural production, as scheduled
under the 1979-83 Kenya Development Plan, places the
industry in a position to contribute significantly to
(a) improving nutritional standards, and (b)
increasing rural incomes and employment. Against the
poverty background described, it was found that
around 13 per cent of the total value of household
consumption on small farms in 1974/75 was accounted
for by fruits and vegetables (table 6.5).
Horticultural produce is an important source of
vitamins in a diet heavily dependent on grains and
beans. Increased incomes arising from expanded
horticultural production on smallholdings would be
equitable in income distribution terms and would also
help alleviate rural poverty. Horticulture is shown
to be an important contributor to employment within
agriculture (Table 6.6) from the standpoint of

existing work opportunities and in its capacity to
create additional ones and help in reducing seasonal
under-employment. Also the important processing,
marketing and input-supply linkages that
horticultural production has with other parts of the
economy mean that any increase in horticultural
output will also have beneficial effects on aggregate
output and employment.

Investment in the horticultural industry in Kenya
would seem to provide at least equal, and probably
higher, rates of return compared with those in the
rest of agriculture. The 1975 Development Plan for
Kenya's horticultural industry saw such investment in
research and extension as receiving high rates of
return and being very worthwhile, when well
directed. But in this regard it is important to note
that the estimated parity ratio for the 1979-83
period indicates that public expenditure on research
and development in horticulture is less than
proportional to the industry's contribution to the
total crop value in agriculture.

Tests of hypothesis

<u>Hypothesis 1</u>
"The development and the performance of agricultural
research systems are conditioned by the national
resource and cultural endowments."
Looking at this wide-ranging postulate from the
standpoint of the national horticultural research
system in Kenya, there is no reason to disagree with
it in general, although specific qualifications can
be put on the ways in which conditioning of
development and performance takes place.

For instance, in Kenya, horticultural research has
developed rapidly over recent years, having emerged
in 1963 from a colonial administered organisation
which still forms the basis of the organisation that
exists today. And while British colonial government
imported horticultural innovations to suit a mostly
small white settler segment of the population, the
process has continued and expanded to meet the
development priorities placed on increased
horticultural production. Thus production techniques

for temperate vegetables, ornamentals and fruit (and for some tropical types), while often adapted to local conditions by national research stations, are nevertheless very much dependent on prototype technical innovation from abroad.

This influence of imported know-how in Kenyan horticulture has great advantages, for above all it has usually been an economically efficient way of gaining rapid improvement in the industry. It can be said to have been complementary and supplementary rather than competitive. However, this kind of transfer of knowledge and, sometimes, physical inputs too, still leaves a crucial role for the national research system to play in adapting and improving imported technology into new local technology.

This aspect, and the overall pattern of development of the horticultural research system and its ensuing performance, will necessarily be conditioned either directly or indirectly by national resource and cultural endowments. It has been shown clearly that the relative shortage of good land in regard to available labour, capital and management places a premium on using the most intensive crops and techniques in horticulture. Furthermore, the supply of qualified research scientists and technicians (depending on the education system), the food tastes of the population, the prevailing economic and political philosophies, and the general welfare of people, all help condition the horticultural research system over time.

It has been stated in the introduction that Hypothesis 1 permits the formulation of specific hypotheses regarding:
(a) the path of technological change;
(b) the economic and political demand for technical change;
(c) the political system influencing the latent demand for technical change.
Without actually formulating hypotheses in these areas, some brief comments can be made based on observations from the study.

<u>The path of technological change</u> The history of horticulture in Kenya is really too short to see

marked secular changes in technological development, as have been identified in Japan and the United States, in the case of biological and mechanical technologies respectively. If any pattern of change is to be discerned it is within a biological (yield increasing) mould from the start, which under early settler conditions stressed labour intensity and gradually adjusted itself to modern economic conditions, where labour and capital inputs (e.g. agro-chemicals and irrigation) were brought more and more into intensive combination. Certainly a broad mechanical technology emphasis cannot be identified. Even in large-scale production, where mechanisation of operations occurs and which accounts for the smaller part of total production in horticulture, one cannot say in general that there is an overriding adoption of mechanical technology. In particular cases such as pineapples and potatoes, practices do conform more to this type of technological emphasis, but in a country where labour is relatively cheap and usually obtainable on a long-term commitment basis (worker-family settlements), there is seldom the economic inducement to push the substitution to the technical limits. And even if there were, the political environment would find it unacceptable where it ran counter to rural development policies embracing increasing welfare of the population and labour employment opportunities.

<u>Demand for technical change</u> The previous discussion, therefore, already touches on the economic and political demand for technical change in horticulture. The economic system, as encouraged by the Presidential, one-party, parliamentary system of government in Kenya, is based on mixed economy principles and institutions. Thus, while the economy is one of free enterprise in the private sector, co-operative and public sector control is invoked in special areas, where it is thought necessary to safe-guard levels of investment and standards in what are considered to be essential services and activities from the standpoint of national interests and welfare.

Horticultural producers in Kenya, whether small or

large, work under the profit incentive even when
organised under institutional structures such as
co-operative irrigation schemes, settlement schemes,
co-operative marketing schemes and scheduled
Horticultural Production Centres. Consequently, the
profit incentive for technical change exists and is
officially encouraged. This translates into an
economic demand for technical change in horticulture
which in theory is governed by supply and demand
conditions in home and overseas markets. The
government can and does use measures to influence the
market conditions, by way of taxes, subsidies, import
tariffs and quotas, regulations, price control and
other policy instruments. Horticulture in general
remains unaffected by controlled prices for certain
crop products. On the other hand it is affected by
price controls (e.g. minimum wage rates) on a range
of agricultural inputs.

Here it is important to note that increased
intensity and higher total output in horticulture
can, and mostly will, involve increased labour
employment, without any concomitant substitution of
labour for capital inputs. Where the latter is found
to be feasible and economically relevant, then it too
can be an additional means of increasing labour
employment. But the scope for it seems to be
somewhat limited at the moment and in the future.

<u>Latent demand for technical change</u> The study has
discussed the Kenyan Government's important
priorities for the horticultural industry as set down
in the development plans. These priorities are
forward looking and have already led to induced
institutional change, e.g. the Horticultural Crops
Development Authority, and to the commissioning of
high-level consultant studies, with the help of
international agencies, for outlining measures which
can achieve stated policy objectives. A good example
of an innovative and far-reaching scheme of an
integrated production-marketing nature for increased
horticultural production in designated areas of the
country is that which will set up Horticultural
Production Centres. These will involve cooperating
groups of growers receiving and making use of central
services and facilities.

310

It is therefore evident that the political system
has to date, and through its likely stability in the
future, been an important influence on the latent
demand for technical change in horticulture. It has
by its policies and actions called for increased
technical change in the present and the future, and
indicated what financial resources can be expected
for the support of research and other activity in the
industry.

<u>Hypothesis 2</u>
"An agricultural research system which is
decentralised with respect to location, management,
and funding tends to be more responsive to the
resource and cultural endowments, and hence more
productive in supplying new technology to farmers,
than a highly centralised system."
The study shows that the public sector
horticultural research stations in Kenya show a mixed
system of organisation and control. Funding is
largely controlled through the Ministry of
Agriculture, even where foreign aid is concerned.
Therefore in this aspect a high degree of
centralisation takes place. Management of research
stations is also highly centralised in the Research
Division of the Ministry of Agriculture, and control
passes down through the Director of the NHS-Thika and
the Director of KARI. The horticultural research
system does however show a large measure of
locational decentralisation. This is manifested in
the siting of stations to serve under very different
environmental conditions, ranging from low-elevation
tropical to high-elevation tropical, and arid to
adequate rainfall areas.
From the evidence provided in the study, locational
decentralisation of horticultural research stations
would appear to be the most satisfactory way of
catering for local technological improvement needs.
No evidence to the contrary has been noted in the
literature on horticultural research in Kenya. The
fact that serious inefficiencies have been observed
in the local research station-extension-farmer
communication network does not argue that a
centralised location research system would achieve

any improvement in this direction. In fact the reverse is probably true in Kenya, where extreme contrasts in soil, altitude and climate give rise to specialised administrative and technical advisory services at provincial, district and divisional levels. Applied research needs to work closely with locally appointed government officials, as well as farmers, and is therefore probably most responsive to local problems when located near them.

Since funding and management of horticultural research in Kenya is highly centralised and always has been, it is difficult to judge the effects of any substantive change in procedures. A move in this direction has begun to take place in the case of KARI, which is not only controlled by the Ministry of Agriculture but also by the Ministry of Livestock Development and Ministry of Natural Resources. However, since its involvement in horticultural research is relatively small, any far-reaching moves to decentralise its funding and management control would affect agricultural research more than horticultural research.

There is the important consideration that the central role of government in formulating and implementing development policy within the resource context of Kenya as a developing country places certain limitations on the extent to which decentralisation can take place. The future may well see these limits being explored more and more and it would be a mistake to assume there were no pay-offs from a more decentralised structure. For instance the history of more favourable resource provision being made through foreign aid to international research institutions in East Africa, prior to the break up of the East African Community, raises questions as to whether greater decentralisation in funding and management might not hold some real advantages, for presumably this favourable treatment would not have taken place if it had been thought at the time that greater national research centralisation was needed.

<u>Hypothesis 3</u>
"An agricultural economy characterised by great

disparity in size (or income) of farms will be less effective in generating a sustained demand for a productive national agricultural research institution than an agricultural system characterised by reasonable equity in farm size distribution."

The estimated Gini ratio measurements of inequality among small farmers with respect to farm size and household income in the six agro-ecological zones of greatest horticultural importance in Kenya, lead to two main conclusions:

(a) the split in zones between low and high inequality in farm size was equal; and

(b) in the case of household income two zones exhibited high inequality and four medium inequality.

Therefore for the horticulturally important zones the variation exhibited in farm size inequalities helps neither confirm nor reject hypothesis 3. Moreover, while it is possible to say that the demand for a productive horticultural research system has been reasonably strong in recent years, this observation is certainly not matched by any finding that there is reasonable equality in farm size or income distributions, which would be necessary not to reject the hypothesis on positive grounds. Any suggestion that an agricultural economy characterised by a mixed pattern of area inequality in farm size or income distributions might be neutral in its effect on generating a sustained demand for a productive national research institution must be put as an hypothesis which in turn needs testing. However, it is obvious that the observations presented cannot test such an hypothesis.

In relation to the demand for horticultural research and technology which is responsive to economic forces, the study shows that Kenyan growers appear to exert little direct pressure on the system. Rather it is left mostly to government officials and organisation to decide what needs doing. It has been shown that horticultural producers have remained underorganised, particularly in comparison with producers of important commodities in the rest of agriculture. This seems to account for them being less effective in bringing their interests in research to bear on the allocation of research resources. Disproportionate funding (underfunding) of horticultural research has been observed. In this regard and to the extent that disparity of farm-size and incomes may affect matters, the issue is probably not so much disparity within the horticultural industry as that between the horticultural industry and other agricultural commodity producers.

Nevertheless, horticultural research has responded well in the last two decades to supplying recommendations for technical innovation. Although the technology developed depends a lot on adapting imported innovations to local conditions, this is in the main an efficient procedure and the system copes reasonably well with such demands. Perhaps the biggest question mark regarding innovation performance in horticulture in Kenya lies over the efficiency with which technical knowledge is transmitted and feedback is received. As one might expect in a developing country with a small-farm population the size of Kenya's, the diffusion of technical knowledge, involving research stations, extension service and growers, has been shown to face serious difficulties and deficiencies. This no doubt adversely affects horticultural activity, since it is very much the smaller part of the total agricultural effort anyway.

NOTES

[1] According to Republic of Kenya (1978), high rates of population increase are forecast for towns and cities, and these correspondingly increase demand. This agrees with the high population rate of increase for Kenya as a whole, approaching 4 per cent in 1980.
[2] These values include cashew nuts, tinned pineapples, and other items: it omits the beans, peas and lentils group.
[3] For a critical evaluation of the induced innovation thesis see Chapter 1 of this volume and Dorling 1982.

[4]   In all the IRS 1974-75 sample frame was
      estimated to include 1.48 million smallholders
      in six provinces, ranging in size from below 0.5
      hectare to 8.0 hectares and above. In fact no
      holdings above 20 hectares were included and the
      percentage included with 8.0 hectares and above
      was only some 3.5 per cent. This small
      percentage occurs as a result of most of the
      over 8.0 hectare holdings being excluded by
      definition.

[5]   Broad agreement on these land prices and wage
      rates was conveyed in an interview with the
      Deputy Chief Valuer and Senior Valuer in the
      Lands Department, Office of the President,
      Nairobi, in January 1981. However, it was
      pointed out that while percentage increases for
      land prices in agricultural areas in Kenya have
      borne a close resemblance to each other between
      1963 and 1981, the price levels themselves can
      be very different.

[6]   This apportionment of the Joint Research
      Services total expenditure to horticultural
      activity is obviously on a very rough basis.
      For instance this total includes Forestry, which
      was not present in the crops development
      category in table 6.12 and on whose constituency
      proportional breakdowns for horticultural
      expenditure have been extended to the Joint
      Research Services total.

[7]   The general commentary on the interview
      procedure and a summary of information gained on
      horticultural research inputs and projects is
      provided in the next section. The list of the
      research stations, government and parastatal
      organisations interviewed is as follows:
      National Horticultural Research Station - Thika;
      Coastal Agricultural Research Station - Mtwapa;
      Pyrethrum and Horticultural Research Station -
      Molo; Kenya Agricultural Research Institute -
      Muguga; Department of Crop Science, Faculty of
      Agriculture, University of Nairobi, Kabate;
      Research Division, Ministry of Agriculture,
      Nairobi; Crops Division, Ministry of
      Agriculture, Nairobi; Natural Resources Section,

Ministry of Economic Planning and Development, Nairobi; Treasury, Office of the Vice-President, Nairobi; Lands Department - Valuation Section, Office of the President, Nairobi; National Council of Science and Technology, Office of the President, Nairobi; CIMMYT, Nairobi; Horticultural Development Authority, Nairobi; Horticultural Cooperative Union Ltd., Nairobi.

[8] The term research in this section implies the broader research and development meaning, whereby new knowledge is created and applied so that it can give rise to improved technology and innovation.

[9] This committee is given the same title as that used by Warui and van Eijnatten (1979) in their paper on extension and research links. Ministry of Agriculture officials often refer to it as the Provincial Research Advisory Committee (PRAC).

[10] Irrigation does not show in this listing. Related research activity in horticulture is difficult to assess because of the diverse contribution in this area from the full range of agricultural research stations and from equipment supply firms.

7 Institutional factors and technological innovations in Vietnam

NGUYEN NGOC LUU

INTRODUCTION

This case study is an attempt to trace the course of
and to analyse the dynamics of the process of
technological change in agriculture in North Viet Nam
prior to its reunification with the South. It will
focus on the interaction between political, economic
and cultural factors that have determined the course,
pace and rythm of the process. Although the study
focuses on the 1954–75 period it will occasionally
include events of the post–1975 period (i.e.
post–reunification) in order to clarify events and
trends that occurred before then.

This study will adopt an "integrated approach",
including in its analytical framework various
economic, cultural, institutional and technical
factors. This will help first to study the process
of technological change in its historical context;
and, second, to probe into the dynamics of the
process. The "integrated approach", as adopted some
time ago by Stavis (1976) and Luu (1979) is indeed
very close to Ruttan's analytical model (as shown in
Chapter 3 of this volume).

Based on the experiences in most Western, non-socialist economies, Ruttan constructed his induced innovation model on the basis of responses to changes in the prices of production factors, among which land and labour are the two most important. Although he rightly pointed out that the major constraints imposed on agricultural development are either an inelastic supply of land, as in the case of Japan, or an inelastic supply of labour, as in the case of the United States, the fact that his model is based on the free-market prices of land, labour and other factors of production renders an induced innovation model inapplicable to socialist countries unless adjusted in some way. The reason is simple: the prices of factors of production – as they exist in a free-market economy – are absent in socialist countries where the economies are centrally planned. Furthermore, in socialist countries, all means of production are supposedly collectively-owned and mobilised to meet the society's needs. The very basic differences in the modes of production mean that any analytical framework designed for one cannot be easily and totally applied to the other without modification.

Apart from that, Ruttan's approach and that adopted here are very much the same. Many hypotheses put forward by Ruttan – as readers may find later – will be borne out by the empirical findings of this study, among which one should specifically mention the hypotheses on the positive effects of a socialised network for scientific and technological research and diffusion, and on the increasing degree of institutionalisation that corresponds with progress towards sustained innovation.

Research on most socialist countries is often handicapped by a lack of the necessary information and statistics. North Viet Nam has experienced a long war, so that many records and documents have been destroyed. The sources remaining are sometimes contradictory or fragmented. Most statistics are also computed and presented in very different ways. All this creates problems in interpreting and generalising even for a relatively small area of study like North Viet Nam, and even during a short

period of time. To overcome this problem, reliance
was placed on statistical figures provided by
official sources or sources close to the North Viet
Namese authorities, whenever they were available.
Other figures, which are used without mentioning
sources, are computed from fragmented information
from newspapers, periodicals, and serial publications
which were accessible, or through direct interview of
individuals in North Viet Nam.

THEORETICAL CONCEPTS

From a theoretical point of view, technological
development has never been seen by the Marxists as an
independent variable, only loosely linked with the
process of overall social change. On the contrary,
the Marxists maintained that the development of
technology would necessarily lead to the development
of productive forces. In other words, there was no
science and technology that existed independently of
production. As part of the superstructure, science
and technology had close links with the economic base
of the society.
The development of production forces in turn was
dialectically related to the social relations of
production. Any significant progress in developing
production forces would necessarily breed
contradictions in the social relations of
production. Therefore, neither of the two should be
left behind with respect to the advancement of the
other. From a practical point of view, any attempt
to build up a new system of social and production
relations should always rely on the possibility of
bringing about appropriate development of productive
forces that were strong enough to support the new
systems of production relations. This theoretical
tenet thus maintained that institutional change and
technological change were organically related to each
other, and the change in the social relations of
production should necessarily be accompanied by the
advancement of production forces. If the task of
revolutionising the social relations of production
was already hard to fulfil, the development of

productive forces appeared to be even more strenuous, particularly for a country with a weak and backward scientific base. Such development was very much influenced by the country's stage of development, the priority that society set for itself, and by the degree to which the political leadership succeeded in mobilising resources.

North Viet Nam represents a case in point. As the communist writers often put it, Viet Nam decided to move forward directly to socialism, bypassing capitalist development. Or more metaphorically: "Viet Nam moved to socialism with its traditional plough".

In two decades of existence as a state independent of its rival, South Viet Nam, North Viet Nam had undergone intensive changes that affected both institutional and technical aspects of agricultural production. From 1954 to 1975, although national liberation always occupied the central position in the struggle, social revolution had its own importance and involved drastic political and economic changes. In the field of agriculture, the revolution had been made concrete in the replacement of the old system of social relations of production by a new one which was constructed on the basis of collective production relations. The traditional private farming system was terminated, and in its place collective farms were set up. Apart from the ideological considerations, it was believed by the communist leaders that the new agrarian system would pave the way to large scale socialist agriculture. Throughout the process of constructing a new mode of production, the leadership had to face various kinds of problems which evolved either from the revolutionisation of production relations or productive forces, or from the combination of these two.

Social revolution in North Viet Nam - of which agrarian revolution was one component - followed Marxist-Leninist guidelines. From a theoretical point of view, although the Viet Namese communists advocated the tenets of the Marxist-Leninist model of economic development, putting high priority on socialist industrialisation, they had to recognise the fact that agriculture would remain the most important sector of the economy for some time. The urgent need for food sufficiency, the need to employ the 80 per cent of the population who lived in the countryside, and the need for a guaranteed supply of raw materials and for capital formation for industrialisation had all drawn the attention of the leadership of North Viet Nam to agricultural development.

Seeing the society in its totality, the Viet Namese communist leaders not only recognised the relationship between technological change and institutional change, but they also emphasised the ideological and cultural dimensions of the revolution. They put forward the theory of three revolutions, namely the revolution of production relations, the technological revolution, and the ideological cultural revolution.

These three revolutions constituted the three organic parts of the socialist revolution. They were interwined, exerted influence upon each other and propelled each other forward. In the perspective of this theory, the Viet Namese communists always let politics take command, and for this reason, revolution in production relations was a sine qua non step which paved the way for the development of productive forces and quickened the pace of technical revolution (Le Duan 1970 and Giap 1978). Once the political forces succeeded in making a step forward in establishing new production relations, scientific and technological innovation was to follow with a view to consolidating the new production relations, and to pushing forward cultural progress. This made scientific and technological development in agriculture in Viet Nam unlike the experience of the Green Revolution elsewhere in Asia, where scientific and technological changes in agriculture took place without institutional changes first having paved the way for them.

Although the revolution in production relations was allowed to take the lead, it was not the intention of the communist leaders to complete the building of the new system of production relations before initiating the technological and cultural revolutions. Instead, the three revolutions were to be undertaken simultaneously, and were meant to support each other.

It is interesting to note that the Viet Namese communist leaders emphasised reliance on the most abundant and precious asset of the country, namely the labour force, and on a good combination of human resources with natural and technological resources. This meant that technological development should have started with, and primarily relied on, human development. In the view of the Viet Namese leadership, labouring people constituted the most essential and basic element of the productive forces in the agricultural sector. As the natural productive force, labour had the great potential of transforming nature and producing wealth for the society. But more important than that, this labour force was at the same time the master and the manager of the socialist production system.

Thus, human development in the agricultural sector occupied the fulcrum where the three revolutions converged. In other words, the development process started with the people, and centred around the development of the people. In this view, technological development in agriculture was not merely aimed at increasing productivity and production but also at the realisation of the potentialities of the people.

In line with this philosophy, research on technological development required the services of the natural, human, and social sciences. It was not merely a matter of finding new seed varieties, designing and manufacturing new or improved farm machinery and implements, etc. - but it touched all aspects of life for the vast mass of agricultural producers and affected the whole dynamics of the national development process. From this viewpoint, the Viet Namese communist leaders tackled the problems related to technological development in its social-cultural totality. They also broke through the rigid and static division of the development process into stages in which only one sector was allowed to develop at the cost of the others; they brought some moderation to the orthodox socialist pattern of socioeconomic development.

In general, the orthodox socialist strategy for development often showed a clear bias in favour of industry. In the early stage of development, industrialisation was often favoured at the cost of agriculture. The development of industry had always been considered the locomotive of national economic development; the development of agriculture was to follow. In other words, at the early stage of national development, resources had to be drained off agriculture and invested in industry. Agriculture had to wait a long time before it could expect any service from industry in return. The experience of most socialist countries proved that this model of orthodox socialist development was detrimental to agricultural development. Agriculture in these cases often lagged behind, or was crippled by an excessive drain of resources for too long a period. The weaknesses of agriculture in almost all socialist countries is not surprising particularly when one looks at the handling of the relations between industry and agriculture, and at the lack of an effective incentive system.

It seems that Viet Nam could never have afforded such an overemphasis on industrialisation at the initial stage of development. Therefore, although the country's communist leadership had continuously adhered to the theoretical principle of giving industry higher priority and putting it in the leading position in national development, in practice the leaders not only avoided exhausting agriculture, but also did their best to support agriculture – particularly the food-crop production. In the case of Viet Nam there had been a gap between the ideological and theoretical positions vis-à-vis the questions of development and the reality and practice. The probable explanation is that if the Viet Namese leaders – as communists – found themselves bound by a set of measures that every proclaimed communist must take in order to remain communist, they also found themselves in the position to work out strategies and adopt measures that could help to respond to the challenge of circumstances. However, although moderation and flexibility could be observed in the way the leadership steered the

development process, reality was actually never able
to affect the Viet Namese communists strongly enough
to stimulate them to re-think development, critically
to review the orthodox model of socialist
development, and to examine its applicability to Viet
Nam. The quick development of heavy industry had
remained very much the top priority. Only when
circumstances would not permit such an ambitious
strategy for development did the Viet Namese leaders
tilt back to agriculture. On the theoretical plane,
heavy industry always retained supremacy, although in
various instances references in official documents
appear to modify this emphasis to some extent (Giap
1978). For example, after having been confronted
with reality, the leadership of the Communist Party
of Viet Nam maintained:

> "Priority should be given to reasonably
> developing heavy industry on the base of the
> development of agriculture and light industry, to
> combining the development of industry and
> agriculture in the whole country and to making this
> combination an agro-industrial structure."
> (Communist Party of Viet Nam 1977)

From the point of view of technological
development, Viet Nam had not followed the normal
path by first developing science and technology in
industry and afterwards turning to agriculture.
Instead, under the pressure of circumstances, Viet
Nam had simultanously undertaken scientific,
technological revolutions in both industry and
agriculture (Giap 1978). In other words,
agricultural development did not have to wait for the
completion of industrialisation for a strong and
supporting heavy industry to come into being first.
Agricultural development had to proceed towards
large-scale socialist production at the same time as
socialist industrialisation. This was because
industrialisation could not be undertaken if it were
not constructed on the foundation of a strong
agriculture. Industry and agriculture were
organically inter-linked and therefore one could not
develop without the support of the other. One could
find similar ideas on the interrelationships of these

two economic sectors in the writings of Mao Ze Dong (selected readings 1971).

But while industry – particularly heavy industry – was still in its infancy, and science and technology were still backward, on what technical basis was agriculture to develop? In other words, at the existing level of development, with what type of science and technology must agriculture start? Where could it draw technological resources for development?

As far as the type of technology was concerned, it seems that the Viet Namese knew pretty well their needs and their capabilities. They were fully aware that with Viet Nam's meagre means, a drive for modernity in a short period was futile and beyond their capacity. Consequently, they adopted a two-pronged policy. Le Duan, the Secretary General of Viet Nam's Government Party, noted that the development of productive forces in Viet Nam could and should at the same time develop gradually according to the law of change from small- to large-scale production, and by leaps directly to first mechanisation and then automation. In combining gradual change with leaps, Viet Nam thus had to rely on the gradual improvement of traditional empirical science and technology in order to bring the existing production forces to a higher level of development, while at the same time it took a short cut and embarked upon quick mechanisation and automation using advanced technology wherever it was appropriate and economical (Le Duan 1970). Thus, the communists in Viet Nam hoped to compensate for industry's weak support of agriculture with the body of empirical knowledge on farming that peasants had accumulated over the centuries. This policy was later elaborated by Vo Nguyen Giap, the former member of the Political Bureau of the Party, who was put in charge of the development of science and technology. He called for adequate attention to be given to the rich and valuable experience of Viet Namese peasants, and he positively evaluated the contribution of this body of knowledge to the development of agriculture (Giap 1978).

This policy of 'technical duality' had been translated into concrete measures in the field of agricultural mechanisation. Lacking highly mechanised machinery, most farm activities in Viet Nam had been practiced manually. For this reason, the quantity and quality of simple, traditional or improved implements were of great importance. There had been a push to improve traditional tools and implements. Giap criticised the error of some local cadres for having over-emphasised modern tools and machinery, and for failing to appreciate the available manual or semi-mechanised farm implements. He was equally critical of the defeatist attitude of technicians who when faced with failure in their attempts to improve traditional tools had turned back to the traditional ones and refused to continue their work. He emphasised:

"We must take the improvement of hand tools and implements seriously because it not only helps to increase labour productivity but also makes implements increasingly suitable for advanced farming techniques and for the conditions of different regions. It also paves the way for the mechanisation of production." (Giap 1978, p. 59)

The choice of science and technology for development was therefore made against the background of the country's reality. It took into consideration the short-term and long-term objectives and the availability of resources at the time the choice was made. The direction of the process of change was definite, however, moving away from intermediate technology and towards advanced technology.

With regard to land and labour as factors of production, the question then arose whether Viet Nam should adopt technology aimed at saving land or at saving labour.

The Viet Namese leadership looked at the limited amount of arable land, which had been under the increasing pressure of population growth, and concluded that intensive farming was a must for Viet Nam. Thus they emphasised the need to save land as a scarce factor of production. Although there was no attempt to introduce labour-saving technology into agriculture in order to release labour for other sectors, the Viet Namese leaders did not overlook

the implications of certain technology on the amount
of labour available at peak periods, and the quality
of labour's performance with regard to important
farming activities. In general, newly introduced
technology should help facilitate intensive farm work
during short periods, such as land preparation,
transplanting and harvesting. With these tasks, the
existing labour force could not even manage to do its
work well and on time, let alone improve it. This
situation in turn hindered the full utilisation of
land resources. Thus, for these tasks there had been
both a shortage of labour and a need to improve the
quality of labour. In these phases of the production
cycle, saving labour could lead to saving land as
well, since the labour replaced by mechanisation
could be used for other farming activities in the
effort to multiply the number of crops sown on each
unit of land. New technology in these cases aimed at
improving the quality of farming activities as well.
The use of tractors in land preparation, for example,
not only helped to complete the task in a short
period so that a second crop could be started on
time, but it also allowed for higher productivity
thanks to deeper ploughing. This policy had its
economic reasons; the country had to utilise fully
its cheap and abundant factor of production - labour
- and other limited resources (financial and
technological) were supposed to be invested in
generating innovations that would have the scope for
the widest possible diffusion.
   On the basis of all the above-mentioned theoretical
conceptions, the Viet Namese leadership designed a
comprehensive strategy for the development of
agriculture. Although the strategy was consistent,
its implementation had always been affected by the
historical circumstances of the country. The whole
process of development had always been disturbed by
protracted war to such a degree that sometimes one
could hardly talk about development, but rather about
survival. The pace and rythm of the development of
productive forces in Viet Nam's agriculture were
determined to a large extent by the war situation.
Still, one could realise that the fundamental
character of the process did not alter: it remained

comprehensive and was guided by Marxist-Leninist
ideology.

CROPPING PATTERN

North Viet Nam's agriculture had all the
characteristics of backward monoculture. Rice
cultivation dominated rural life and monopolised
almost all available resources in the countryside.
Activities in the countryside were organised and
centered around the production of this crop. Before
North Viet Nam's "Green Revolution", the Viet Namese
peasants grew rice in the same manner, using the same
techniques that their forefathers had used centuries
ago. As a consequence, agricultural production
stagnated or grew at a lower rate than the
population. Millions of smallholders worked
desperately on tiny plots ranging from 0.18 to 0.36
hectares in size in order to eke out a bare
subsistence. Millions more found themselves tilling
the land of others without any guarantee of having
enough rice to eat for more than three months per
year. Under such conditions, no one had the
incentive to innovate, or even if they had, no means
would have been available to undertake innovation.
This contributed to the irony of the Viet Namese
history; a strong nation existed on a weak economic
base.

The structure of crops in North Viet Nam was
characterised by the two major rice crops sown on
large areas of the country's cultivated land.

The Chiem was the rice crop of the cold and dry
season. It was sown in the 10th and 11th months and
harvested in the fifth month of the lunar calendar.
Seedlings of this rice crop were transplanted in
January, the coldest month of the year. The cold
slowed down the growth of the plants and
unnecessarily extended the growing period of the rice
stalks. Rice plants of this crop suffered from all
the problems arising from unfavourable climatic
conditions and gave low yields (1 to 1.1 tonnes per
hectare).

The Mua was the rice crop of the rainy season. It

was sown in the fifth and sixth month and harvested in the 10th and 11th months of the lunar calendar. Under the traditional level of farming, even if the Mua enjoyed far more favourable climatic conditions, its yield remained very low: 1.2 to 1.3 tonnes per hectare.[1]

While the Chiem had often been threatened by drought and lack of water, the Mua had often been damaged by flood, water-logging and typhoons.

Under optimal conditions (availability of water, enough land and other capital), agricultural production in North Viet Nam was undertaken according to the calendar in table 7.1

The traditional calendar shows two areas of overlapping: the two peak periods are the fifth to sixth months and the 10th-11th months. During these periods, urgent and highly labour-intensive farm activities have coincided, such as: harvesting, threshing, drying, storing, ploughing, puddling for land preparation, preparing the seedling fields, seed treatment, and sowing. All these activities had to be completed within one to two months. The pressure of time obviously created labour shortages and negatively affected the quality of the farm work. In many areas, owing to lack of labour and bad organisation, peasants failed to complete the work in time and therefore suffered from damage or lost the momentum for a good harvest. In this situation, there was no reason even to consider further multiple cropping or intensified farming.

This system of monoculture with the two rice crops bore all the characteristics of backward agriculture. Apart from the heavy dependence on natural conditions and unstable and low yields, the Viet Namese agricultural authorities saw in this system of cultivation structural weaknesses to be overcome. Their view on Viet Nam's rice cultivation system was interesting in the sense that it did not point to the lack of modern input as the primary reason for stagnation but to the irrational crop structure and the imbalance between the different branches of agriculture. They noted the following weaknesses:

1. There was a lack of crop rotation which might

Table 7.1
Viet Nam's traditional agricultural calendar

Crops sown	Lunar month											
	1	2	3	4	5	6	7	8	9	10	11	12
Chiem	weeding	protection			harvest					sowing		transplanting
Mua					sowing	transplanting	weeding, protection			harvest		

have helped to improve the fertility of the soil. Indeed, under pressure of food shortages, the Viet Namese peasants had no way to move away from the monoculture of rice. Rice occupied the supreme position in agricultural production, covered the largest percentage of the country's cultivated area and employed the great bulk of the labour force. As the calendar shows, rice farming activities extended throughout the year and, as a consequence, rural life was synchronised to the rice farming cycle. This system on the one hand created a disequilibrium between rice and other food crops and, on the other hand, it meant an almost total lack of division of labour in agriculture in general, and in rice cultivation in particular. It did not leave room for the development of specialised techniques in rice and other crop cultivation. Every peasant family was involved in the cultivation of rice; other subsidiary crops received only minor attention. Every grower also used the same techniques, and performed the same tasks at the same time. For rice cultivation in Viet Nam, there were thousands of production units (the families) doing exactly the same thing at any one time of the year.

2. As a consequence of monoculture, there had been not only disequilibrium among food crops, and between food crops and industrial crops, but also a neglect of animal husbandry. This had resulted in an impaired agriculture that had not only failed in supplying enough foodstuffs for a balanced diet, but also in establishing mutual supporting relations between cultivation and husbandry. Animal husbandry, as experience had shown, made optimum use of by-products or residues in rice production and processing. In return, it could supply an important quantity of manure that could be used as organic fertiliser. This mutual support could not be exploited for the benefit of agricultural development unless an equilibrium between the two branches was established.

3. Besides, the monocultural system had not
   rationally utilised natural resources and
   exploited fruitfully the country's natural
   conditions. One of the notable examples has been
   the Chiem crop in Viet Nam: the long cropping
   season limited land use for other crops and had
   taken up the largest position of the cultivator's
   time. It started at the end of Autumn and
   extended over Winter and Spring. The occupation
   of the land for so long a period and the demand
   for so much labour in exchange for an unstable
   and low-yield harvest had made this crop
   uneconomical and undesirable. Its long
   vegetative development had made the agricultural
   schedule so tight that no sooner had the previous
   crop been finished than the peasants had to start
   work on the next crop. Thus, the existence of
   such a long period of intensive work had been
   detrimental to the two most important elements of
   good farming: timing and the quality of the
   operations. Timing was not only vital for sowing
   and transplanting, but also for other farm
   operations of the production cycle (Cua 1980).
With all these weaknesses, the monocultural system
of rice farming failed in ensuring food sufficiency,
and cyclical famine occurred. Efforts to develop
Viet Nam's agriculture were therefore geared towards
breaking this claim of backward production and
modernising forces using technologies that had been
well tailored to meet the economic, cultural and
technical conditions of the country.

INSTITUTIONAL REFORM AND TECHNOLOGICAL INNOVATION

On the clear practical plane, the leadership of North
Viet Nam clearly pursued a two-pronged policy: on the
one hand they wanted to break the vicious circle of a
backward and monocultural system, and institute in
its place a more balanced cultivation structure
whereby diversification and specialisation could be
realised; on the other hand, they sought to intensify
the rice cultivation in order to increase the
country's yearly food production.

The implementation of this policy at the beginning
of the modernisation campaign ran into the following
problems:
(a) the inelasticity of land as a scarce production
    factor. In the mid-1950s, little arable land
    could be brought under cultivation unless the
    country could afford massive financial and
    technical investments so that industrial and dry
    crops could be introduced into the mid- and
    highlands;
(b) a very tight farming schedule that would not
    allow for multiple cropping unless new technology
    was introduced that would enable the schedule to
    be rearranged;
(c) a weak and atomised peasantry which had little or
    no capacity to absorb new technology and to
    finance new inputs. Living barely at the
    subsistence level, the peasants could hardly
    afford to take the risk of innovation. This lack
    of receptiveness was a consequence of the
    peasants' destitute socioeconomic situation: a
    high rate of illiteracy, widespread malaria and
    trachoma, and undernourishment due to low incomes;
(d) most serious of all was the newly established
    socialist government's inability to support a
    "take-off" in agricultural production. Its lack
    of investment resources meant a lack of any
    substantial support from outside the agricultural
    sector. Any agricultural development thus had to
    be financed mostly by the agricultural sector
    itself. In other words, it relied on a
    mobilisation of intra-sectoral resources.
For all these reasons, North Viet Namese leaders
emphasised in the first place institutional change in
the hope that they would liberate the traditional
productive forces from the constraints imposed by
"feudal" and colonial social forces. They also hoped
that without financial and technical support in the
early stages, it would be possible to boost
production through agrarian reform that would give
the peasants incentive to produce, and would improve
the organisation of the agricultural labour force.
This would buy time for agricultural sciences to
develop and produce some useful and practical

discoveries, and for industries to reach the stages of being able to supply some agricultural outputs.

During the first stage of innovation, which was dominated by the transformation of production relations, technological changes resulted largely from the propagation of traditional or improved technology rather than from the adoption of new, advanced technology. Innovation during this period had very much the personal character that Ruttan believes to be the common character of any country's initial stage (Chapter 3 of this volume).

Land reform

By 1954, when the Government of the Democratic Republic of Viet Nam returned to its capital after nine years of resistance against French colonialism, it had committed itself to an intensive class struggle in the countryside that aimed at an agrarian revolution. The campaign had been launched in 1953 when the Lao Dong Party (which was later to become the Communist Party of Viet Nam) and the resistance Government granted their all-out support to the landless and land-poor peasants in their fight to overthrow the landlord class and redistribute agricultural land. The communist-led regime's regaining control over the northern half of Viet Nam opened up the development dimension of the extremely political land reform campaign. Although the land reform campaign remained very much influenced by political necessities after 1954 (i.e. preparation for re-unification, establishing an effective Party-controlled state apparatus, etc.), the Party's Political Bureau stated in September 1956 that land reform must continue in peacetime because it was essential for economic recovery and development. In the resolution passed by the Political Bureau, North Viet Nam's communist leaders described the reconstruction and development of agriculture as the fundamental issue at the very moment that a programme for rapid large-scale industrialisation appeared to be impossible (Phuong 1968).

After theoretically assuming full control of North Viet Nam, the Communist leaders allowed the class struggle in the countryside to go ahead in full swing in order to complete land reform. Since the great majority of the landless and land-poor peasants actively participated in the class war, the Lao Dong Party and the Government of North Viet Nam gradually found the whole situation getting out of their control. Many mistakes were made owing both to the peasants' jealous fighting for a piece of land and to the Party's attempt to purge from its ranks members with rich-peasant and landlord backgrounds. Although the Party later admitted frankly that a number of middle and rich peasants had been unjustly treated as landlords or rural despots, it could not have been more pleased to see all potential opposition (inside and outside the Party) crushed by the mass movement, and to see a large number of its local-level cadres with rural bourgeois backgrounds purged. Despite all the upheaval, the Party and the Government drove the campaign to its completion in 1957, after having spent one year rectifying the excesses that had been committed.

In the wake of the land-reform campaign, North Viet Nam's countryside had a relatively egalitarian system of property. Although it did not bring about a drastic change in production it obviously succeeded in establishing a more equitable distribution of incomes in the countryside. The fact that 77 per cent of rural households benefited from the campaign, and that around 90 per cent of North Viet Nam's peasants were then owners of farms of roughly 1 hectare in size, and that after Land Reform there has been constant demand for innovation on the part of peasants, bore out de Janvry's hypothesis that an agricultural economy characterised by great disparity in farm size or income would be less effective in generating sustained demands for innovation (and peasants' participation, we might add). This held particularly true for hydraulic innovation, the kind of technical innovation that North Viet Namese leaders planned to implement alongside the reform campaign.

Besides, although some concern for reconstruction and development was shown during the last and most drastic stage of the land reform campaign, it became quite clear that the political aims of the campaign had gone much further than mere economic objectives. Perhaps the most significant achievement was the improvement of the landless and land-poor peasants' socio-political situation in the countryside. With the class-struggle campaign, there was a clear shift in the leadership away from landlords and rich peasants and towards the poor peasants.

This forged a loyalty towards the regime which undertook further institutional reform and technological innovation in the subsequent years. This loyalty eventually became an all-out support of poor peasants for the communist agricultural policies in Viet Nam in the late 1950s. This made agricultural innovation a joint venture between the communist leadership and the mass of poor peasants, instead of a partnership between the well-off modernising farmers and agri-business enterprises, as was the case in non-socialist neighbouring countries. In Viet Nam, the Communist Party and the State played the modernising role, in a sense occupying the position of agri-business and the modernising farmers in non-socialist countries. The achievements in national development in general and in agriculture in particular thus depended largely on the concrete terms of the relations between the Party and the state on one side and peasants on the other. In other words, they depended largely on the incentive system that the policy established in order to mobilise peasants' participation.

On the basis of such institutional change and peasant support, could North Viet Nam's leadership succeed in promoting substantial technological innovation? In order to answer this question, it is first necessary to examine the state's control over economic factors during the period in question.

First, as the leading agent of change, the state emerged after nine years of resistance with virtually no means to activate and influence economic agents for the sake of development. During the resistance, the infant state apparatus, which had ruled for barely a year, had absolutely no chance to improve its own organisation or gain much experience in running a country, let alone to develop that country. The only skill that North Viet Nam's communist cadres were equipped with was the ability to mobilise the people and organise them to fight a protracted war. This skill in political economy for a war of liberation was not necessarily suited to the new tasks. The lack of capable cadres to run an economy that had been distorted by the war of liberation resulted in the paralysis or poor performance of state organisations, or state-run enterprises. It was reported that 70 out of 131 administrative services handed over to the communists came to a virtual standstill (Le Chau 1966). Under such circumstances, there was little the state could do to intervene in the operation of the economy.

In the field of science and technology, there was no infrastructure whatsoever for research and training. Before 1954, there was virtually no regular training in agronomy and agricultural economy in North Viet Nam. The few agronomists the whole country possessed were mostly trained in Paris or Saigon. They made up the handful of privileged intellectual bourgeois who often switched to political or administrative professions. When the country was divided in July 1954, most of these elements departed for South Viet Nam. The few other agronomists who had joined the resistance came back to Hanoi and found themselves working empty-handed. The meagre and primitive laboratory facilities of the University of Hanoi were partly shipped to Saigon and partly destroyed at the departure of the French for the South.

The reconstruction of the national schooling system seems to have been much easier at the primary and secondary levels than at the university level. However, the effort of North Viet Nam to develop technical education during this period was noteworthy. In 1955, there were only 2,800 students in middle-level technical schools, and 1,200 in universities. Three years later, the figures rose to 8,300 for technical schools, and 3,700 for universities (General Office of Statistics 1978). This rapid increase in the number of trained personnel could only benefit the subsequent period.

In view of all the advantages (psychological and

political) and disadvantages (lack of infrastructure, skilled personnel, etc.) only one type of technological innovation could be undertaken: the development of irrigation.

When peace was restored in July 1954, the war had destroyed almost all the major irrigation systems in North Viet Nam. At least eight major hydraulic systems had been bombed, leaving 250,000 hectares without any irrigation. Dozens of other smaller irrigation systems had been damaged. Altogether, some 320,200 hectares of land had inadequate water supplies (Tri 1967). If the estimate was correct, it meant that all cultivated land in North Viet Nam in 1954 was without irrigation; since in pre-war 1939, the total irrigated area of North Viet Nam had only been 326,000 hectares (Le Chau 1966).

The achievements in reconstructing hydraulic systems were impressive. Through the newly re-organised peasant associations, the mobilisation for repairing dykes, dredging canals and ditches, and building dams successfully brought thousands of peasants to the fields. It was reported that within a year, more than 8 million work days were spent to repair hydraulic systems and to strengthen 2,750 kilometres of flood-protection dykes. By 1955, almost all damaged systems had been repaired and the areas they serviced were brought back under some sort of irrigation. Besides, this achievement improved the irrigation of 241,400 hectares of land that were planted with the Chiem crop (Le Chau 1966 and Khoa 1964).

Since 1955, a more comprehensive plan for water control has been prepared and carried out. On the average, it was possible to mobilise peasants to move 5.7 million cubic metres of earth each year.

There were several reasons that accounted for such a successful mobilisation of the peasants. Firstly, it was very easy for the poor peasants who had just been allotted a piece of land to recognise how vital such hydraulic works were for the exploitation of their plots. Development of the water system obviously enhanced the stability and the increased yields of the land. The most important psychological factor was that the peasants were sure that they

themselves would feel the full benefit of that
increase. Second, mobilisation on such a scale could
only be done if there were grass-root organisations
where genuine participation could take place.
Peasant organisations and local village government
seem to have met this qualification. During the
land-reform campaign, the mass line was proclaimed,
and the principle of "relying on the poor and
landless peasants, united with the middle peasants"
served as the guideline for reforming peasants'
organisations and village administration. As a
result of the application of this principle,
two-thirds of the members of peasant associations'
executive committee were land-poor and landless
peasants; the rest were middle peasants. These
committees were in charge of implementing land reform
in villages by mobilising peasants to participate in
the class struggle. In this capacity, peasant
associations gathered together the great bulk of poor
peasants who were eager to join the associations in
order to fight for and defend their interests. It is
thus possible to say that peasant associations really
had a popular base, and could mobilise peasants. The
very presence of these associations and their
participation in the repair and construction of
hydraulic works obviously contributed to the
successes in this field. One can say that the
mobilisation of peasants for hydraulic work inherited
the positive psychological effects of land reform and
the organisation that made the mobilisation possible.

The second support for agricultural production was
the Government's drive rapidly to supply the hundreds
of thousands of new peasant owners with farm tools.
With 2,104,138 rural households consisting of
8,323,636 individuals having benefited from land
redistribution, one can assume that many of these new
family farms were badly in need of farm tools.
During the land reform campaign 1,846,000 farm tools
and 106,448 draught animals were confiscated from the
landlords and distributed to beneficiaries. But the
average of less than one piece of farm equipment for
each family was far from sufficient. The efforts of
the North Viet Namese Government to solve this
problem could not yield the same dramatic results as

hydraulic works. By the time the communists took over the North, its stock of fuel was only 30,000 litres, and most manufacturing plants stayed idle because the French had taken away or sold stocks of raw materials and had stripped the plants of all machinery (Limbourg 1956). Faced with this problem, the agricultural authorities turned to local manufacturing workshops, particularly to the newly formed co-operatives of blacksmiths and other craftsmen.

Despite all the lack of inputs, new investments and farm tools, the peasants could already increase production because of the improved irrigation system. Sources near the communist authorities later claimed increases over the figures of 1939 of some 24 per cent in area sown with rice, 43 per cent in production per head, 36 per cent in yield per hectare, and some 13.25 per cent in rice consumption (Le Chau 1966 and Tri 1967).

The recovery of rice production was achieved almost totally by a production system of tiny holdings. It was the outcome of the peasants' struggle to alleviate the threat of famine, and to improve their diet. Acting in the capacity of small producers and relying on their own means, peasants were able to surpass the pre-war production level. However, with an average holding of about 1 hectare in size, small peasant owners could not be sure of a stable source of income at the existing level of production. At best they could only maintain a bare subsistence; there was no question of producing surpluses in order to support the urban areas and contribute to an accumulation of capital for national development. The fact that tiny farms were suited to family farming using low-level farming technique in a densely populated area like North Viet Nam demonstrated that land reform could bring about a more equal distribution of wealth and income but could not lead to agricultural development. The small producers living at the subsistence level were hardly able to defend their holdings let alone afford new technology and bear the risk of innovation.

In conclusion, during the land reform campaign between 1954 and 1957, institutional reform boosted production and created some psychological and organisational advantages for development. However, it could not initiate large-scale technological innovation. The breakthrough had to wait for the further development of the infrastructure for building up science and technology, and for the establishment of social and production organisations that could help peasants to afford new inputs and to absorb the risk. These two conditions were to some extent established in the subsequent period of Viet Nam's rural transformation.

The collectivisation of agricultural production

After completing land reform in 1957, and strengthening state and Party control over the economy in North Viet Nam, the communist leaders concluded that the short period for national democratic revolution had ended with some success and they decided to embark upon a process of building socialism (Chink 1976).

Apart from ideological considerations, there were practical reasons for collectivising agriculture in the late 1950s and early 1960s.

First, in the eyes of the Viet Namese leaders and planners, the subsistence peasantry that land reform had created was neither able to obtain secure social economic welfare for itself, nor could it positively contribute to the development of agriculture. Although a poor peasant had enough land to produce food for his family, and had been freed from the exploitation of landlords, his economic position remained very precarious. The small savings he could accumulate during a good harvest year were so meagre that they could scarcely tide him over for a year with a bad harvest. He ran the risk of going bankrupt and again being caught in the debt trap (Vien 1967). The economic viability of the tiny family farms in Viet Nam was extremely fragile: in one year out of three, the harvest was a serious loss. The two years 1958 and 1959 bore this out: in 1959, Truong Chinh, the then Party Leader who was put in charge of rural transformation, reported to the National Assembly that after just one year of bad harvest (1958), many peasants had had to sell their lands; and a small number of wealthier peasants had begun to enlarge their holdings (Truyen and Vinh 1964). The post-reform economic situation of different categories of peasants had initiated a debate at a meeting on Exchange Labour and Agricultural Production Co-operatives held in Thanh Hoa on 12 May 1958. The whole question of rich and middle peasant economy and the generation of capitalism was raised. In the aftermath of this debate, a highly analytical work by Tran Phuong shed some light on the rural political economy (Phuong 1960). His work mentioned the results of the survey constructed by the Central Committee for land reform in 3,104 villages. It was revealed that in the above-mentioned villages, there were 26,604 rich peasant households, and out of that number 10,259 households had been able to retain their lands and property and their economic power had not been hurt by Land Reform. Besides, while pointing at the process of middle-peasantisation Tran Phuong observed: "... Among the category of poor peasants who had received land and other means of production, not everybody turned out to become middle peasants. On the contrary, for a number of persons, poverty gradually ate up their small newly acquired means of production. Stratification came about as the unavoidable trend of the small peasant economy ..." The underlying reason for the stratification inherently lay in the fragmented private plots of lands of the small peasants. It looks as though the small peasant operated at will like a little king in his own kingdom, but in reality, the small peasants were one by one gradually crushed under their private plots, under their 'freedom of impoverishment'. Seizing the opportunity that was thus offered, better-off middle peasants started to practice hoarding, to manipulate the rice market, and to advance credits like rich peasants. The well-off strata in the countryside also continued to hire wage labour illegally.

Thus, after the land reform campaign, the communists in Viet Nam were faced with the re-emergence of what they called a 'feudal agrarian structure' where exploitation through usury, low wages and land rent was possible. The loose control over the free market allowed such a trend to develop. As zealous socialists, the communists in North Viet Nam could not afford a reversal of the situation.

From a technological point of view, North Viet Namese leaders advocated that given the existing production system and technical level of farming, production and productivity had reached a plateau.

In terms of labour requirements, with an average farm size of around 1 hectare, it was estimated that in the mid-1950s each crop in North Viet Nam required about 400 work days per hectare. This work day requirement meant a constant shortage of labour during peak periods of transplantation, harvest, and land preparation. Intensive farming, an objective set by the Viet Namese communists, obviously intensified the labour requirement. If the number of crops sown on one unit of land was to be increased and better farming practices were to be used, then larger labour input was unavoidable unless some degree of mechanisation was introduced. Thus technological innovation in agriculture at the existing stage of development in North Viet Nam clearly required an increase in work days per unit of land. It can be estimated that if all planned innovation took place, the requirement would increase to 600 or 700 work days per year (Nguyen). In the absence of mechanisation, family farms in North Viet Nam could not supply that amount of labour input (it is necessary to note that the employment of wage labourers had been banned because it was considered as a form of exploitation). Thus, agricultural development in the communists' view necessarily demanded a re-organisation of the rural labour force and a rationalisation of the use of the limited resources. As communists, the solution they had in mind was co-operativisation. The socialist production co-operatives – as the communist leaders envisaged – would allow concerted and better organised efforts with respect to hydraulic works, better organisation and division of labour, and full utilisation of the limited number of farm implements and draught animals. With this in mind, the North Viet Namese leaders launched the campaign of collectivisation. The process was undertaken along three guidelines: (a) peasants had the right to join or to withdraw from co-operatives at will; (b) co-operatives operated for mutual benefit; and, (c) they would be managed under democratic principles.

Although the peasants were in principle free to join or not to join, North Viet Nam's leaders very skilfully applied a two-pronged strategy to group peasants into co-operatives.

On the one hand, a campaign of persuasion was meticulously conducted. The authorities set up many model co-operatives in the provinces. These pilot collective farms received substantial financial and technical support from the Government so that their success was guaranteed and the peasants would be convinced. Thousands of seminars, observation tours and demonstration sessions were organised to propagandise the superiority of co-operatives. Besides, the promise of a supply of inputs at cheap prices, of credits and of favourable tax rates all made joining co-operatives attractive to peasants.

On the other hand, the control over the sale of basic commodities, including rice, and the almost total monopoly on agricultural inputs gave the Government strong leverage on the peasant economy. A peasant who wanted to stay outside co-operatives could have run the risk of having difficulties in marketing his produce, and in purchasing necessary inputs (Chalian 1968).

Apparently, throughout the whole process of collectivisation, the peasants' freedom to join or not was respected. A number of peasants reportedly withdrew from co-operatives, particularly when these co-operatives were badly managed. However, the total number of withdrawing members never surpassed 1.6 per cent of the country's total number of members of co-operatives.[3]

Apart from giving peasants the right of "free choice", the communist government took care to undertake the process of transformation very slowly. The pace and rythm of collectivisation were determined by many factors; among them the war, and the lack of managerial and technical skills among the personnel were the most dominant.

Throughout the whole period of collectivisation, the process was undertaken with great caution and pragmatism.

First, a rudimentary and non-socialist form of co-operation was developed on the basis of the mutual help that had traditionally been practised among peasants. During the resistance, the Government had encouraged peasants to form mutual-aid or labour-exchange teams to solve the problems of labour shortages. After 1954 attempts were made to reach a higher degree of co-operation in the context of these teams in the form of a more rational plan for farming practices among members. The leadership, although encouraging peasants to join teams, saw this form of co-operation as an exercise through which peasants could later be introduced to higher forms of co-operation. By the first of May 1958, 245,000 teams had reportedly been organised.

In the same year, the co-operativisation process began with the formation of low-level co-operatives. Land was pooled for cultivation. In return, owners received rent from the co-operatives proportional to the amount of land they had contributed. As a rule, the rent was not to exceed 25 per cent of the income from the land, and the owners were responsible for land taxes. The cultivation was done in accordance with a common plan drawn up by the co-operative. The income of the co-operative – after rents, production costs and taxes had been deducted – would be divided among the members. A certain percentage of the co-operative's income might be reserved for re-investment. An elected executive committee was in charge of the management of the co-operative. Again, in order to prevent the domination of better-off peasants, it was decided that two-thirds of the committee's membership should be poor peasants, the remainder middle peasants. The speed at which co-operativisation took place can be seen in the rapid growth in the number of co-operatives. In one year alone (from 1958 to 1959) the number of co-operatives grew from 4,800 to 27,400.

The two subsequent years were for strengthening

management. By 1961, when low-level co-operatives had demonstrated that they could survive, the leadership made the final step towards collectivisation. Co-operatives that were too small were merged and upgraded to make larger, high-level co-operatives in which land and other production factors became collectively owned. Crop yields and other income were deposited to the account of the co-operative while the members were paid according to their contribution of work days. A low-level co-operative could be converted to a high-level one only when it had proved that it could achieve some success and could demonstrate the advantages of co-operation to its members.

Close examination of the process of collectivisation allows one to distinguish three elements which determined its pace:

1. The collective spirit of the peasants: there had been a clear policy of relying on the traditional co-operative spirit of peasants living in a neighbourhood. The trust and solidarity that existed among peasants living close to each other was used as the basis for the new forms of co-operation. This largely facilitated the managerial tasks of co-operatives in the early years. The authorities had therefore deliberately avoided creating co-operatives that encompassed very large numbers of families. The number grew from 14 families per co-operative (on the average) in 1957 to 127 in 1967. To make peasants accept a new mode of production was not an easy task. Alongside educational campaigns, the leadership had tried hard to prove to them that each step taken would definitely bring further advantages. This was difficult as North Viet Nam had only very limited resources for investment. The expected support from industry had been very limited mainly because it was still in its infancy. The achievements of ten years of industrialisation were later destroyed (after 1965) by American bombing. The co-operatives then not only lacked support from industry, but also had to operate with an extremely serious shortage of inputs. There were periods when

setbacks were seen under the disguised form of private farming. However, in the main, collective agriculture survived thanks to the Government's successful appeal for patriotism and sacrifice. In addition – at least from the point of view of the majority of poor peasants – the more difficult the situation was, the more they were convinced that their economic situation was more vulnerable with small-scale and private farming. This element, combined with the state's control of rice trading and the supply of agricultural inputs, had resulted in a peaceful collectivisation. The Government appears to have patiently waited until the peasants' consciousness was firm enough before taking any further steps.

2. Collectivisation had been undertaken with serious consideration given to co-operative management. Since poor and middle peasants had been put in leading positions in the co-operatives, the need to provide them with managerial skill had become vital. Planning, bookkeeping, and organising and combining production factors had been problematic. Moreover, the complicated tasks of quantifying individual labour by applying a system of work points had also required skill and consensus. In brief, the question of training executive committee members to run co-operatives as business enterprises had taken time. This had made management a criterion for converting a low-level co-operative into a high-level one. However, the problems of management seemed to be still far from having been solved. In the past, there had been repeated campaigns for the improvement of management.

3. Although advocating large-scale agricultural production, the North Viet Namese seemed to have taken into account the fact that productive forces has not been strong enough for really large-scale farming. An overextension of the area of each co-operative would have created more problems than it would have solved. The leaders had deliberately avoided forming too large co-operatives, as this would have made the

combination of production factors and the rational organisation of labour uncontrollable for the still-weak management bodies of the co-operatives. Statistics showed that in the ten years from 1957 to 1967 the average acreage of co-operatives rose from 25 to only 74 hectares (Ching 1969).

With all these factors, collectivisation had not been carried out in one fell swoop, but gradually. Table 7.2 confirms this observation.

It is interesting to note that the ideal of a socialist economy had also been pragmatically compromised by the existence of a family economy. As a rule, each family was allowed to occupy and farm a small plot of land. The total amount of land reserved for private plots was not to exceed 5 per cent of the co-operative's total cultivable land.[4] The plot ranged on average from 270 to 300 square metres. The ownership of these small plots had also been collectivised in the campaign for high-level co-operation. However, peasants retained the right of possession and could enjoy the full yields of the plots. Transfer of these plots became impossible (Lai 1967). The existence of a family economy side by side with the co-operative sector had been seen as a historical necessity for the transition to socialism. At the beginning, however, cadres in some areas tended to suppress the family economy by strict planning that minimised the peasants' liberty to exploit their plots. This appeared not only to have been unpopular but also to have affected the peasants' enthusiasm for the work in the co-operative. This mistake was severely criticised. The Party stressed that the two sectors of the rural economy should be handled properly so that they were not only complementary but also mutually promoting. Nevertheless, some guidelines were laid down to orient the production of family plots in accordance with the development of each co-operative and with the state plan. This moderate policy was believed to have a positive effect. In the situation of Viet Nam, while collective agriculture could not generate sufficient income for co-operatives' members, family plots served as a supplementary source of income for

Table 7.2
Production relations in agriculture, 1957–67*

	1957	1958	1960	1964	1967
Agricultural producers	45	4,820	41,400	31,900	23,550
Co-operatives (in numbers)					
– low level	42	4,800	37,000	16,390	5,511
– high level	3	20	4,400	15,510	18,039
Collectivisation ratio	0.03	4.7	85.8	86.8	93.7
Co-operative share in cropland (per cent)	–	–	76	–	90

*Source: Theodore Bergmann (1975).

them. Thanks to this policy, one could see secondary production and handicrafts developing in rural areas. This contributed greatly to the production of consumer goods. Incomes from this sector had increased the peasants' purchasing power and had allowed them to improve their standard of living.

The above account of the progress of collectivisation through time implied by no means that the process was a linear and smooth transition from small-scale private farming to large-scale socialist farming. The establishment of a socialist mode of production in a backward country that was at war, like Viet Nam, appeared to have been a far more tortuous path, full of ups and downs, twists and turns. The most significant contradiction from both theoretical and practical points of view was the contradiction between "advanced" production relations (socialist) and the weak and backward production forces.

In 1958, the argument put forward by the leading figures of the Party was that despite the low degree of development of the productive forces, Viet Nam should embark on the building up of socialist production relations. In other words, the change in production relations was the key to higher productivity; only on the basis of new production relations would technical revolution succeed. In the late 1950s, there was obviously a debate within the Lao Dong Party leadership; and not every person who had taken part in the debate was enthusiastic about making a break with the orthodox pattern of socialist development by cutting too short the period of bourgeois democratic revolution, and quickly moving towards socialist production relations without waiting for the productive forces to develop further. It seemed that ideological considerations and the revolutionary zeal of the Lao Dong Party leaders predominated, and precluded an objective weighing the merits of the two options. A writer very close to the Viet Namese leadership summarised the arguments put forward in the debate:

"Unanimity was reached on the necessity of agricultural co-operation, but opinion differed as soon as one had to decide the right moment for

starting the campaign. The Soviet Union was already producing tractors and farm machinery when the first Kolkhoses were created. Some inferred from this that Viet Nam should wait until conditions were ripe, i.e. until the country could make tractors -- in other words, until heavy industry has been built up ...
We could not wait until we had developed industry before launching agricultural co-operation. Small individual production, with its rudimentary techniques, was hardly able to insure simple production ... Under such conditions, how could we speak of enlarged production; how could we provide the food and raw materials necessary for industrialisation." (Vien 1967)

In the first three years of collectivisation, the semi-socialist character made co-operatives not especially at odds with the backward productive forces. But in the 1960s, when the agricultural authorities had pushed forward the socialisation of the agricultural means of production by converting low-level co-operatives into high-level, socialist ones, the socialist production relations had become much more difficult to reconcile with the low technical level of the productive forces. The hasty push towards further collective farming at the beginning of the 1960s was obviously frustrated by the inadequate material and technical base of the co-operatives. Production at the co-operative level ran into trouble. In 1960, Le Duan had noted that it would have taken a rather long time to transform the production relations in agriculture, and "due to the fact that our production was still low and based on manual labour, and that collective production was not developed yet, the co-operatives could not be immediately brought to a higher level. As long as production was not raised, the shifting of co-operatives to a higher level was impossible" (Le Duan 1976). Here, Le Duan wanted to point out that high-level co-operatives had already achieved high production and thus convinced peasants. He also returned somewhat to Lenin's assertion that without a good material and technical base, socialist co-operatives could not survive.

Thus, after several years of experience, the Viet Namese leaders found themselves caught in a dilemma: they wished to establish socialist production relations in order to boost the development of productive forces, and thus increase production. But through experience, they had learned that unless the productive forces were strengthened enough to generate a material and technical base, socialist production relations would be scarcely able to withstand the problems of production. In other words, they found it was impossible to move very fast with production relations, leaving productive forces far behind. Another dimension of the problem was also discovered: the managerial problems. As a consequence of these, the conversion of co-operatives was slowed down considerably. In 1960, only 12 per cent of the co-operatives were collectives. The figure rose to 36 per cent for 1966.

In 1968, when almost all low-level co-operatives had already been transformed into collective farms, the contradictions between socialist production relations and weak productive forces with weak management remained far from being resolved.

The effects of the foundation of co-operatives on employment were contradictory. On the one hand, co-operatives allowed North Viet Nam to transfer a large part of its rural labour force to serve the war, while production was slowly increased. It looks as though co-operatives succeeded in organising and fully utilising the available manpower in the countryside during the period in question (1958-75). On the other hand, however, close examination reveals that the mobilisation of peasant labour faced serious problems. Le Duan revealed that the average number of work days that each member worked for a co-operative was only 130 days. The reason for this failure lays in the contradiction between the economics of the co-operatives and of the private plots. Members were supposed to work full-time for the co-operatives, while in return receiving an amount of rice for family consumption and some cash income. In general, peasant co-operative members could only draw about 50 or 60 per cent of their

income from the co-operatives. The family plots
(which made up only 5 per cent of the cultivated area
of the co-operatives), despite their secondary role,
on the other hand generated the rest of the peasants'
income, even though they were not allocated together
with any other resources (not even with the time to
exploit them). This made the family plots compete
strongly with co-operatives for labour. Many members
of co-operatives preferred working on their own plots
to serving co-operatives. The resulting lack of
incentive to work harder for the collective sector
not only negatively affected the co-operative's
performance, but also stood as an obstacle to the
intended technological development.

But from an organisational point of view, the
co-operatives also brought some advantages for the
development of technology in agriculture. The
evolution of the interrelationships between
technological and institutional innovation is
historically presented in table 7.3.

First, as a large-scale and collective enterprise,
a co-operative could far more easily plan production
and allocate resources in order to respond to the
demands of intensive and modernised farming. It made
specialisation of labour within a co-operative
possible, so that all peasants no longer had to
perform the same activity. As a member, a peasant
could be assigned to specialised work that entitled
him to a fair share in the co-operative's income.
This stimulated the development of technical skills
and made room for specialisation. It was within the
context of co-operatives that one could see teams
emerge that were specialised in hydraulic works, seed
or green manure production, as well as workshops for
the manufacture, repair and maintenance of farm
implements. These groups or workshops made the
transfer of technology to the co-operative easier and
faster than it would have been to atomised private
peasants somewhere else. Besides, collective farming
opened up the possibility for exchange of
experiences, and propagation of traditional
techniques which families had kept secret for
generations. The propagation of Azolla as a green
manure was a case in point.

Table 7.3
Institutional reform and technological innovation, 1953–75

Technical innovation	1953–57	1958–60	1961–65	1966–75
Institutional reform	Land reform (LR) Rectification of LR's errors Reform of Peasant Association Labour exchange groups	Labour exchange groups Low-level co-operatives	Transformation of low-level co-operatives to high-level co-operatives Co-operatives' managerial reform Enlargement of co-peratives	Campaign for the improvement of co-operative management Contract system and abolition of contract system
Hydraulic	Repair of damaged systems Strengthening dykes Construction of small projects	Expansion of existing systems Construction of more small projects and some medium and large projects	Continued expansion and some mechanisation Local hydraulic stations Formation of hydraulic teams in co-operatives	Reconstruction of rice fields of continued expansion and mechanisation Improved quality of water conservancy
HYV and agricultural calendar	Selection and better production of seeds Search for short-termed rice varieties (Nam Nink) Better treatment of seeds	Better seed production and seed treatment Local seed stations Improved resistance of seeds Continued search for new varieties	Co-operatives' seed field and special teams New high-yield and dwarf varieties: Trân Châu Lún	Adaptation of IRRI's varieties Establishment of spring crop with NN8, Trân Châu Lún Rearrangement of agricultural calendar: multiple rotating cropping
Alternative sources of nutrients	Intensive use of night soil and animal manure Better techniques of procurement and utilisation	Better techniques of application Research on Azolla and Sesbania	Propagation of Azolla and Sesbania Search for better techniques with regard to Azolla and Sesbania	Wide and established use of green manure Establishment of green manure crop in agricultural calendar Co-operatives' specialised teams

Second, as a collective enterprise, a co-operative could afford to set aside resources for the development of technologies. Since their foundation, the co-operatives had contributed considerably to the expansion of irrigation systems, and to the setting up of local experimental stations and of special fields for seed production. Their contributions in terms of land, materials and labour largely made up for the investment at the grass-roots level. It was far easier for a co-operative than for private peasants to reserve some land for the expansion of the hydraulic system, and to construct the special fields used for seed production. One of the achievements of co-operatives in paving the way for technological development was their success in reconstructing the rice fields. Land was consolidated and divided into large plots well linked to the hydraulic system so that water could be brought in or out at will. Better roads were also constructed to facilitate the circulation of vehicles, tractors and machinery. All these innovations obviously surpassed the capacity of an individual and could only be implemented in the framework of co-operatives.

Third, as an enterprise larger than the family farm, a co-operative was better able to absorb the risk involved in the adoption of new technology. Similar to China, Viet Nam waged a peaceful campaign of collectivisation. The reason for this smooth transition was the highly valued security that the co-operative offered to poor peasants. By joining a co-operative, the peasant surrendered his ownership of land and the absolute private control over his produce; but in exchange he received protection from the co-operative if the harvest was lost. It had always been said in Viet Nam that: "with land reform the poor peasants regained their dignity; with co-operatives they got security". Co-operatives' own savings, plus state assistance, ensured this security. This protection removed the peasant's reluctance to innovate, which was rooted in his fear of risk. In other words, with the co-operative as the production and income-distribution unit, the economic basis of so-called peasant conservatism was

destroyed. It was true in the case of Viet Nam that
the peasant co-operators were far more receptive to
innovation than the private peasants.

The real weakness of the agricultural co-operatives
in Viet Nam did not result from the peasant's strong
attachment to private ownership of land, which is
often overemphasised by researchers unfavourable to
collectivisation. In fact it lay in the incompetence
of the co-operative's managing cadres.

The management of a co-operative as a production
and financial unit was a crucial issue during the
whole period of technological change. In a private
farming system, technological change results entirely
from the response of capitalist or peasant farms as
economic enterprises. They respond to the changes in
the price of production factors, as Ruttan has
explained. With private ownership of production
factors and with maximisation of profit as the
guiding principle, these farms operate merely as
economic enterprises of owners.

The production co-operative in socialist countries
was quite different. In principle it was supposed to
serve three interests simultaneously; that of the
society, of itself, and of its members. Being at the
same time an economic and a social political unit,
the co-operative's functions shaped both production
organisation and the distribution of incomes. The
complex of functions, the enlarged size of farms, the
large number of members, and backward but large-scale
production – all these made the management of a
co-operative a difficult task for the inexperienced
and poorly trained managerial cadres.

Moreover, not only did co-operatives have to be
managed, but technological development also had to be
planned and undertaken while there was a serious lack
of managerial skills among the co-operatives'
executives. The strict political leadership of the
Communist Party could not compensate for the lack of
skilled manpower in planning, organising and
bookkeeping, etc. The policy of enhancing the
socio-political position of the poor peasants by
entrusting them with the leadership of co-operatives
was obviously an achievement of mass politics. But
the very weaknesses of the newly instituted rural

leaders aggravated the managerial problems, and resulted in economic costs that the State and the co-operatives had to pay. As one writer in Viet Nam observed, many of the co-operatives' leaders could hardly read and write when they assumed leadership (Cuc 1979). Yet they were supposed to run enterprises covering dozens of hectares, employing large numbers of people, and to deal with all the problems that evolved from planning and implementing technological development in agriculture at the co-operative level. Although the official literature recognised the success of co-operatives in ensuring production during wartime, it often pointed to the managerial weaknesses as the reason for the co-operative's failure in becoming the stronger, modernised production unit that the leadership had expected.

The impact of managerial weaknesses on agricultural development in general, and on technological innovation in agriculture in particular, was significant.

The Viet Namese political leaders and planners seem to have believed not only that collective farms were indispensable for agricultural modernisation, but also that they should be _large_ and _socialist_. There has therefore been a constant drive for these two characteristics.

But in the circumstances of North Viet Nam, it often happened that the further agriculture moved towards large-scale socialist production, the more was technological innovation frustrated by managerial weaknesses.

From 1960 onward, North Viet Nam's agricultural authorities undertook several campaigns to improve the management of co-operatives. A five-year study-and-work programme was designed in order to equip cadres with some skill in running co-operatives. Although training is always valuable, the fact that the co-operatives' cadres had to participate in such a long training programme meant that it consumed a lot of time and energy that could have been spent on quicker modernisation. To a large extent, the level of management at co-operatives had always lagged behind with respect to the expected speed of innovation. Just to run the co-operative's everyday affairs, the cadres already had to perform complicated tasks such as drawing up alternative plans for unexpected circumstances (drought, flood, bombing), planning production and mobilising labour, monitoring members' work and distributing incomes on the basis of work done. Added to these, modernisation of the productive forces entailed hundreds of more difficult tasks involving planning and implementation; it was not surprising that a number of cadres failed to cope with problems encountered.

By the time efforts to improve co-operative management had produced some results, the war spilled over into North Viet Nam, and the bombing put new constraints on the process of development. The war situation after 1964 broke down the Party's ideological control and disturbed the economic relationships between co-operatives and the State, and among the co-operatives themselves. The war not only devastated the results of North Viet Nam's ten years of industrialisation, depriving agriculture of the little technical support and supply it had, but it had also taken the most able cadres and workers away from the co-operatives. It was reported that between 1965 and 1975, women made up more than 60 per cent of the co-operatives' workforce. The breakdown in the Party's control and the crumbling management caused collective farming to degenerate into a disguised form of private farming with the _Khoan_ system (contract system) so that collective planning for production and collective farming were virtually no longer practised. Instead, the land was contracted to a family or group of families in return for a quota of produce delivered to the co-operatives, and in turn handed over to the State. Later, sensing that collective production was at risk, the Party condemned the contract system, accusing it of opening the way to capitalism. The Party then launched a campaign to reassert the ideological basis of the policy for cooperation.[39] Still, managerial problems could never be totally resolved. As late as 1974, an editorial in _Nhan Dan_, the Party's official newspaper, revealed that in the previous ten years not only had the cultivated area declined, but the amount of co-operative land being used illegally by private individuals was nearly equal in size to the area allotted to private plots (Nhan Dan 1974). In other words, nearly 5 per cent of the co-operatives' land had been under illegal private exploitation. This situation suggested that not only had the co-operatives failed in controlling their resources, but they were also unable to make full use of the land at their disposal. The matter was not so simple as merely to involve reasserting the co-operative's control over the land; the question had to be examined whether co-operatives really had the capacity fully to utilise all the land they possessed. Land-use policy after 1978 showed that the lack of an effective incentive system and the weak technical capacity of the co-operatives required the authorities to allow – or even to encourage – peasants to cultivate the land that remained beyond the capacity of the co-operatives.

Although the managerial problems have never been totally overcome, co-operatives succeeded surprisingly well in meeting the challenge of producing crops during wartime. They proved able not only to ensure a constant food supply for the North, but also to contribute to the support of the war effort in the South. Despite this achievement, co-operatives had never become the strong and modernised production units that the leaders of North Viet Nam had expected. Perhaps it was the lessons of experience that led to the drastic reform of the functions of co-operatives in 1981. With this reform, a far more flexible formula for co-operation was applied; the activities of co-operatives were reduced to planning, supervising and supporting. The three most labour-intensive activities in rice production – namely transplanting, weeding and harvesting were assigned to families or groups of families. A far more attractive incentive system was also instituted; land was farmed out to families according to their labour capacity. They were supposed to grow the crop and meet all the technical requirements that were agreed upon with the co-operative. The co-operative and the peasant contractor agreed upon an amount of produce to be handed over. This was often reasonably low and the peasant was free to do what he wished with the rest. This two-pronged reform, combining a reduction in the co-operative's administrative work and a new material incentive, appears to have yielded good results.

Throughout the period from the foundation of co-operatives in 1958 to the present day, the co-operatives have passed many tests. They survived and contributed positively to the technological developments in agriculture. They could also have done better, however. Now that the war has ended, a critical review of the whole experience is very much needed in order to work out new adjustments that will help to utilise fully the available resources of a reunified Viet Nam.

INTERSECTORAL LINKAGES AND TECHNOLOGICAL INNOVATION

For a country with a centrally planned economy like Viet Nam, the State played an extremely crucial role in handling the inter-relations between economic sectors so that technological innovation in agriculture could be promoted and achieved. In the place of the market mechanism, in Viet Nam the State played the coordinating, mobilising role.

If productivity was to be increased, agriculture would need biological, chemical and mechanical technologies. Industry had obviously been expected to supply agriculture with chemical and mechanical inputs. The construction of hydraulic works, for example, required the products of heavy industry (steel, cables, cement, etc.) and the machine-building industry (motor pumps, electric generators, etc.). The mechanisation of agriculture required a substantial supply from the machine-building industry as well, in terms of tractors, motors, machinery and implements. Similarly, from the chemical industry agriculture needed chemical fertiliser and insecticides. Thus, the support of industry for a technological breakthrough and for sustained innovation in agriculture was one of the indispensable conditions for an increase in productivity. On the other hand, industrialisation could only be achieved on the basis of a stable agriculture. One North Viet Namese scholar insisted that socialist industrialisation should take the transformation of agriculture as its starting point (Lai 1979). This meant that the development of industry in its initial stage should be directed to serve the transformation of agriculture and industrialisation should start with those branches that are linked to agriculture. In the 1960s, the Party reiterated this policy and repeatedly insisted that:

"In industrial development, emphasis should be put on producing the means of production in order to develop agriculture; farm implements, simple and improved means of transport, veterinary products, machinery for agricultural processing, etc. .. with a view to increasing labour productivity and to improving the working conditions of peasants."

The choice of priority between industry and agriculture, and between different branches within the agricultural sector had changed through time, and had been reflected in investment policy.

The period between 1953 and 1957 witnessed a strong bias towards industrialisation: investment in industry was double the amount of investment in agriculture. Agriculture's share in total public investment was only 13.3 per cent, while for industry it totalled 27.7 per cent. Of the total investment in industry, 65.6 per cent was allocated to branches that produced capital goods, notably to mechanical industry. Because of the lack of technology, investment in the chemical and fertiliser industry remained low.

Due to the prevailing weaknesses in the scientific and technical infrastructure, most of the resources allocated to agriculture during this period were channelled to the labour-intensive development of small hydraulic systems. This branch received 10.9 out of the 13.3 per cent of investment in agriculture.

Although the period 1957-60 witnessed an increase in investment in agriculture in terms of value, the bias towards industry increased during this time. On the basis of 1957 prices, investment in agriculture had increased 157.1 per cent by 1960. The share of hydraulic works in agricultural investment tended to grow at a lower rate, by only 35.6 per cent.

In industry, investment in capital goods producing branches increased sharply (201 per cent). Of these branches, the increase in the electricity-generation industry was 74.7 per cent, the steel industry 280 per cent, and the mechanical industry 49 per cent. With limited resources, and on the basis of extremely low investment, the rates of increase might appear to be high (19.5-fold for non-iron metallurgy, 8.9-fold for the chemical industry, for example), but investment in real terms in those sectors or branches was definitely small and inadequate.

While the total amount of investment in agriculture increased, its share in the total investment in fact declined. From 13.3 per cent of total investment, it went down to 11.9 per cent. Consequently the share of investment in hydraulic development also decreased to only 6 per cent, compared to 10.9 per cent during the previous period. The increased bias against agriculture during this period coincided with the formation of the co-operatives. The decreases in investment could be explained by the ongoing process of re-organisation of agriculture. Industry, on the contrary, by that time had been firmly socialised, and the North Viet Namese leaders intended to move quickly towards development of heavy industry.

The period of the first Five Year Plan (1961-65) was characterised by a return to a more balanced investment policy. In terms of value, investment in agriculture and industry continued to increase. But agriculture nearly doubled its share during the first period - on the average, agriculture received 20 per cent of the total investment. Meanwhile, investment in some agricultural-supporting industries went down: investment in the chemical and fertiliser industry declined to 11.6 per cent, while the share of the chemical industry remained stable at 6.4 per cent (Tri 1967).

The increase of investment in agriculture at this time seemed to have been a clear attempt to boost the production of the co-operatives. Once peasants had been induced to join co-operatives, financial and technical support were very much needed in order to make co-operatives viable. This was particularly true when the North Viet Namese leaders planned to convert semi-socialist co-operatives into fully collective socialist co-operatives. How much of this investment could be channelled into promoting technological innovation, and how much was spent just to keep the collective farms from disintegrating? Those questions could not be answered due to the lack of statistical data. Nevertheless, on the basis of available statistics and information, some qualitative observations could be made about investment policy.

First, agricultural investment went mostly into the development of hydraulic works. This included financial support for the construction of large projects, for purchasing water pumps, electric generators for water stations, etc. Another part went to the development of seed-production stations in the provinces and districts. The remaining part of invested resources went towards the purchase of tractors and machinery for state farms.

With regard to the development of hydraulic systems, there had been consistency in the investment policy: the costs of the construction projects had always been shared by the State and the co-operatives. State investment was mostly in the form of subsidies for large and medium-scale systems, or in the form of supplying scarce material (cables, valves) and machines (pumps, generators). State assistance to small projects was mostly provided in the form of loans to the co-operatives that benefited from the development. This credit supply allowed the co-operatives to purchase the machines and material needed for their projects. The co-operatives in their early years had few resources and the State could not provide all of the much-needed investment to develop the water system. The 6 per cent share for hydraulic development during this period was rather low compared to what was needed for establishing an extensive and effective hydraulic network. It was the great labour contribution of the co-operatives that compensated for low investment in this branch.

The second largest amount of investment in agriculture was reserved for setting up seed experimentation and seed production in several regions and provinces. These stations contributed very positively to the selection of good traditional rice varieties, and to improving and propagating them in the areas where the stations were located. They not only contributed a great deal to the development and popularisation of new seeds at regional or provincial levels, but also helped the co-operatives to set up their new seed-producing fields by providing training and technical advice and by supplying them with good seeds to be reproduced. Some of the development of the new seeds was done at state farms. In this way state farms were expected to spearhead the technological breakthrough in agriculture. But it seems that state farms in North Viet Nam had been more successful as industrial crop producers than as pioneers in technological development. Part of the reason was that state farms were mostly located in the midlands or highlands and operated under rather different climatic and topographical conditions than most co-operatives. Any technology developed there would have required another long process of testing and adaptation in the areas that intended to adopt it.

Second, during this First Five Year Plan period, state support for strengthening the technical base of co-operatives was indispensable, particularly since the co-operatives were in their infancy. The weaknesses of the newly converted socialist co-operatives in terms of production and management made a number of co-operatives unable to maintain a sound level of production that would ensure economic security for their members. In 1963, for example, more than 40,000 households withdrew from co-operatives. Some explained that they withdrew in order to consolidate the family economy first, and then would rejoin the co-operatives when they had achieved economic stability for the household. The weak economic base of the co-operatives was also reflected in the lack of fixed capital and working capital. It was reported that the value of fixed capital per peasant member of a co-operative during this period was only 40 to 50 dong. The lack of working capital was also reported as serious.

Industrial support for agriculture during this period in terms of supplying the means of production steadily increased at the rate of 25 to 30 per cent per year. It totalled about 25 per cent of the total value of the production of heavy industry (Tao).

Late in 1964, the war in South Viet Nam spilled over to the North, and the American bombing of North Viet Nam put technological development in North Viet Nam's agriculture into a new phase. From that year on, one can observe on the one hand an intensive process of agricultural modernisation and on the other hand a highly troubled state trying to invest in and give industrial support to agriculture. The war really dealt a blow to the ongoing agricultural development in North Viet Nam. The highly emphasised modernisation of the means of production was very much disturbed.

The war brought about important changes in the country's economic life. Industry was the first sector hit by the war. Bombing obliterated the achievements of North Viet Nam's ten years of industrialisation. Most of the country's industrial infrastructure was attacked and destroyed. Very little could be saved except some movable machinery and equipment. The direct consequence for agriculture was that it lost all support from the central, heavy industry.

To minimise the effects of the bombing, production was dispersed alongside a process of de-urbanisation. The leaders of North Viet Nam adopted a policy for decentralising economic planning as well as economic activities. Since 1965, regional economy had become more and more important. This had been particularly true for regional industry because it had had to make up for the interruption in the production of central industry. The shift in emphasis had been reflected in sharply increased investment in this sub-sector. During the period between 1965 and 1968, for instance, the share of regional industry in the total industrial investment increased five-fold. In addition, regional industry also switched its production more to capital goods:

in 1972, 62 per cent of investment was for the production of capital goods, compared to 15.5 per cent during the period 1961–65. The development of regional industry had obviously benefited the agricultural sector. Located closer to agricultural production, the burgeoning regional industries were more sensitive to the needs of agricultural modernisation and therefore more able to respond quickly and appropriately to agriculture's demands for inputs. In 1964, the production of regional industry totalled 48.7 per cent of the total industrial production of North Viet Nam. The percentage increased in subsequent years, and an important portion of this production was capital goods for agriculture.

Emphasis was put on the production of:
- mechanised or semi–mechanised farm implements: hydraulic pumps, electric and diesel motors, small threshing and milling machines, etc.;
- improved farm tools such as ploughs, insecticide sprayers, and other manual equipment;
- means of transport;
- chemical fertiliser (mostly phosphate, apatite) and insecticides;
- construction materials;
- electric generators.

The demand resulted in the rapid establishment of small factories and workshops in townships or districts. At the end of 1968, more than 200 mechanical factories were operating in several provinces, offering 22.1 per cent of their output to agriculture. Alongside these machine and tool–building factories, another 1,029 industrial enterprises, employing 123,618 workers and technicians, supplied 58 per cent of the country's machinery, 38 per cent of its electrical motors, 35 per cent of its transportation means and all of the country's simple and improved farm tools (Tuan and Lai 1976). The most interesting development was that the production of machinery and farm implements had been scaled down to village and co–operative levels. Of the above–mentioned enterprises, many were set up in the co–operatives and by the co–operatives themselves. It was reported that in the midst of the first year's bombing, 450 of the co-operative's workshops were using diesel or electric machines for their production (Tao).

Despite the intensity of the war, investment in developing agriculture's infrastructure continued to increase steadily. Compared to 1965, it had increased by 12 per cent in 1966, 33 per cent in 1967, and 51 per cent in 1968.

The co-operatives' consumption of electricity gives some idea of the progress in the mechanisation of irrigation and farming. Compared to 1960, the amount of electricity supplied to co-operatives increased by a factor of 12.4 in 1966 and 13.7 in 1968. This shows that the number of co-operatives that became equipped with some machinery continuously increased. In 1966, only about 10.9 per cent of all co-operatives possessed generators and some agricultural machines; in 1967, the figure rose to 19.8 per cent and in 1968 to 29.9 per cent (Nien Giam 1970). While the war drew off a large portion of the rural labour force, the development of biological technology in the late 1960s made a higher degree of intensive farming possible. Multiple cropping was being practised more widely and led to a sharp increase in the demand for labour. This accentuated the need for some mechanisation of farming practices. The plan at the time was to set up for each co-operative a machine depot equipped with three water pumps, six threshing machines, one milling machine, carts and a large number of improved ploughs, insecticide sprayers, and improved rakes (Hoc Tap 1968). The implementation of this policy led to a quick increase in the number of machine depots in the co-operatives. In 1960, there were only 1,000 such small stations; in 1968 the number increased to 6,500 and in 1969 to 8,783. The rapid increase in the number of machine depots shows that regional industry, despite its technical weaknesses and its lack of energy and raw materials, could contribute fairly well to this development. Close examination of the technical level of the equipment delivered to machine depots revealed that they were small in size and built with intermediate technology. According to Viet Namese planners, these depots were to be the foundation for the long process towards full-scale and proper mechanisation. Despite their modest size and low technical level, the equipment depots helped the co-operatives to compensate for the lack of labour during wartime, and even allowed them to take some steps towards modernised agriculture.

In this context, North Viet Nam's real "Green Revolution" had started in the early 1970s on the basis of previous advances in hydraulic works and in biological technology. What made the Viet Nam case different from others was the fact that the Green Revolution in this country was preceded by fundamental structural changes; it was not a direct import of a technological package from foreign countries but rather a slow build-up of indigenous technological resources.

THE BUILD-UP OF A TECHNOLOGICAL INFRASTRUCTURE

As Ruttan rightly suggests (Chapter 3 of the volume) the failure of many countries to bring about sustained development in agricultural technology often results from their failure to set up an efficient system of generating the type of technology that the countries need at their particular stage of development. What Ruttan called 'system' should not be interpreted merely as an R & D system. Instead, one could give it the meaning of an infrastructure that includes the institutional and non-institutional elements that serve as the basis for a technological "take-off". This infrastructure includes not only prestigious scientific institutions where proper scientific research is undertaken, but also far more simple but indispensable elements for a country's technological development, such as the achievements in eradicating illiteracy among peasants, and the grass roots technical organisations where new technology is made known to peasants, and tested, adapted and popularised with the participation of peasant users. In other words, one should look at the whole set of prevailing conditions in which technology is developed rather than looking merely at the existence of a few top scientific institutions.

Research and scientific organisations

Research in the field of agricultural technology had
always received high priority in North Viet Nam. As
conceived by the leadership, research in agriculture
in North Viet Nam has followed two lines: one aimed
at studying and improving the body of empirical
knowledge among peasants with a view to fully
exploiting this body of centuries-old know-how; the
other emphasising the study, adaptation and
dissemination of new advanced technologies that had
been developed in Viet Nam or in other countries.
Research along both lines had remained very much on
the practical level. It had always been directed to
the needs of production. In Viet Nam's situation,
what was sorely needed was a realistic and
comprehensive research policy that aimed at resolving
the problems that existed at the country's particular
stage of development. Taking the country's stage of
development into consideration drew the attention of
the authorities to the existing resource endowments
of the country, and helped to avoid overly ambitious,
unrealistic research. Research in Viet Nam,
therefore, was production-oriented instead of being
aimed at extending the limits of theoretical
knowledge. With such a production-oriented research
policy, science and technology had to involve both
analysis and synthesis. On the one hand, basic
research on the various factors involved in the
production process had to be undertaken in order to
find ways to maximise the effects of innovation on
each of these factors; on the other hand,
synthesising research had to bring data from all
research areas together in experimentation that was
applicable to production.
Such a combined research policy required a
comprehensive organisational system that could
fulfill the tasks along both lines, being at the same
time analytical and synthetic and developing at the
same time traditional and modern technologies.
At the central level, the highest body for policy
formulation and research coordination was the State

Commission for Science and Technology that was founded in 1956. In its early years, the Commission was in its infancy and had almost no resources. Yet, it was supposed to assume the heavy responsibility of managing all scientific and technological activities, including those of the social sciences. Ten years later, the social sciences broke away and were put under the administration of the State Commission for Social Sciences.

The State Commission for Science and Technology served as the nucleus of a network of scientific and technological organisations that spread throughout all regions of North Viet Nam, from the centre out to the local levels. It had the status of a ministry and worked closely with the science and technology departments of the ministries. The fact that the Commission had the same status as a ministry, and that each ministry had its own department for science and technology, made the relations between the Commission and these departments unclear. The departments received a major portion of the resources for research and development. They were under the direct control of the ministries. Besides, since research in Viet Nam had been production-oriented, one may assume that each ministry had the final word on its own research, and that the power of the Commission would have been reduced to drawing guidelines and monitoring the research undertaken by various organisations. The coordination of research done by universities, independent institutes and ministerial science and technology departments has remained a bit of a mystery to the outsider down to the present day.

From an organisational point of view, the Commission had local offices just like the ministries. Each province had an office for science and technology that came under the central office of the Commission. This provincial office worked as the intermediary between the central office and the research and experiment stations located in a number of agricultural co-operatives. Organisations at these three levels were strongly linked with each other (Dap 1970 and Sigurdon 1982).

The existence of a number of research and

experiment stations spread throughout the country gave Viet Nam's system for scientific and technological research a base at the grass-roots level, and made research closely allied with production. This greatly facilitated the testing, adaptation and popularisation of improved or imported technology. These research and experimentation stations, situated in co-operatives, have contributed very positively to the development of agricultural technology in Viet Nam in the last two decades. It was in these stations that the discoveries of higher research institutions were tested, adapted, selected and popularised. It was here that theories and practice, experience and new knowledge were brought together; that new or improved technology was confronted with the problems of production and was scrutinised by peasants equipped with a rich body of empirical knowledge. It was also in the context of these stations that researchers and technicians had a chance to hear the ideas and needs of the peasants, to test the peasants' response to the prospect of adopting a new technology. The Viet Namese agricultural authorities had rightly paid great attention to these stations. Their attention was justified because it was reported that the role of peasants in checking research results and in the final testing and adaptation of technology contributed greatly to the completion of many research projects. Their contribution not only shed light on the technical aspects of the research, but also pointed out the economic implications of adopting newly developed or imported technology (Dap 1970). It was also at these stations that two-way communication between peasants and scientists took place. Scientists and technicians meeting and working with peasants at the co-operatives' stations obviously meant that modern science and technology were brought down to the production level at the same time as the centuries-old experiences of peasants were passed upward to research and development institutions. This achievement constituted what Vernon W. Ruttan terms the socialisation of research and diffusion, and views as a favourable condition for induced technological innovation.

At the top level, Viet Nam's structure for science and technology bore a strong resemblance to the Soviet model. The reason for this was not only that Viet Nam received substantial scientific and technological assistance from countries of the Soviet bloc, but also that this was its only source of assistance. In the early 1950s and 1960s, the experiences of China were studied meticulously. Due to the great similarity between China and Viet Nam in terms of crop pattern and cultural background, the Chinese experiences were useful in the effort to modify the Soviet model and make it suitable for the conditions of North Viet Nam. One can say that whereas the Soviet Union and Eastern European countries provided Viet Nam with scientific knowledge and hardware, China offered a grass-roots organisation model that brought science and technology closer to production. Since the end of the war in 1975, Viet Nam had been trying to diversify its contacts. It had sought contact with Western scientific circles, but the international political atmosphere had frustrated these attempts so that the Soviet Union and Eastern European countries had continued to be Viet Nam's major source of supply and inspiration, even though the leaders realised clearly that isolation from the international scientific community would not do Viet Nam any good at all.

Education and training

There is no doubt that the rhythm and pace of technological development depended heavily on the existence of a body of capable and active scientists, technicians and cadres who were responsible for the development and dissemination of technology at all levels. Fully aware of this pressing need, the North Viet Namese leaders had always emphasised the importance of education and training for improving the level of science and technology as well as the receptivity of cadres and peasants. The efforts to develop the country's human resources for technological innovation had been made within the framework of the combined technological revolution and cultural ideological revolution.

During the anti-French resistance period, the "Ministry of Agriculture" had had under its auspices one agricultural middle school, one research institute and one veterinary institute. In the war situation, they had existed only in name. In 1954, these institutions returned to Hanoi and started working from scratch. Despite the lack of almost all facilities, the first class of the Agronomic Middle School graduated in 1955.

In 1965, the High School of Agronomy was founded and two years later it merged with the Institute of Agronomic Research to become the Central Institute of Agronomy. Since then, the process of institution building for the development of technical education has continued. In 1960, the first three regional middle schools of agronomy were opened in order to supply the demands for skilled personnel for agricultural development in localities with varying physical and climatic conditions. A middle school of agricultural machinery also came into being in preparation for the introduction of tractors to rice production.

1961 marked a more comprehensive phase in the field of scientific and technical education for agricultural development. In a meeting held at the end of March 1961, it was reported that between 1951 and 1961 the Central Institute of Agronomy had focused on the following ten subjects (Nghiep, undated):

(a) soil survey, solutions for the problems of erosion;
(b) fertiliser and crop response to fertiliser;
(c) selection and improvement of good traditional rice varieties;
(d) plant protection;
(e) aspects of and techniques needed for rearranging the agricultural calendar;
(f) improvement of farm tools, design of simple agricultural machinery;
(g) fodder;
(h) animal husbandry;
(i) prevention and treatment of animal diseases;
(j) basic survey and geological cartography.

In the same year, in order to pave the way for agricultural development during the period of the First Five Year Plan (1961-64), the most significant step towards a modern peasantry was taken with the foundation of agronomic schools in all regions. Since then, the extensive network of primary and middle schools of agronomy had continued to expand. They drew from co-operatives the young and active members and trained them in techniques of seed production, plant protection or water control. Some trainees later worked on experimentation at provincial or district levels, but most of them returned to resume their production work.

At the highest level, in 1963, the Central Institute of Agronomy was subdivided into the Institute of Agricultural Research, the College of Agronomy and the College of Forestry. Four years later, another College of Agronomy was founded. Since the expansion of agriculture in the midlands and highlands posed new technical problems, the College of Agronomy for the highlands was opened in 1970 in order to train cadres in dry crops and industrial crop production. Many students at this school came from ethnic minorities.

By 1970, after 15 years of continuous development of training institutions, North Viet Nam had an extensive network of agronomic schools consisting of three colleges of agronomy, 35 middle schools of agronomy, 113 primary schools of agronomy and 12 institutes of agricultural machinery. The numbers show that despite the war situation, an increasing amount of resources was invested in agronomic education, and in training agricultural mechanics.

Alongside technical education, general education also made impressive strides. First, the achievement in eradicating illiteracy among peasants was remarkable. Nowadays, Viet Nam's literacy rate is among the highest in the Third World. After 25 years of general educational development, a World Bank report noted that the rate of adult literacy in reunified Viet Nam was as high as 87 per cent in 1981.[6]

With the development of general education, institutions for higher education have increased in number and diversified in disciplines. From a single university in Hanoi in 1954, Viet Nam now has four comprehensive universities, 68 colleges, and 263 higher vocational schools. The number of research institutions has also increased considerably; apart from the National Science and Technology Centre, Viet Nam at present has some 75 research institutes under ministerial control. They are staffed by 20,000 researchers and technicians, 800 of whom have a doctoral degree or its equivalent. Of these 75 institutes, 17 are specialised in agriculture, and some other institutes of natural sciences also do research that serves agriculture.

The network of general and technical education has trained several hundred thousand managerial and technical cadres in all fields. In agriculture alone, it has been reported that from 1955 to 1970, nearly 99,000 technical cadres passed their training courses (table 7.4).

Table 7.4
Number of agricultural trainees (1955–70)[a]

High level technical cadres	8,100
Middle level technical cadres	40,977[b]
Low level technical cadres	49,786
Total	98,863

[a] Source: Pham Cuong 1970.
[b] of which 8,582 graduated from central government schools and 32,395 from regional schools

The 8,582 cadres who graduated from the central government schools later worked in the state sector: in state farms, in agricultural achools, in research and experiment stations at local levels. It was only the 90,763 cadres at the middle and low levels who worked in co-operatives. If one compares this figure with the number of middle and low level technicians working in agricultural production units in other Third World countries, one sees that the achievement of North Viet Nam was remarkable, particularly in view of its war situation. But if one compares it with the intended degree and speed of technological development in agriculture, one realises that the body of technical cadres for co-operatives and local experiment stations still appears to fall short in terms of both quantity and quality. It has been observed that if the authorities planned to provide each co-operative with just four technicians - one in agronomy, one in animal husbandry, one in pisciculture, and one in agricultural machinery - then the number of technicians to be trained would far surpass the capacity of Viet Nam's training institutions.

The method of training was a combination of study and work. This system prevented the trainees from becoming detached from physical work and from production activity, directing peasants instead of working together with them. Besides, as noted above, most trainees were recruited from the young and active members of co-operatives. After training, they were sent back to their co-operatives to work in production. This made training an investment in co-operatives and tended to prevent the phenomenon of trainees changing profession after their studies. A survey of the post-training jobs of 456 graduates of the Hai Hung Provincial Middle School of Agronomy revealed that despite the war's tremendous demand for manpower, most graduates came to work in the same field in which they had been trained: 68 per cent continued to work in the co-operatives, while 12.7 per cent worked in provincial agricultural service. Altogether, more than 80 per cent of the graduates continued to work in agriculture (Cuong 1970).

Even with the growth in the number of R & D institutions and the expansion of education, the technological infrastructure in Viet Nam was still weak. Although the quality of the scientific and technological work done during the war was relatively good, it is obviously not up to Western standards. The reason could be found in the lack of resources and facilities, particularly laboratory facilities. At the present time, Viet Nam can allocate only 0.3 per cent of its GNP for science and technology. This small amount (200 million dong) must be shared by all research institutions. As noted above, research in Viet Nam is mostly production-oriented, so that the bulk of these financial resources go to research institutes in various ministries (from two-thirds to three-quarters of the total amount). Although agriculture always gets high priority, its share in this modest amount of money is small and obviously not enough. The lack of financial resources and research and teaching facilities is also an obvious handicap to the formation of a strong contingent of researchers and technicians. Up to the present day, the group of scientists and researchers who have been trained abroad occupy the leading role and continue to spearhead scientific work. Together with the improvement of research and training facilities, improvement of the quality of locally trained researchers and technicians remains a task for the years to come.

NORTH VIET NAM'S GREEN REVOLUTION

Technological innovation in North Viet Nam's agriculture has gone through different stages that were determined by institutional and cultural endowments, as Ruttan rightly pointed out in his conceptual analysis (Chapter 3 of the volume). Ruttan's analytical framework is based on the prices of the main production factors, the most important among them being land and labour. Although in the case of Viet Nam there were no prices for these two elements (at least not in the sense that there were in countries with free market economies), the availability of land and labour has certainly been the independent variable of which technological innovation was partly a function. Throughout the whole process of agricultural modernisation, the supply of land as a natural resource remained inelastic, and the supply of labour fluctuated. For one period, there was an excess of labour

(1954-64); for another period (1965 onward) there was a labour shortage. Meanwhile, other resources grew short. Technological innovation in Viet Nam was not a response to the change in prices of a single production factor. Instead, it was a conscious attempt to adopt a comprehensive strategy for bringing agricultural production to a more advanced stage through sustained innovation. North Viet Nam's technological package for its "Green Revolution" was not much different from that adopted in other countries. It included better irrigation, high-yield and early maturing varieties, and fertiliser. However, as one will see below, although the package might look the same, its contents were of a different nature.

Irrigation

"Water, fertiliser, labour, seeds" is the traditional folk-saying that has been invoked by the Viet Namese to characterise the priorities in developing rice cultivation. The saying still holds true, for water is the prerequisite for wet rice cultivation. The development of an extensive and effective system for water control was thus a must for the development of agriculture. Water control was of cardinal importance because the natural water supply in North Viet Nam is very unevenly distributed. In the May-October period, the South-east Monsoon brings 85 per cent of the annual rainfall. The water level of the major rivers were high at this time and often caused floods. The North-east Monsoon season covering November-April is dry, with only 15 per cent of the annual rainfall. Crops were threatened by drought during this period, particularly in the rain-fed areas. Given this situation, the development of an irrigation system to bring water to a larger area of rice fields, and to protect them from floods and drought, was vital for stable production. Furthermore, both multiple cropping and the adoption of high-yield varieties required better irrigation. This explains the great emphasis that the North Viet Namese authorities placed on the development of the country's hydraulic system.

The period 1954-60 witnessed great efforts to mobilise private peasants to take part in the development of the traditional irrigation system. Mass mobilisation was the approach to the problem. It helped the State to draw a large part of the needed resources from the local government and the peasant community. The little that the State invested was mainly for large projects aimed at flood prevention and some drought control. The peasant contribution in repairing the damaged hydraulic systems and in digging or dredging canals and ditches constituted the largest investment. New constructions were mostly small in size. The improved hydraulic system helped to some extent to alleviate the effects of natural calamities during this period.

Although the development of hydraulic works was broadened in scope, the same investment policy continued to be implemented between 1961 and 1965, with the first Five Year Development Plan. The system became much more extensive. Old systems were expanded and new ones constructed. With the formation of the co-operatives, it was easy to reserve more land for construction, and labour could be mobilised and organised on a regular basis. The policy was to combine water supply with drainage to link small projects to middle-sized and large projects in order to establish a network.[7] In mobilising production, North Viet Nam linked the movement of co-operativisation with the development of the hydraulic system. The two campaigns were made mutually supporting: co-operativisation allowed greater mobilisation of the peasants' contribution and constituted a big push for further developing the hydraulic system. In return, development in hydraulic works helped the newly founded co-operatives to wage a more effective struggle against calamities, and thus demonstrated the superiority of collective production. The agricultural authorities of North Viet Nam were then fully aware that without strong technical support to increase productivity and ensure production, co-operatives were vulnerable and peasants could easily lose faith in them. They therefore

concentrated efforts and available resources on
developing a hydraulic network.[8] The State, for
its part, contributed some capital, equipment and
technical advice, while it was the co-operatives that
bore the cost of labour and construction. The
financial contribution of the co-operatives during
this period was small because they were still weak
and could not yet ensure reasonably stable
production, let alone accumulate and invest.

In this five-year period, there had been not only
an expansion of the scope of the hydraulic network
but also a clear improvement in the quality of water
control. Although most co-operatives were still weak
and still feeling their way towards survival, the
creation in a number of co-operatives of specialised
brigades in charge of water control signified clear
institutional and qualitative progress. With the
creation of these brigades, water supply became a
controlled technical element in the production of
rice, and a sine qua non element for subsequent
innovation.

By the end of the first Five Year Plan, North Viet
Namese agriculture had 56 important hydraulic systems
completed, including old projects that had been
repaired, and 213 hydro-electric stations that fueled
water-control stations. All these systems served to
increase the irrigated area in 1965 to 2,418,000
hectares. The most significant achievement was that
the area under permanent irrigation (i.e. where water
could be brought in or out of the fields at will at
any moment) increased from 29 per cent to 53.5 per
cent of the total irrigated area under rice, and from
33.4 per cent to 60.5 per cent for the irrigated area
under dry crops.

Improved and expanded irrigation systems allowed
the total cultivated area to increase by 16 per cent
between 1960 and 1965. Of the area newly brought
under cultivation, 5.6 per cent went to rice, and
66.5 per cent went to dry crops. These figures
showed a clear change in crop pattern. In rice,
thanks to better irrigation, some 30,000 hectares
could be sown to two crops, and could reach a yield
of 5 tonnes per hectare per year.

After 1965, hydraulic works were no longer limited

to the construction of dams and reservoirs, and the digging of canals and ditches as before, but a new course was embarked upon: the overall reconstruction of rice fields. This campaign involved levelling the lands on a large scale, and consolidating and reconstructing the fields with a view to reorganising the water-supply network and to constructing roads for tractors and other agricultural machinery. It sometimes involved refilling existing canals and ditches and digging new ones, removing old roads and constructing a broader and better road network. A number of water systems were therefore abolished or altered. This made some of the investments during previous periods a waste of resources, and reflected a lack of long-term planning in the construction of the hydraulic network.

While efforts were made further to develop the water control network, the bombing annihilated a large part of North Viet Nam's achievement in this field. Bombing from 1965 to 1972 seriously damaged flood-protection dykes, destroyed dams and reservoirs, and blew up hydro-electric and pumping stations. Communist sources claimed that between 1965 and October 1972, American bombers caused 783 breaches in the flood-control dykes, destroyed 70 water reservoirs and attacked and destroyed several hydraulic systems. Despite the problems caused by bombing, the North Viet Namese authorities claimed that by 1969 the subsequent reconstruction of the fields and the hydraulic network had by and large been quite speedy and successful.

In 1965, double cropping with high-yield varieties of rice was practiced on only 30,000 hectares; by 1972, irrigation allowed co-operatives to sow two rice crops, yielding 5 tonnes per hectare per year, on some 380,000 hectares. A more than 12-fold increase was achieved in seven years under heavy bombing.

After 1972, the hydraulic network continued to be enlarged and improved. It had been claimed that by 1974, water conservancy could already ensure irrigation for 94 per cent of the total area sown to two rice crops, but prevention of water-logging could only be ensured on 83 per cent of the area under Mua crops. The achievement was impressive. However, with the use of motorised and electrical machinery limited, water conservancy still had to rely a great deal on labour. Its technical efficiency before 1975 was still below the standard of modern agriculture.

Since the reconstruction of the fields, the technical quality of irrigation has largely improved. Water conservancy no longer meant only a struggle against natural calamities and a supply of water for irrigation. With the more extensive hydraulic network and with better technical cadres specialising in this field, North Viet Nam could take a more "offensive" approach to agricultural production. Hydraulic-network development was mobilised to contribute to the progress of other factors. Good water supply contributed to the improvement of the soil by affecting microbiological processes, fertiliser absorption, aeration of the soil, temperature of the field, structure of the soil, erosion, acidity, salinity, etc. Advanced irrigation could no longer merely rely on experience. When production was to be modernised, better control over water was indispensable. For this reason, North Viet Nam established a wide network of hydraulic stations in order to study water availability and effects of water on crops under different conditions at different times.[62] If put to work to maximise other factors, it was reported that good irrigation could increase the yield of rice from 10 to 30 per cent, of cotton from 25 to 27 per cent, and of maize from 10 to 34 per cent. In order to achieve this, members of a co-operative's hydraulic brigade worked closely with technicians at the local hydraulic station and received from them information and technical advice. The improvement in the technical quality of irrigation resulted in the application of several different techniques, such as rotating irrigation, continuous irrigation and integral irrigation. Together, technical cadres and the co-operative's hydraulic brigade searched for the optimal irrigation coefficient and fixed the economic norms for the use of water.

The success in developing the hydraulic network in North Viet Nam owed a great deal to agricultural

co-operatives. Through co-operatives, the State was
able to successfully mobilise millions of man-hours
and could thus considerably reduce its own
investment. Without this transfer of a large part of
the costs to the peasantry it would have been
impossible for North Viet Nam's Government to
undertake such an extensive programme of water
conservancy.

The spring rice and the new agricultural calendar

Of the two traditional rice crops in North Viet Nam,
the Chiem (rice crop of the dry season, sown in the
ninth-tenth month and harvested in the fifth month of
the subsequent years) had been the weak link in the
country's rice production cycle. It required too
many resources and produced only low yields. The
Chiem crop occupied the rice fields for six to seven
months and demanded larger labour requirements while
its average yield could never be higher than 1.1
tonnes per hectare. This crop was not only
uneconomical, but also hampered any attempt to
promote multiple cropping.
   In the eyes of the Viet Namese agronomists and
agricultural economists, the presence of the Chiem
crop in the agricultural calendar represented an
obstacle on the road to modern agriculture. As early
as the 1950s, they had already conducted research on
the various technical and economic aspects of this
crop. The results of their research raised the
possibility of replacing this crop with one that had
a shorter growing cycle. Once this replacement could
be effected, the whole agricultural calendar could be
rearranged in order to make room for multiple
cropping and crop diversification. What was sorely
needed for this was a rice variety that met the three
following requirements: short growing cycle,
resistance to cold, and high yield. This new variety
would then be adapted to late sowing, in December –
January. When all these requirements had been met, a
new crop called Lua xuan (spring rice) could be
established. Spring rice could then be harvested
only four months after it had been sown, whereas the
Chiem crop took six to seven months to mature. The

search for new varieties and other means of establishing the new crop began in the early 1960s. It took several years before a technological breakthrough permitted a rearrangement of the agricultural calendar.

North Viet Nam's adoption of new high-yield varieties had depended on the country's achievements in other fields of agricultural research (hydraulic, mechanisation) as well as on the development of new rice varieties. It was not an abrupt step of merely adopting a new "technical package" imported from foreign countries. Instead, the adoption of new varieties came as a result of the technical ground-work laid in the previous periods.

In the late 1950s and early 1960s when plant-breeding techniques had not yet been developed, North Viet Nam took care to select a number of good traditional varieties of rice and to adapt them to different soil and climatic considerations before popularising them more widely. The major efforts in the early period were geared towards ensuring the availability of good seeds. These efforts resulted in the establishment of seed-production stations in different areas, and later in government help in setting up special fields for seed production in the co-operatives. In these special fields, seeds were reproduced with maximum technical care in order to avoid degeneration of quality. Moreover, research led to the application of better techniques of seed treatment before sowing. In the late 1950s and early 1960s, besides the old varieties, two new ones were propagated: Nam Ninh and Tra Trung Tu. They were tested and showed good results when sown as a spring crop at the existing level of farming and with the same supply of other inputs. However, these two varieties bore all the features of traditional varieties; long stalks, long and bending leaves, limited response to fertilisers, and fragility. Their ratio of paddy/straw could never surpass 0.35, and the yield seldom exceeded 4 tonnes per hectare. Despite all these limitations, Nam Ninh and Tra Trung Tu were sown for some time because they proved able to give the highest possible yield when a rather small quantity of fertiliser was used.

Moreover, their labour requirements were more manageable for the newly established co-operatives. The advantage came from the capacity of these varieties to compete with weeds, thanks to their speed in reaching a relatively high ratio of leaf surface in a short period.

The studies in the early 1960s on the traditional varieties' biological development and fertiliser response revealed that premature vegetative growth of the rice plant precluded a good yield. This premature and excessive growth of the stalk was caused by an excessive dose of a nitrogenous nutrient. This discovery led to a technique for controlling the unnecessary growth of the rice stalk. The application of nitrogenous fertiliser was interrupted and the rice fields were allowed to dry up during the period when the stalk of the rice plant grew. Fertiliser and the plant's strength were thus saved for the development of the ears and grains. This technique greatly helped to increase the yields of traditional varieties. But in order to increase the yield, one must not only find a way to make the rice able to respond positively to high doses of fertiliser. The discovery of a dwarfing gene and its adoption in plant breeding was a breakthrough in the biological technology for rice cultivation. Whereas dwarf varieties were developed in IRRI in the Philippines, Viet Namese agronomists bred short-stalked rice varieties such as Tran Chau Lun, TH2. These new varieties could reach far higher ratios of leaf-surface per square metre of rice field (from 8 to 10 per square metre). This led to a high rate of photosynthesis that resulted in better development of the ears and grains. The ratio between paddy and straw of these varieties reached 0.5 (Tuan 1973).

With these two dwarf rice varieties available, testing began in the second half of the 1960s on the establishment of the spring rice crop. At the same time, Viet Namese agronomists learned of the achievements in rice breeding at IRRI. Starting with the 20 bags of IR8 rice seeds that had been given by Fidel Castro, the Viet Namese agronomists began the process of hybridising with local varieties in order to increase their resistance to pests and to the climatic conditions of Viet Nam. This resulted in Nong Nghiep 8, the Viet Namese version of IR8. With Nong Nghiep 8, yields could reach 5 tonnes per hectare per crop. In 1970, the peasants in Thai Binh province harvested their first crop of Nong Nghiep 8.

The development of new high-yield varieties had been conducted alongside research to increase the resistance of these new varieties to the cold winters in North Viet Nam. The conclusion was reached that in order to have good harvests, the timing should be such that the rice blooms in late April for the Chiem, and in October for the Mua. The creation and then adaptation of the new varieties in order to establish a short-term spring crop during winter appeared to be difficult. It was only in 1970 that the new varieties suitable for a spring crop were finally created and propagated. Among them, Nong Nghiep 8 was the most important and most widely adopted.

In 1970, spring rice was sown on 18 per cent of the area that had been under the Chiem crop, in 1971 on 58 per cent and in 1972 on 64 per cent. The average yield of spring rice in the whole of North Viet Nam in the early 1970s was 3.2 tonnes per hectare. A new era of rice cultivation started in North Viet Nam in 1972: the spring rice quickly replaced the Chiem and released land for the cultivation of other crops. Since spring rice is transplanted in February, the land was freed from October to February, allowing the introduction of a new crop. As a result, the agricultural calendar could be rearranged and land resources could be more fully utilised. This entailed a whole process of rationalising land use and reorganising farm activities and labour. Since 1972, the new agricultural calendar in North Viet Nam has included several alternate formulae for multiple cropping:

(a)   better land preparation + azolla + spring rice + sesbania cannabica + mua rice;
(b)   azolla + spring rice + azolla + mua rice;
(c)   azolla + spring rice + azolla + fast-growing rice (CS1 variety).

But multiple cropping raised a very serious problem

that could not be easily solved in the situation of
North Viet Nam in the early 1970s: the problem of
replenishing the soil with nutrients in order to
permit multiple cropping with sustained high
yields. The above formulae give an indication of
North Viet Nam's solution to the lack of chemical
resources: the intensive use of the country's
available resources, namely full utilisation of
organic fertilisers.

Alternative sources of plant nutrients

As mentioned in the first part of this study, the
Viet Namese authorities right at the beginning
envisaged animal husbandry as the second major branch
of agricultural production. They planned for a
balanced development of cultivation and animal
husbandry. There were two reasons for this policy.
First, the planned improvement in the diet of the
growing population required a larger supply of
protein-rich foods. Second, once it was
well-organised, animal husbandry would help to solve
the problem of fertilising the soil by supplying an
important quantity of manure.
In the context of a strategy aimed at intensive
farming, the lack of fertiliser posed a serious
problem. Unless it could be solved, intensive
utilisation of land meant a complete exhaustion of
nutrients every couple of years. Given the country's
war situation and financial weakness, it has been
impossible, even up to the present day, to import
substantial quantities of chemical fertiliser. In
1960, for the whole of North Viet Nam, the Government
could supply only 111.3 tonnes of nitrogenous
fertiliser, some 60,000 tonnes of phosphate
fertiliser, and no insecticide at all. In 1971, on
the average, for each hectare under two rice crops,
366.9 kg of chemical fertiliser was available. Of
this amount, only 93.87 kg was nitrogenous fertiliser
(urea).
Not only was the amount of chemical fertiliser used
for each hectare of land far below the requirement,
but the nutrient content of the chemical fertiliser
used was low (Luu 1982).

Faced with this problem, North Viet Nam chose to direct its efforts towards developing alternative sources of fertiliser, namely animal and green manure. In the very first months of the co-operatives' existence, the agricultural authorities urged co-operatives to build pigsties for the systematic rearing of pigs, and encouraged peasant members to rear a couple of pigs on their private plots of land. The Government sent cadres to help build the pigsties and to train the co-operatives' cadres in pig-rearing and the peasants in the techniques of animal husbandry and manure procurement. Special care was given to the hygienic aspects of procurement and compost-making in order to prevent the bad side-effects of the use of pig manure. In the early 1960s, it was planned that for each hectare of cultivated land, one pig would be reared;[9] later, in 1980, the Party and the Government called for rearing two to four pigs per cultivated hectare (Cuc 1979 and Le Duan 1976). The authorities emphasised that the amount of organic manure procured from pig rearing at the existing stage of the country's development could not be provided by chemical fertiliser. It was said to be good not only for improving the soil but also for fixing other nutrients as well. To give an idea of the importance of this source of manure for the agriculture of North Viet Nam, it was necessary to point out that since the mid-1960s, pig manure had constituted 60 per cent of the total nutrients provided for rice cultivation. Thus, this source of nutrients had still been in short supply and was unable to meet the growing demands of intensive rice cultivation. In order to ensure high yields for a hectare under intensive cultivation, 15 to 20 tonnes of pig manure was needed per crop. But in the late 1960s and throughout the 1970s, pig rearing had only been able to provide from 6 to 10 tonnes for each hectare. The fact that North Viet Nam's planners linked the number of pigs reared to the area sown to rice showed that they obviously intended to continue to rely on this source of nutrients for many more years.

Since the date the Americans started to bomb North Viet Nam, the population had continuously suffered from a shortage of protein-rich food. Pig rearing for meat therefore became the most attractive way for peasants to earn cash income. As a result, while the herd of pigs reared on state farms and co-operatives grew, the number of pigs reared on private plots also increased considerably. In the 1970s, the number of pigs kept in private pigsties could be estimated at nine times larger than in the collective sector. This situation points out the difficulties in procuring all pig manure from private pigsties for fertilisation. With a total of 6,304,100 pigs in 1973, the quantity of manure that could be procured can be computed as follows;
- 6,304,100 pigs = 1,260,820 animal units (five pigs = one animal unit);
- amount of animal manure: 1,260,820 x 7 metric tonnes per unit = 8,825,740 metric tonnes; and
- available as fertiliser; 8,825,740 x 70 per cent = 6,178,018 metric tonnes.

Even if one adds to this amount other supplies of organic fertilisers such as night soil and manure from other branches of animal husbandry, North Viet Namese agriculture's demands for fertiliser still could not be met. The agricultural authorities therefore sought to develop the country's other major source of plant nutrients, namely green manure.

Together with animal manure, green manure had been strongly emphasised since the early 1960s. Research had been undertaken in order to develop fully the sources of supply of nitrogenous nutrients. Starting from traditional practices, scientific research had made a technological breakthrough in producing and using sesbania cannabica and azolla as major sources of nitrogen-rich fertilisers.

Sesbania, known in Viet Nam as Dien Thanh, was grown by seed or by cuttings. In the delta areas, it was sown in February, March and April; in the coastal areas, it was sown a bit later. The Viet Namese had four ways to cultivate sesbania for manure:
1. The easiest and most economical way was to sow sesbania along the roadsides. Several rows on each side can be sown. With this method, one

needed 0.5 to 0.8 kg of seeds for each kilometre of road. When the plants reached a height of one to 1.2 metres, they could be cut, yielding 1 to 1.5 tonnes of sesbania leaves and stalks that could be used as fertiliser;

2. Sesbania could also be sown in the seedling fields. In January or February, after rice seedlings had been transplanted from 40 to 50 kg of sesbania, seed was sown to 1 hectare of seedling fields. After three months, one could harvest from 5 to 10 tonnes of green manure from each hectare;

3. One could also sow sesbania between rows of dry crops such as sweet potato and maize;

4. Peasants in Viet Nam also cultivated sesbania in the rice fields between the rows of rice or sowed sesbania when they drained the water from the fields before the harvest. This method replenished the soil with several tonnes of nitrogenous manure before the new crop started. Research on sesbania in Viet Nam had revealed that 1 tonne of sesbania was equivalent to 5 kg of pure nitrogen. Viet Namese agricultural scientists, however, were obviously more interested in azolla, a floating algae, as a source of nitrogenous nutrients. Viet Nam's experience with azolla was a good illustration of the fact that traditional technology could be developed by scientific research, and that a local natural resource could be fully mobilised to replace chemical fertiliser, an input the country could not afford. It also demonstrated the Viet Namese ability to find an alternative way of increasing yields without overloading the land with chemicals, and without disturbing the eco-biological balance of the fields.

Originally, azolla had been used for centuries as green manure in a small area in Thai Binh province (in the three villages La Van, Bung, and Bich Du). Wealthy families in these villages reserved land to cultivate azolla, which they sold as manure. They monopolised production and kept the techniques secret. Very early, scientists recognised the potential of this algae and tried to break through the secrecy. Since the 1950s, there has been consistent research on azolla. Early studies revealed the techniques of production and utilisation. Altogether, Viet Namese scientists studied three categories of azolla, consisting of 28 varieties of algae, in order to select the most suitable varieties for various climatic and topographic conditions. They discovered that the local variety of azolla of La Van village showed greater resistance to cold and reproduced faster in winter, and therefore was the best variety for cultivation in the fields sown to spring rice. On the other hand, green azolla and the variety that Viet Namese scientists collected from Bangkok were more productive in warm weather. They were therefore suitable for fertilising the fields sown to the Mua crop, and for the preparatory fertilisation of the fields sown to the spring crop.

Azolla reproduced very quickly. Each day its weight could increase from 20 to 35 per cent. In one week, azolla might double the surface area of water it covers, and in three months it can multiply 170,000 times. Thanks to this reproductive capacity, azolla grown in 1 hectare of land could provide from 200 to 300 tonnes of manure each year. It had been estimated that during the Mua crop, if from 5 to 10 per cent of the land could be reserved for two months to cultivate azolla, peasants could gather enough green manure for the whole area that would be sown to spring rice.

The value of azolla lies in its richness in nitrogenous nutrients. Dried azolla contained 3 to 5 per cent nitrogenous nutrients. A layer of azolla covering 1 hectare gave 10 tonnes of green matter that contained 25 kg of nitrogenous nutrients. This amount of green manure helped to increase the yield from 10 to 25 per cent. If 20 tonnes of azolla manure was applied to 1 hectare before transplanting, it could provide some 50 per cent of the nitrogen requirement for a yield of 5 tonnes. Apart from being a good fertiliser, azolla was also a good source of animal feed. It was rich in protein (13 per cent of its dried matter): in one month on one hectare, azolla could produce from 540 to 720 kg of

protein. For this reason, it was used to feed pigs
and could be substituted for 50 per cent of the rice
bran pigs required (Thuyet and Tuan 1973).

Traditionally, Viet Namese peasants only turned
their attention to azolla after they had transplanted
the seedlings of the Chiem rice crop in December or
January. The azolla was first allowed to multiply
for a couple of months in its seedling fields before
being introduced into the rice fields. This meant
that it matured and became manure only at the end of
the rice stalk growth. From the point of view of
fertilisation technique, the manure became available
too late in the cycle of the rice plant and thus lost
some of its effectiveness.

After research, the cultivation of azolla was
rescheduled and resources were allocated for its
development. It became an integral part of the whole
agricultural calendar. In order not to take up land
needed for rice cultivation, co-operatives grew
azolla in the seedling fields that had been left
fallow to rest. These amounted to some 10 per cent
of the area reserved for rice cultivation in each
co-operative. Another method was to grow azolla in
the rice fields between two rice crops. The
replacement of the uneconomical Chiem crop by the
spring crop allowed the co-operatives to make use of
the fields from December to January or February to
grow azolla or other dry crops. As the agricultural
calendar described above indicates, azolla and
sesbania as green manure came to occupy established
positions. Other costs involved in growing azolla
were not high. Some P_2O_5 and K_2O was used to
stimulate the algae development and to fix the
nitrogen in the air. For each kilogram of P_2O_5
used for azolla cultivation (at the rate of 30 kg per
hectare of ordinary soil), 2.2 kg of nitrogenous
nutrients could be obtained. On poor soil, K_2O
gave better results if a dose of 5 kg of K_2O was
applied to 1 hectare every five days (Thuyet and Tuan
1973; Nghiap 1980).

Taking azolla seriously and relying on it as a
major source of nitrogenous nutrients, the Viet
Namese agricultural authorities decided to start
courses to train technicians and specialised members

of co-operatives in the techniques of producing and processing and using Azolla. In most of the co-operatives, besides the hydraulic groups, fertiliser-making groups were also formed, and their members were thoroughly trained. These groups were made up of young peasants who after finishing seven years of general or agricultural education received training on azolla in three to five month regular courses. Through these courses, the difficult techniques of azolla use were widely propagated. The speed of popularisation was remarkably high: in 1961 the new methods were introduced; in 1964 it was grown on a surface area of 140,000 hectares; by the early 1970s it had spread to some 700,000 to 800,000 hectares.

The extensive use of animal manure as fertiliser, and the development of biological and chemical technology for growing and using green manure on a large scale had made Viet Nam an interesting case. Apart from China, Viet Nam was perhaps the only country that was undergoing its "Green Revolution" with almost no import of technology or material from outside. In this sense, Viet Namese technological development was endogenous, as Ruttan maintained while describing his model of induced innovation.

CONCLUSION

Agricultural modernisation and productivity

On the theoretical plane, North Viet Nam's communist leaders repeatedly emphasised that the country's long-term aim was large-scale socialist agricultural production with full-scale electrification and mechanisation, and with full application of biological and chemical technology.

On the practical plane the objectives were set with realism and policies were tailored to suit the country's stage of development and resource endowments. Although increases in both land productivity and labour productivity were sought, it seems that the concrete situation of the country had made it fairly impossible to increase labour productivity in the 20 years of agricultural modernisation. The first reason for this failure was that the rate of growth in agricultural production during the period had always been more or less equal to the rate of population growth. The direct consequence of this was that North Viet Nam could not produce a net increase in its surplus and thus remove the threat of food scarcity. Besides, the very socio-political system dictated that everyone was entitled to work and to earn some income. This put pressure on the co-operatives to employ every active member of the rural labour force, which not only put a check on labour productivity but also prevented an increase in the income that members drew from working in co-operatives. The effort to employ an increasing number of member peasants, while co-operatives' earnings were not increasing at the same pace, resulted in a low rate of remuneration. It was estimated that the payment per member for one work-day ranged from only 0.6 kg to 1 kg of paddy rice. This had negatively affected labour productivity since little incentive was provided by the income generated from co-operatives. Secondly, despite all the remarkable achievements in chemical and biological technologies, North Viet Nam had still missed the massive adoption of modern inputs that would have brought about a productivity "take-off". The quantity of plant nutrients applied to each hectare of land had remained far below that of modern agriculture in Japan, for example. Thirdly, the need to support the war of liberation by providing food for a large armed force, and to supply cheap food for the urban areas, had been added to the zealous attempt to bring about quick capital formation in agriculture. All these demands had consumed most of the gains of modernisation, leaving little over for the members of the co-operatives. It was reported that during the period 1965-68, co-operatives on the average retained as much as 72.8 per cent of their revenues for the funds of accumulation (Luu 1982). The fact that peasants earned only 36.2 per cent of their incomes from co-operatives obviously made them anxious to direct their effort to other more profitable pursuits.

While labour productivity seemingly stagnated, land productivity clearly increased. With more intensive use of land resources and with the wide adoption of biological technology, land productivity had risen steadily. However, bringing under cultivation the less fertile lands that were newly claimed in the midlands and highlands had undoubtedly brought down the country's average increase in yield. For the area where modernisation had been undertaken, the yield per unit of land doubled or nearly doubled. If one compared the average yields in North Viet Nam with the yields in neighbouring countries where the Green Revolution was launched with considerable imports of modern inputs from foreign countries, the yields in North Viet Nam had been either equal or slightly higher. This underlined the real economic costs of the alternative paths towards agricultural modernisation under different socioeconomic systems. In North Viet Nam, high yields per unit of land were achieved without much imported chemical fertiliser or other inputs. The achievement was therefore not paid for with foreign assistance or with earnings from foreign trade. North Viet Nam did not depend at all on the supply of agricultural inputs from foreign countries. If one also takes into account the damage to the environment and disturbance of ecosystems as long-term economic costs of modern agriculture, one can appreciate even more the achievement of North Viet Nam. The intensive use of organic sources of nutrients helped to increase land productivity while preserving the structure of the soil, stimulating microbiological processes in the soil, and avoiding the pollution of water with chemicals. Naturally, the intensity of the use of labour partially accounted for the achievement. High yields were partly the result of intense care in transplanting, weeding, protecting, etc. But appropriate and low-cost technologies accounted for the other part of the success. North Viet Nam's agriculture at the end of the war in 1975 looked less 'modern' than that in other countries if one counts the number of tractors and machines and the quantity of chemical nitrogenous fertiliser it utilised. But if one looks at the technical base of the co-operatives and local units,

and at the real costs involved in increasing yields,
one can say that North Viet Nam's agriculture was
relatively modern in its own right - and in its own
way.

Agricultural modernisation and employment

In principle, agricultural co-operatives were
supposed to employ all active members of the rural
workforce and provide them with some income. This
adds another problem to the lack of statistical data
and makes it impossible to assess the impact of
modernisation on the generation of employment.
Instead, one can make some general remarks on the
qualitative change in the rural labour force.

Firstly, there had been clear improvement in the
outlook and receptiveness of the North Viet Namese
peasants. Alongside the eradication of illiteracy,
peasants had become increasingly acquainted with
science and technology. Nowadays, the most
widely-read literature in the countryside of North
Viet Nam was the simple but highly useful Journal of
Elementary Sciences. Through this periodical, not
only was agricultural technology introduced, but
other knowledge and techniques were popularised as
well. In North Viet Nam today, science and
technology in a modest sense have become part of the
peasant's life.

Secondly, there had been a sharp increase in the
degree of specialisation, something that had not
existed before. Besides, the labour force had been
more evenly distributed among the various economic
activities in the countryside. Statistics gathered
from the Dinh Cong co-operative, one that was
advanced, showed that in the early 1970s, there was a
transfer of labour from agricultural activities to
rural industries. In 1970, 71 per cent of
co-operative members worked in farming and 4.8 per
cent in animal husbandry; in 1977, only 50.6 per cent
were still in farming and 9.6 per cent in husbandry,
whereas the number of members who had shifted to
industries and handicrafts had increased to 24.3 per
cent (Luu 1982). At the district level, one could
also notice a migration of labour from cultivation to

husbandry and to rural industry and handicrafts. Within co-operatives emerged specialised members working in hydraulic teams, seed-production teams, fertiliser and compost-making teams, animal-husbandry teams, etc. This diversification and specialisation of the labour force marked a qualitative change in the peasantry.

To conclude, one can say that with all its achievements, North Viet Nam's agriculture still did not reach the stage where the increase in land productivity gave way to an increase in labour productivity. The special view of the leaders and the country's special circumstances have put agricultural modernisation on a course of change along which alternative resources and methods had been used. These resources and methods, despite their limitations, had helped the Viet Namese to obtain some of their objectives. They had constituted the strength as well as the weakness of the experience.

<u>NOTES</u>

[1] In Viet Namese literature, the Chiem crop is often called the fifth month rice crop or Winter-Spring crop, and the Mua crop is called the tenth month crop. In order to avoid confusion arising from the use of several names, they are hereafter referred to as the Chiem and the Mua.

[2] In China, with more intensive farming, the work-day requirement of 1 hectare sown to rice in 1958 was 525 (computed on the basis of statistics given by Dawson 1979).

[3] In 1963, for example, there were more than 40,000 rural households which withdrew from co-operatives, but with the consolidation of co-operatives, the number decreased in subsequent years.

[4] Resolution of the Sixth Plenum (II) of the Central Committee of Lao Dong Party, April 1959.

[5] Resolution of the Tenth Plenum of the Central Committee of the Lao Dong Party, 1964.

[6] World Bank's annual report 1981. The rate of
    literacy in North Viet Nam alone was higher that
    the rate for the whole reunified Viet Nam. This
    was due to the lower rate of literacy in South
    Viet Nam at the date of reunification.
[7] Resolution of the Fifth Plenum (III) of the Lao
    Dong Party, July 1961.
[8] See _Political Report_ to the National Assembly of
    Prime Minister Pham van Dong, April 1965, and
    Binh 1973.
[9] Resolution of the 19th Plenum of the Lao Dong
    Party.

8 Summary and conclusions

IFTIKHAR AHMED and VERNON W. RUTTAN

INTRODUCTION

Both the conceptual and empirical chapters of the volume point to the need for continued assessment and analysis of the role and performance of the research and extension institutions concerned with agricultural technology. Methodologically, current emphasis on quantitative techniques of analysis focusing on management aspects of these institutions may result in the neglect of the necessary and complementary political economy research which is important for both policy analysis and action. The recognition of the non-transferability of agricultural technology, even across limited geographical areas, has increased the importance of adaptive and applied research at the national level. This, in turn, focused attention on the potential of and limits to international transfer of agricultural technology.

This concluding chapter attempts a synthesis and overview relating to a set of theoretical and analytical issues important for policy-making and project formulation. The conceptual and case study chapters have covered the following broad categories of issues: (a) the empirical verification of the theories of induced technological and institutional innovations and their interrelationships; (b) the effect of farm-size distribution on the demand for a productive national agricultural research system; (c) the effect of decentralised research on the supply of new technology to farmers; and (d) biases in agricultural research and extension programmes and their implications for poverty alleviation. Finally, the chapter identifies areas for future research.

INDUCED TECHNOLOGICAL INNOVATION

The empirical case studies confirm the existence of a general process of induced technological innovation based on national factor endowments, although the form and intensity of this process was not uniform among countries. This was as valid in a poor country like Bangladesh with underdeveloped indigenous scientific research and a predominance of imported technology, as it was in Kenya where the population explosion and limits to land expansion led to a rational institutional response of a labour-using bias (e.g. inducement towards biological as against mechanical technology) in the path of technological change. However, the experience of India reveals a regional bias in programmes of agricultural technology generation. As was noted (Chapter 4), the regional allocation of research resources did not bear any relation to the corresponding regional contribution to national agricultural output. The influence of socioeconomic factors and political forces on agricultural research resource allocation is reviewed later.

The major criticism of the theory of induced innovation was its critical reliance on the market mechanism for its validity. The case studies have established its validity in both free market and mixed economies, countries with different degrees of rural factor market imperfections and where agricultural technology generation is overwhelmingly a public sector activity. The real test of the

induced innovation theory was its relevance to a
centrally-planned economy where the means of
production are collectively owned and free market
prices of factors of production (especially land and
labour) cannot be observed. It was clearly seen from
the case study on Viet Nam (Chapter 7) that the
nature and direction of technological innovation was
influenced by the availability of land (largely
inelastic) and labour (fluctuating) as Viet Nam's
agriculture passed through stages with differing
institutional and cultural endowments.

INDUCED INSTITUTIONAL INNOVATION

As had been noted in Chapter 1, agricultural research
and extension, being an exclusive public sector
activity, was considered as a step towards
socialisation of research at the macro level. The
alteration of the national priority from primarily
extension to research was assumed to be the work of a
process of induced institutional innovation. Higher
priority assigned to adaptive (and to a lower extent
applied) research compared to basic research,
particularly in India (Chapter 4), could be regarded
as a rational response by agricultural research
institutions within the public sector because
adaptive research responds more closely to the
localised and divergent agro-climatic and
socioeconomic environments. In contrast, the
stochastic nature (risk and uncertainty of outcome)
and longer gestation period of basic research make it
unattractive to research personnel. The significance
of adaptive and applied research is more dramatic at
the grass-roots level in Viet Nam (Chapter 7) where
the co-operatives became the centres of adaptive and
applied research and where scientists mingled freely
with peasants.
There is evidence of technology-induced
institutional innovations simultaneously at the
national and grass-roots levels. In the case of
Bangladesh (Chapter 5), the macro level institutional
innovations for delivery of modern inputs (chemical
fertilisers, HYV seeds, irrigation pumps, etc. by

Bangladesh Agricultural Development Corporation and the Bangladesh Water Development Board) and provision of supportive cheap institutional credit (by a network of the Agricultural Development Bank and Farmer Co-operative Societies) were induced by the technological innovation (bio-hydro-chemical)[1]. This technology also induced the intensification and widening of a network of co-operatives at the grass-roots level[2]. Similarly, in Kenya (Chapter 6), the establishment of the Horticultural Crops Development Authority was a rational public sector institutional response at the macro level, and the parallel formation of co-operating groups of small-scale growers under an integrated production-marketing scheme known as the Horticultural Production Centres was an institutional innovation at the grass-roots level.

The process of technology-induced institutional innovations is most dramatic in its effect at the grass-roots level and is as convincingly valid in a centrally-planned economy as in a free-market system. Following the collectivisation of Viet Namese agriculture, the co-operatives virtually served as grass-roots technical organisations where new technology was made known to peasants and tested, selected, adapted and popularised with the participation of the peasant users (Chapter 7). The process of technology-induced institutional innovation is extended a step further owing to specialisation of the workforce with diversification when the co-operative membership is disaggregated into yet smaller peasant organisations, such as the hydraulic teams, seed production teams, fertiliser and compost-making teams, animal husbandry teams, etc.

Bangladesh (Chapter 5) provides an interesting example of technology generated by farmers' own informal R & D (e.g. locally improved rice variety) which was then picked up by the country's formal R & D institution for further refinement and blending with imported varieties.

INDUCED TECHNOLOGICAL AND INSTITUTIONAL INNOVATIONS:
INTERRELATIONSHIPS

The existence of causal interrelationships between
induced technological and institutional innovations
have been noted for both a centrally-planned economy,
Viet Nam (Chapter 7), and the free-market system, the
Philippines and Indonesia (Chapter 1).

The process and path of institutionally induced
technological innovation in Viet Nam was the
following. Collectivisation of agriculture was the
single most institutional reform which set in motion
a chain of developments: (a) as a large-scale and
collective enterprise, a co-operative could easily
plan production and allocate resources in order to
respond to the demands of intensive and modernised
farming; (b) as a collective enterprise, a
co-operative could afford to set aside resources for
the generation of technologies; and (c) as an
enterprise larger than a family, a co-operative was
better able to absorb the risk involved in the
adoption of new technology.

As noted in Chapter 1, changes in institutional
arrangements in both the land rental and labour
markets in the Philippines were induced when
disequilibria between marginal returns and marginal
costs of these two factor inputs emerged as a result
of changes in relative factor endowments (land/man
ratio) and technical change in agriculture. A
similar institutional basis of labour use induced the
bias and direction of technological change in
Indonesian agriculture.

FARM-SIZE DISTRIBUTION AND DEMAND FOR INNOVATIONS

In Chapter 3, it was hypothesised that an economy
characterised by great disparity in farm size or
incomes would be less effective in generating a
sustained demand for a productive national
agricultural research system than an agricultural
system characterised by a reasonable equity in farm
size distribution. The multiple regression model
using inter-state cross-sectional data from India

(Chapter 4) clearly showed that investment in agricultural research is <u>negatively</u> related with the inequality in farm size distribution (Gini coefficient of operated area). On the other hand, the Kenyan case study (Chapter 6) is neither able to confirm nor reject this hypothesis since the cross-sectional data (Gini ratio of farm size distribution) was limited in size.

DECENTRALISATION AND TECHNOLOGY GENERATION

In Chapter 3, it was hypothesised that an agricultural research system which is decentralised with respect to location, management and funding tends to be more responsive to the resource and cultural endowments, and hence more productive in supplying new technology to farmers, than a highly centralised system.

This hypothesis is qualitatively supported by the finding that a decentralised research system in Bangladesh (Chapter 5) is likely to be more responsive to the needs of various regions and different farmer groups since the socioeconomic, climatic and agronomic conditions vary over different parts of the country and across different farm size groups. Evidence from Kenya (Chapter 6), supports the hypothesis from the locational point of view.

Institutional innovation (formation of co-oper -atives), coupled with massive decentralisation of agricultural research right down to the grass-roots level, has successfully induced the generation and diffusion of technological innovations in Viet Nam. As has been observed (Chapter 7), the existence of a number of research and experimental stations located in co-operatives spread throughout the country gave Viet Nam's system for scientific and technological research a base at the grass-roots level, making research closely allied to production. Such a decentralised structure greatly facilitated the testing, adaptation, screening and popularisation of improved or imported technology. The location of research and experimentation within the co-operatives permitted a blending of theories and practice and of experience and new knowledge. These also serve as centres for vigorous cross-fertilisation of new ideas and practical experience of production among scientists and peasants.

INSTITUTIONAL BIASES AND POVERTY ALLEVIATION

The existence of the three major forms of bias identified in Chapter 1 have been empirically confirmed by the case study chapters, often with important implications for policies and programmes of poverty alleviation.

Extension bias

The extension bias revealed by the international cross-sectional data is confirmed by two individual country case studies. For example, in the early 1980s, nearly 70 per cent of India's total research and extension expenditure was on lab-to-land type of extension (Chapter 4). This is also confirmed by both traditional extension programmes and institutionally innovative ones exclusively responsible for the delivery of critical technological inputs and working capital in Bangladesh (Chapter 5). Both of these developments constitute a rational public sector response to the changing needs for a widening of the strategy and thrust of conventional extension programmes.

Commodity bias

The bias against food crops (Chapter 1) in national agricultural research and extension programmes is confirmed by the case study from India (Chapter 4). For example, the staple cereal, rice, receives about 7 per cent of the research funds, while its contribution to national agricultural output is more than one-third of the total. Similarly, wheat, another important food item which constitutes over a quarter of the national agricultural production, is allocated about 14 per cent of the total research resources.

A neglect of the food crops like cassava, sweet potatoes and field beans has failed to transfer the benefits of technology to the poor in the form of lower prices given that the demand for these basic carbohydrates (including rice, wheat and maize) and vegetable proteins are relatively unresponsive to changes in income and prices (Chapter 3).

The choice of commodity emphasis in agricultural research could favour the incomes on small farms, for example, in Latin America, through a choice of beans (an important source of protein) grown on small holdings as against beef produced in huge ranches. At the same time as low-income consumers, small farmers and landless workers would gain from a choice in favour of beans relative to beef (Chapter 3). The research emphasis on horticultural crops is intended to benefit the poor of Kenya as these crops are directly produced and consumed on small holdings (Chapter 6).

The bias in favour of traded agricultural commodities, as opposed to non-traded ones, is clearly a consequence of the political and economic power of the commodity constituencies, as in the cases of sugar in Colombia and maize in Argentina (Chapter 1).

Among the neglected food crops, the position of root crops is acute since the longer gestation period and the public sector's tendency to value short-term gains serve more highly as a disincentive to research on these items as compared to rice, wheat or maize, for which research has been in progress for many years (de Janvry and Detheir, 1986).

Public sector bias

The Third World public sector shoulders almost the entire responsibility of agricultural research and extension (Chapter 1). However, the inter-state data from India revealed that regions with higher income levels were able to attract greater research attention (Chapter 4). The recent shift in relative priorities from extension to research in the Third World has been accompanied by a shift towards adaptive research, as is confirmed by the Indian study.

Urban bias is found to affect adversely the level of agricultural research and extension activity (Chapter 1). For example, in India (Chapter 4), the level of research activity is <u>negatively</u> related to the agriculture-industry terms of trade; it is also <u>inversely</u> related to the socioeconomic disparity between the urban and rural population.

That the level of public sector research and extension is influenced by the socioeconomic and political power of the regional and commodity constituencies is confirmed by the Indian inter-state data (Chapter 4). The level of agricultural research activity in India is <u>positively</u> related to the ratio of rural-urban literacy rates; it is <u>inversely</u> related to the unemployment rate. Similarly, eradication of rural illiteracy in Viet Nam played a significant role in both technology generation and diffusion (Chapter 7).

METHODOLOGICAL INSIGHTS

The case studies reveal that it is possible to verify empirically the set of hypotheses concerning induced technological and institutional innovations by means of both historical and cross-sectional data. These also show that it is possible to arrive at similar conclusions by means of an econometric analysis of quantitative data and through qualitative analysis where adequate statistical data is not available. The volume (Chapters 1 and 2) also reveals that the price-induced technological and institutional innovations model has its limitations, although the set of empirical case studies in the second part of the volume were able to use it as an analytical framework. However, there are some research gaps, described in the next section, that still remain to be filled.

Some of the country case studies (notably Bangladesh, Kenya, India and Viet Nam) make some effort to establish the validity of the induced innovation model at the micro level by extending the analysis to farm-level data.

AREAS FOR FUTURE RESEARCH

There are a number of areas that have not been adequately reflected in empirical research on induced innovation. Two areas which can be considered as particularly relevant are (a) the integration of demand induced and factor endowment induced technical change and (b) the integration of models of induced technical change and international trade.

Demand and factor induced technical change

The model of induced innovation employed in this book has drawn primarily on the Ahmad-Hayami-Ruttan-Biswanger models of induced technical change in which the direction of technical change is guided by relative resource endowments, often interpreted through relative factor prices. There is also the Griliches-Schmookler tradition of demand-induced technical change in which the rate of technical change is influenced by the rate of growth of demand.

Binswanger (1978) has suggested that the two approaches can be linked by introducing the concept of a research production function. In the proposed model the growth of product demand increases the marginal value product of resources devoted to research, thereby increasing the optimal level of research expenditure. The larger research budget results in an inward shift of the innovation possibility function, graphically visualised as the envelope of unit isoquants corresponding to the alternative technologies that can potentially be developed for a given research budget.

In this combined model technical change is guided by the innovation possibility function through changes in relative factor prices, while the above function itself is induced to shift inwards by the growth in product demand. So far no one has attempted to specify and estimate such a combined model. If such a model were developed and estimated, it would add to our capacity to allocate resources in agricultural research among commodities. It would be particularly valuable in a developing country context where the role of growth in commodity demand is changing rapidly in response to economic growth.

Technical change and international trade

In the standard Heckscher–Ohlin–Samuelson model of international trade, differences in resource endowments are the primary determinants of trade – each country is expected to export its abundant or factor-intensive products. In the theory of induced innovation, the path of productivity growth is directed towards releasing the constraints on growth imposed by resource endowments. To the extent that technical change can release the constraints on growth resulting from inelastic factor supplies, the power of the differential factor endowment explanation for trade is weakened (Jones 1970). And to the extent that trade can release the constraints of factor endowment or growth, the theory of induced innovation may lose part of its power to the direction of bias in productivity growth. Different factor endowments induce different production functions and initial factor endowments are in turn modified by this new production function. Once the technologies are allowed to differ among countries, the factor price implications of the standard trade model are weakened. Yet these two bodies of literature have not yet been adequately integrated (Cheng 1984).

When it does become possible to build empirical models that incorporate the factor endowment implications of both induced innovation and trade theory, it will provide a more powerful tool for analysing both the choice of technology and decisions about investment in research and development in response to the constraints imposed by resource endowments. For the smaller developing countries the issue of when to borrow rather than develop the technology appropriate to their resource endowments represents an important economic decision.

SUMMING UP

A general process of induced technological
innovations based on variations in national factor
endowments is observed for a range of countries at
different levels of development, research and
extension capacities and economic systems. Despite
the fact that a rational public sector response to
changing national factor endowments critically hinges
on price signals, the general validity of the induced
technological innovation theory remains unaffected in
situations where market mechanisms are non-existent
(e.g. Viet Nam) or, where they exist, are distorted
primarily through factor market imperfections (e.g.
Bangladesh, the Philippines, India and Kenya).
The assumption of research and extension
responsibility in the Third World by the public
sector is interpreted as a rational public sector
response to the stochastic and public goods character
of agricultural technology generation. Similarly, a
recent shift in priorities from extension to
research, and to adaptive research as opposed to
basic or applied agricultural research, are also
considered as components of the induced institutional
innovation processes.
Complete take-over of agricultural research by the
public sector has also been described as
socialisation of research which has been observed
when research is decentralised right down to the
grass-roots level (as in Viet Nam).
Technological change and variations in relative
factor endowments have induced institutional
innovations through parallel institutional
developments simultaneously at the micro and macro
levels (e.g. Kenya and Bangladesh) and in both market
(e.g. the Philippines, Bangladesh, Kenya and India)
and non-market (Viet Nam) contexts. Of course, the
phenomenon of induced institutional innovation is
most dramatically demonstrated at the grass-roots
levels in the context of both market (Indonesia and
the Philippines) and non-market (Viet Nam) economies.
The demand for new agricultural technologies cannot
be sustained in countries characterised by
inequalities in the agrarian structure (particularly

land ownership distribution, as in India). Decentralisation of research not only ensures a continuous generation and utilisation of agricultural technologies appropriate to the needs of the highly localised soil and climatic conditions and diverse socioeconomic environment, but also permits the complete integration (as in Viet Nam) of participatory research with production.

The commodity and public sector biases of agricultural research and extension are clear manifestations of the socioeconomic and political power of the interested social groups based on their regional, sectoral and commodity constituencies.[3] The current neglect by the national (and international) programmes of research and extension of agricultural commodities, particularly subsistence food crops produced and consumed by the rural poor, has done little to alleviate mass poverty in rural areas or reduce the level of widespread malnourishment. Instead, a disproportionately large amount of research resources has been poured into commodities (particularly export) based on the socioeconomic and political muscle of these commodity constituencies.

In the light of the above, it is hardly surprising that the existence of urban bias has tended to depress the level of agricultural research activities in the Third World. Agricultural research intensity is increased only where the power of perception, articulation and organisation of the rural population is enhanced. Finally, a rational public sector response is observed in the character and composition of national extension programmes, e.g. lab-to-land extension in India and through institutional innovations like a network of nationwide input delivery systems in Bangladesh.

<u>NOTES</u>

[1] More recently, abject poverty and landlessness contributed to yet another institutional innovation, the Grameen Bank, which provides collateral-free loans for both individual and

collective production enterprises of this specific (large and growing) target group.

[2] These "Comilla" model co-operatives succeeded in boosting incomes and ouput, but were a strain on both national funds and administrative resources. Moreover, these co-operatives have not been fully successful in mobilising savings, recovering public loans and utilising installed irrigation capacity. More significantly, the benefits of the co-operative movement accrued mainly to the large farmers, which was contrary to the objective of protecting and supporting the resource-poor small-scale farmers (see Chapter 5).

[3] Preference for export crops is also related to pressures on the national trade balance and/or to the unfavourable world capital markets which have provided little room for adjusting to trade imbalance and basic needs via borrowing and official aid flows.

References

Abdullah, A.A., et al, 1976: "Agrarian structure and the IRDP - preliminary considerations", in <u>The Bangladesh Development Studies</u> (Dhaka, BIDS), vol. IV, no. 1).

ADC and IRRI, 1980: A summary of the recommendations made at the conference on the <u>Communication Responsibilities of the International Agricultural Research Centres</u>, Agricultural Development Council and International Rice Research Institute, Philippines.

Ad Hoc Committee on Maize and Pasture Research, 1980: <u>Report</u>, Ministry of Agriculture, ibid., Nairobi.

Agarwal, B., 1983: "Diffusion of rural innovations: some analytical issues and the case of wood burning stoves", in <u>World Development</u>, vol. 11 (4).

Agribusiness Council, 1975: <u>Agricultural initiative in the Third World: a report on the conference: Science and agribusiness in the seventies</u>, Lexington Books, D.C. Heath and Co., Lexington.

Agriculture Canada, 1981: <u>Agriculture abroad</u>, Ottawa, vol.XXXVI, no.3.

Agricultural Information Centre, 1979: <u>Fruit and vegetable technical handbook</u>, Ministry of Agriculture, Nairobi.

Agro-Economic Research Section (AER), 1979: <u>Costs and returns survey for Bangladesh, 1978-79 crops</u>, Ministry of Agriculture, Dhaka.

Ahmed, B.U. and Coward, E.W., Jr., 1977: "Village technology and bureaucracy: irrigation development in Bangladesh", in <u>Journal of the Bangladesh Academy for Rural Development</u>, vol. 7 (1), July.

Ahmed, I. 1980: "Technological change, agrarian structure and labour absorption in Bangladesh rice cultivation", in <u>Employment expansion in Asian agriculture</u>, ILO-ARTEP, Bangkok.

__________ 1981: <u>Technological change and agrarian structure - a study of Bangladesh</u>, ILO, Geneva.

Almond, G.A. and Coleman, J.S. (eds.), 1960: <u>The politics of the developing areas</u>, Princetown University Press, Princetown, New Jersey.

Anderson, K., 1979: "Public agricultural research investment in developing countries: a politico-economic theory", contributed paper to The Seventeenth International Conference of Agricultural Economists, Banff (Canada), 3-12 September.

Apter, D., 1965: <u>The politics of modernisation</u>, University of Chicago Press, Chicago.

Arndt, T., Dalrymple, D., Dana, G. and Ruttan, V., 1977: <u>Resource allocation and productivity in national and international agricultural research</u>, University of Minnesota Press, Minneapolis.

Arnon, I., 1975: <u>The planning and programming of agricultural research</u>, FAO, Rome.

Asaduzzaman, M., 1979: "Adoption of HYV rice in Bangladesh", in The Bangladesh Development Studies, Dhaka, vol. VIII, no. 3.

Auckland, A.K., 1980: "Jute breeding in Bangladesh: a review", Paper prepared for the Advisory Group Meeting on the Use of Induced Mutations for Improvement of Oil Seeds and other Industrial Crops, Vienna, IAEA, November 17-21.

Aurora, G.S., and Morehouse, W., 1974: "The dilemma of technological choice in India: the case of the small tractor", in Minerva XII (4), October.

Ayer, H., 1970: "The costs, returns and effects of agricultural research in Sào Paulo, Brazil", Ph.D dissertation, Purdue University, Lafayette.

Ayer, H. and G.E. Schuh, 1972: "Social rates of return and other aspects of agricultural research: The case of cotton research in Sào Paulo, Brazil", in American Journal of Agricultural Economics, No. 54.

Bal, H.K. and Kahlon, A.S., 1978: "An Econometric Model of Returns to Investment in Agricultural Research in India", ICAR Technical Bulletin.

Bangladesh Bureau of Statistics (BBS), 1981: Report on the agricultural census of Bangladesh 1977 (national volume), Dhaka.

Bangladesh Rice Research Institute (BRRI), 1977: Workshop on ten years of modern rice and wheat cultivation in Bangladesh, Dhaka.

Bardhan, Pranab K., 1978: "On Labour Absorption in South Asian Agriculture, with particular reference to India", in Labour Absorption in Indian Agriculture, International Labour Office, Bangkok.

Barker, R., 1979: "Establishing priorities for allocating funds to rice research", in Rural change: the challenge for agricultural economists. Proceedings of Seventeenth International Conference of Agricultural Economists (Banff), Gower Publishing Company Ltd., Aldershot.

Barletta, N.A., 1970: "Costs and social benefits of agricultural research in Mexico", Ph.D dissertation, University of Chicago.

Basu, A.R., 1979: "SIRTDO: the experiences of a technical institute in appropriate technology", in Appropriate technology, vol. 6 (2).

Bauer, L.L. and C.R. Hancock, 1975: "The productivity of agricultural research and extension expenditures in the Southeast", in Southern Journal of Agricultural Economics, vol. 7.

Beckford, G., 1972: Strategies for agricultural development: comment, Food Research Institute Studies, Stanford University.

Bell, C., 1972: "The acquisition of agricultural technology: its determinants and effects", in Journal of Development Studies, vol. 9, no. 1, October.

Bell, M., 1979: "The exploitation of indigenous knowledge, or the indigenous exploitation of knowledge: whose use of what for what?", in IDS Bulletin, vol. 10 (2), Institute for Development Studies, University of Sussex.

Belshaw, D., 1979: "Taking indigenous technology seriously: the case of inter-cropping techniques in East Africa", in IDS Bulletin, vol. 10 (2), Institute for Development Studies, University of Sussex.

Benor, D. and Harrison, J.Q., 1977: Agricultural extension: the training and visit system, World Bank, Washington, D.C., May.

Bergmann, T., 1975: Farm policies in socialist countries, translated by Lux Furtmüller, Lexington Books, Massachusetts.

Bernstein, H., 1979: "African peasantries: A theoretical framework", in _Journal of peasant studies_, vol. 6 (4).

Beveridge, W.I.B., 1957: _The art of scientific investigation_, Vintage Books, New York.

Bhaduri, A., 1973: "A study in agricultural backwardness under semi-feudalism", in _Economic Journal_, vol. 83, no. 329, March.

Biggs, A.G., 1978: _Horticultural education and development in Kenya_, Department of Crop Science, University of Nairobi, Kabete, mimeo.

Biggs, S.D., 1981: "Institutions and decision making in agricultural research," in F. Stewart and J. James (eds.): _The economics of new technology in developing countries_, Francis Pinter, London.

__________, 1982: "Agricultural research: A review of social sciences analysis", (Ottawa, Canada), Report to the International Development Research Centre (IDRC), _Discussion Paper No. 115_, School of Development Studies, University of East Anglia, Norwich.

__________ 1982a: "Generating agricultural technology: triticale for the Himalayan Hills", in _Food Policy_, vol. 7, February.

__________ 1983: "Monitoring and control in agricultural research systems: maize in northern India", in _Research Policy_, vol. 12.

__________ 1983: "A farming systems approach: some unanswered questions", paper for a session of the Second International Course for Development orientated Research in Agriculture, Wageningen, Netherlands.

Biggs, S.D. and E.J. Clay, 1981: "Sources of innovation in agricultural technology", in _World Development_, vol.9, no. 4.

Biggs, S.D. and Grosvenor-Alsop, R., 1983: "Appropriate agricultural technology: a review of governmental and non-governmental programmes", in <u>Manual report</u>, no. 15, School of Development Studies, University of East Anglia, Norwich.

Binswanger, H.P., 1978: "Induced technical change: evolution of thought", in H.P. Binswanger and V.W. Ruttan (eds.): <u>Induced innovation: Technology, Institutions and Development</u>, John Hopkins University Press, Baltimore.

_______________ 1978a: "The microeconomics of induced technical change", in H.P. Binswanger and V.W. Ruttan: <u>Induced innovation: Technology, Institutions and Development</u>, John Hopkins University Press, Baltimore.

_______________ 1978b: <u>The Economics of Tractors in South Asia</u>, Agricultural Development Council, New York, and International Crop Research Institute for the Semi-Arid Tropics, Hyderabad.

Binswanger, H.P. and Ruttan, V.W., 1978: <u>Induced innovation: technology institutions and development</u>, John Hopkins Press, Baltimore.

Blaikie, P.M., 1981: "Class, land-use and soil erosion", in <u>ODI Review</u>, no. 2.

_______________, 1983: <u>The political economy of soil erosion</u>, Longmans, London.

_______________, 1983a: "The political economy of soil erosion" in T.O'Riordan and R.K. Turner (eds): <u>Progress in resource management and environmental planning</u>, John Wiley and Sons Ltd, Chichester, vol. 4.

Blase, M.G., 1973: <u>Institution building: a source book</u>, Michigan State University, East Lansing.

Blase, M.G. and A. Paulson, 1977: "The agricultural experiment station: An institutional development perspective", in <u>Agricultural Science Review</u>.

Blaug, M., 1980: The methodology of economics, Cambridge University Press, Cambridge.

Bo Nong Nghiep: Bao Cao Tong Ket Khao Hoc Ky Thuat Nong Nghiep Ba Nam 1956- 1960/Report on the Scientific and Technical Achievements in Agriculture between 1956-1960, unpublished.

_______________, 1978: Giao Trinh Cay Lua/Teaching material on rice, Hanoi.

_______________, 1980: Tuyen Tap Cac Cong Trinh Nghien Cuu Khoa Hoc Va Ky Thuat Trong Nong Ngheip/Selected research on agricultural sciences and technologies, vol.1: cultivation; vol.2: animal husbandry, Nong Nghiep, Hanoi.

Boserup, E., 1965: The conditions of agricultural growth, Aldine, Chicago.

_______________, 1965a: Conditions of agricultural growth, the economics of agrarian change under population pressure, George Allen & Unwin Ltd, London.

_______________, 1981: Population and technology, Basil Blackwell, Oxford.

Bourdieu, P., 1977: Outline of a theory of practice, Cambridge University Press.

Boyce, J.K. and Evenson, R.E., 1975: National and international agricultural research and extension programmes, Agricultural Development Council, New York.

Brammer, H., 1980: "Some innovations do not wait for experts: a report on applied research by Bangladesh peasants", in CERES, vol. 13 (2).

Brass, P.R., 1982: "Institutional transfer of technology: the land-grant model and the agricultural university at Pantnagar", in R.S. Anderson et al. (eds): Science, politics and the agricultural revolution in Asia, Westview Press, Boulder, Colorado.

Bredahl, N. and W. Peterson, 1976: "The productivity and allocation of research: U.S. agricultural experiment stations", in <u>American Journal of Agricultural Economics</u>, vol. 58.

Bredahl, M.E., Bryant, W.K. and Ruttan, V.W., 1980: "Behaviour and productivity implications of institutional and project funding of research" in <u>American Journal of Agricultural Economics</u>, vol. 62 (3), August.

Brett, Crowther, M.R., 1983: "Research and duty: a review of agricultural research policy by Vernon W. Ruttan", in <u>Food policy</u>, vol. 8 (2), May.

Brokensha, D., Warren, D.M. and Werner, O. (eds.), 1980: <u>Indigenous knowledge systems and development</u>, University Press of America, Lanham.

Bruce, J., 1980: "Implementing the user perspective", in <u>Studies in family planning</u>, vol. II, no. 1, January.

Bui Huy Dap, 1960: <u>15 Nam Dau Tranh Cai Tien Ky Thuat O Vietnam</u>/15 years struggle for technical improvement in Vietnam, Su That, Hanoi.

__________, 1964: <u>Cay Lua Mien Bac Viet Nam</u>/Rice in North Viet Nam, Nong Thon, Hanoi.

__________, 1978: <u>Lua Viet Nam Trong Vung Lua Nam va Dong Nam Chau A</u>/The Viet Namese rice in the rice cultivation areas in south and south east Asia, Nong Nghiep, Hanoi.

__________, 1982: "La recherche agronomique", <u>Etudes Vietnamiennes</u>, no.27, 1970; and Sigurdon, J.: <u>Vietnam's science and technology: a tentative description of structure and planning</u>, University of Lund, Institute of Policy Studies (mimeo).

Bunting, A.H., 1979: "Science and technology for human needs, rural development and the relief of poverty", <u>Occasional paper</u>, International

Agricultural Development Service, New York.

Bunting, A.H. and Curtis, D.L., 1968: "Local adaptation of sorghum varieties in Northern Nigeria", Proceedings of the reading symposium on agroclimatological methods, UNESCO, Paris.

Burbach, R. and Flynn, P., 1980: Agribusiness in the Americas, Monthly Review Press, New York.

Burch, D., 1980: "Overseas aid and the transfer of technology: a study of agricultural mechanisation in Sri Lanka", in Development research digest, no. 3, Spring.

Busch, L. (ed.), 1981: Science and agricultural development, Allanheld Osmun and Co., New Jersey.

Busch, L. and Lacy, W.B., 1981: "Sources of influence on problem choice in the agricultural sciences: the new Atlantis revisited" in L. Busch (ed.): Science and agricultural development, Allanheld Osmun and Co., New Jersey.

Busch, L. and Sachs, C., 1981: "The agricultural sciences and modern world system" in Busch, L. (ed.): Science and agricultural development, Allanheld Osmun and Co., New Jersey.

Busch, L., Lacy, W.B. and Sachs, C., 1980: "Research policy and process in the agricultural sciences: some results from a national study", in Report: RS-66, Department of Sociology, Kentucky Agricultural Experiment Station, University of Kentucky, Lexington.

Byerlee, D. and Collinson, M., et al., 1980: Planning technologies appropriate to farmers: concepts and procedures, CIMMYT, Mexico.

Byerlee, D. and Hesse de Polanco, E., 1986: "Adoption of technological packages: evidence from the Mexican Altiplano", in American Journal of Agricultural Economics, vol.68, no.3, August.

Byres, T.J., 1972: "The dialectic of India's Green Revolution" in <u>South Asian Review</u>, vol. 5.

Cancian, F.A., 1979: <u>The innovator's situation: upper-middle-class conservation in agricultural communities</u>, Stanford University Press, California.

Central Bureau of Statistics, 1979: <u>Statistical abstract</u>, Ministry of Economic Planning and Community Affairs, Government printer, Nairobi.

Chalian, G., 1968: <u>Les Paysans du Nord Vietnam et la Guerre</u>, Maspéro, Paris.

Chambers, R., 1980: "Understanding professionals: small farmers and scientists", <u>IADS Occasional Paper</u>, International Agricultural Development Service, New York.

Chambers, R. and Howes, M. (eds.), 1979: "Rural development: whose knowledge counts?", in <u>IDS Bulletin</u>, vol. 10 (2), Institute of Development Studies, University of Sussex.

Chaparro, F. et al., 1981: "Research priorities and resource allocation in agriculture: the case of Columbia", in D. Daniels and B. Nestel (eds.): <u>Resource allocation to agricultural research</u>, International Development Research Centre, Ottawa.

le Chau, 1966: <u>Le Vietnam socialiste: une économie de transition</u>, Maspéro, Paris.

Cheng, L., 1984: "International trade and technology: A brief survey of the recent literature", in <u>Weltwirtschaffliches Archiv</u>, vol. 120.

CIMMYT, 1979: CIMMYT Eastern Africa Economics Programme 1979. A Report on the Zoning of Central Province, Zambia, into Recommendation Domains to Facilitate the Planning of Appropriate Research and Development Programmes, <u>Report No. 4</u>, CIMMYT, Nairobi.

Clark, N., 1980a: "Organisational aspects of Nigeria's research system", in <u>Research policy</u>, vol. 9.

__________, 1980b: "The economic behaviour of research institutions in developing countries – some methodological points", in <u>Social Studies of Science</u>, vol. 10, (1) February.

Clay, E.J., 1982: "Technical innovation and public policy: agricultural development in the Kosi region, Bihar India", in <u>Agricultural Administration</u>, vol. 9.

__________, 1983: "Food aid: an analytic and policy review", Institute of Development Studies, University of Sussex, Brighton.

Cleaver, H.M., 1972: "The contradictions of the Green Revolution", in <u>Monthly Review</u>, vol. 24 (2), June.

Cline, P.L., 1975: "Sources of productivity change in United States agriculture", Ph.D dissertation, Oklahoma State University.

Collinson, M., 1981: "Understanding small farmers", in <u>Agricultural Administration</u>, vol. 8 (6) November.

__________, 1982: "Farming systems research in Eastern Africa: the experience of CIMMYT and some national agricultural research services, 1976–1981", <u>MSU International Development Paper No. 3</u>, Department of Agricultural Economics, Michigan State University, East Lansing.

__________, 1980: <u>A review of the build-up of a farm economics capacity in the agricultural and livestock research services of the Government of Kenya</u>, CIMMYT Eastern African Economics Programme, Nairobi, mimeo.

Communist Party of Viet Nam, 1977: Resolution of the Fourth Party Congress, in Vietnamese (Hanoi), Su That.

Coursey, D.G., 1982: "Traditional tropical root crop technology: some interactions with modern science", in <u>IDS Bulletin</u>, vol. 13 (31).

Crawford, E. and Thorbecke, E., 1979: "Number of households, population, Kenya, 1976", in D. Ghai, M. Godfrey and F. Lisk: <u>Planning for basic needs in Kenya</u>, ILO, Geneva.

Cummings, R.W. Jnr., 1981: "Strengthening linkages between agricultural research and farmers: An overview", Draft for discussion only. Prepared for forthcoming DAC workshop on linkages between agricultural research and farmers, March 30, mimeo.

Dahlberg, K.A., 1979: <u>Beyond the Green Revolution: the ecology and politics of global agricultural development</u>, Plenum Press, New York.

Dalrymple, D.G., 1979: "The adoption of high-yielding grain varieties in developing nations", in <u>Agricultural History</u>, vol. 53, no. 4, October.

__________, 1980: "Development and spread of semi-dwarf varieties of wheat and rice in the United States: An international perspective", in <u>Agricultural Economic Report</u>, no. 455, Office of International Cooperation and Development, USDA with USAID, Washington, D.C.

Daniels, D. and Nestel, B., 1981: <u>Resource Allocation to Agricultural Research</u>, Report no. IDRC 182e, International Development Research Centre, Ottawa.

Darling, H.S., 1976: "Memorial to cotton research", review of M.H. Arnold (ed.) "Agricultural research for development: The Namulonge contribution", in <u>Nature</u>, vol. 264, November 11.

Dao The Tuan, 1973: "Les nouvelles variétés de riz", in <u>Etudes Vietnamiennes</u>, no.38.

Dasgupta, B., 1977: "India's green revolution", in <u>Economic and Political Weekly</u>, vol. XII, nos. 6, 7 and 8, February.

Dawson, O.L., 1979: <u>Communist China's Agriculture: its development and future potential</u>, Praeger, New York.

Dinh Thu Cuc, 1979: "Qua Trinh To Chuc Lai San Suat Theo Huong San Suat Lon Xa Hoi Chu Nghia O Mot So Hop Tac Xa Nong Nghiep Tien Tien Vung Trong Lua", in Vien Su Hoc, <u>Nong Dan Vietnam Tien Len Chu Nghia Xa Hoi</u>/The Vietnamese peasantry advanced towards Socialism (Hanoi), Khoa Hoc Xa Hoi.

Dinham, B. and Hines, C., 1982: <u>Agribusiness in Africa</u>, Earth Resources Research Ltd., London.

Doan Trong Truyen and pham Thanh Vinh, 1964: <u>Building an independent national economy in Vietnam</u>, FLPH, Hanoi.

Dorling, M.J., 1979: <u>Income distribution in the small farm sector of Kenya</u>, Department of Agricultural Economics Report, University of Nairobi, Nairobi, mimeo.

_____________, 1982: "National research systems and the generation and diffusion of innovations: the horticultural industry in Kenya", ILO mimeo. World Employment Programme research working paper, Geneva, restricted.

Dorling, M.J. and Mutlu, S., 1985: "Commercialisation of new and indigenous technology in Jordanian agriculture", ILO World Employment Programme research working paper WEP 2-22/WP. 150, Geneva, mimeo.

Drilon, Jr., 1977: <u>Agricultural research systems in Asia</u>, SEARCA, Languna, Philippines.

le Duan, 1976: <u>Cach Mang XHCN o Viet Nam</u>/The Socialist Revolution in Viet Nam, Su That, Hanoi, vol.1 and 2.

_____________, 1979: <u>The Vietnamese Revolution: fundamental problems, essential tasks</u>, Foreign Languages Publishing House, Hanoi.

le Duan and Pham Van Dong, 1975: <u>Vers une agriculture de la grande production socialiste</u>, Foreign Languages Publishing House, Hanoi.

Dumont, R., 1971: "Problèmes agricoles en République Démocratique du Vietnam", in Chesneaux, J. et al. (ed.): <u>Tradition et révolution au Vietnam</u>, Anthropos, Paris.

Duncan, R.C., 1972: "Evaluating returns to research in pasture improvement", in <u>Australian Journal of Agricultural Economics</u>, vol. 16.

Duong Binh, 1973: "L'hydraulique agricole", in <u>Etudes Vietnamiennes</u>, no.38.

Durost D.D., and G.T. Barton, 1960: <u>Changing Sources of Farm Output</u>, USDA Production Research Report No. 36, Washington D.C.

Eckholm, E.P., 1976: <u>Losing ground: environmental stress and world food prospects</u>, W.W. Norton and Co. Inc., New York.

Eckhaus, R.S., 1955: "The factor proportions in undeveloped areas", in <u>American Economic Review</u>, Vol. XLV.

Eisenstadt, S.N., 1966: <u>Modernisation: protest and change</u>, Prentice-Hall.

Eklund, Per, 1983: "Technology development and adoption rates: systems approach for agricultural research and extension", in <u>Food Policy</u>, vol. 8 (2), May.

Elz Dieter (ed.), 1984: <u>The planning and management of agricultural research</u>, The World Bank, Washington D.C.

Esman, M.J., 1978: "Research and development organisation", Centre for International Studies, Cornell University, Ithaca, New York, mimeo.

Evenson, R., 1968: "The contribution of agricultural research and extension to agricultural production", (Chicago), Ph.D dissertation, University of Chicago.

__________, 1969: "International transmission of technology in sugarcane production", Yale University, New Haven, mimeo.

__________, 1974: "The Green Revolution in recent development experience", in _American Journal of Agricultural Economics_, vol. 56 (2), May.

__________ 1975: "Technology generation in agriculture", in L.G. Reynolds (ed.): _Agriculture in development theory_, Yale University Press, Chapter 8, New Haven.

__________ 1978: "The organisation of research to improve crops and animals in low income countries" in T.W. Schultz (ed.): _Distortions of agricultural incentives_, Indiana University Press, Bloomington.

__________ 1981: "Research evaluation: policy interests and the state of the art", in _Evaluation of agricultural research_, University of Minnesota, Agricultural Experiment Station.

Evenson, R. et al., 1978: "Risk and uncertainty as factors in crop improvement research", _IRRI Research Paper Series No. 15_, March.

Evenson, R. and D. Jha, 1973: "The Contribution of Agricultural Research Systems to Agricultural Production in India", in _Indian Journal of Agricultural Economics_, vol. 28.

Evenson, R.E. and Kislev, Y., 1975: _Agricultural research and productivity_, Yale University Press, New Haven.

Evenson, R.E., Waggoner, P.E. and Ruttan, V.W., 1979: "Economic benefits from research: an example from agriculture", in _Science_, no. 205, September.

Falcon, W.P. and C.H. Gotsch, 1966: <u>Agricultural Development in Pakistan: lessons from the Second Plan Period</u>, Economic Development Series no. 6, Harvard University.

Farmer, B.H., 1979: "The Green Revolution in South Asian ricefields: Environment and production", in <u>Journal of Development Studies</u>, vol. 15 (4), July.

Feder, E., 1976: "McNamara's little green revolution: World Bank scheme for self-liquidation of Third World peasantry", in <u>Economic and Political Weekly</u>, vol. X (15).

_____________, 1977: "Agribusiness and the elimination of Latin America's rural proletariat", in <u>World Development</u>, vol. 5, nos. 5-7.

_____________, 1979: "Strawberry imperialism: an enquiry into the mechanisms of dependency in Mexican agriculture", <u>Research Report</u>, Institute of Social Studies, The Hague.

Feder, G. et al, 1981: "Adoption of agricultural innovations in developing countries: a survey", <u>World Bank Staff Working Paper no. 444</u>, February, Washington, D.C.

Feder, G., Just, R.E. and Zilberman, D., 1985: "Adoption of agricultural innovations in developing countries: a survey", in <u>Economic Development and Cultural Change</u>, vol. 33.

Fishel, W.L. (ed.), 1971: <u>Resource allocation in agricultural research</u>, University of Minnesota Press, Minneapolis.

Fitzgerald, D., 1981: "Exporting the land-grant model: the Rockefeller Foundation in Mexico, 1943-53", University of Pennysylvania, Philadelphia, draft.

Fleury, J.M., 1980: "Seeds: patent pending", <u>IDRC Report</u>, vol. IX, no. 1, April.

Fortmann, L., 1981: "Indigenous knowledge systems in development", review of Brokensha et al. (eds.): <u>Rural development participation review</u>, vol. II, no. 3.

Frank, A.G., 1981: "Third World agriculture and agribusiness", in <u>Crisis in the Third World</u>, Heinemann, London.

Franke, R.W. and B. Chasin, 1980: <u>Seeds of famine: ecological destruction and the development dilemma in the West African Sahel</u>, Allanheld Osmun, Montclair, N.J.

Gathee, J.W., 1980: <u>Farm system economics: fitting research to farmers' conditions</u>, National Horticultural Research Station, Ministry of Agriculture (Thika), mimeo.

G.B. Pant University, 1981: <u>Maize on-farm research project 1980: report</u>, G.B. Pant University of Agriculture and Technology, Uttar Pradesh, March.

George, S., 1976: <u>How the other half dies: the real reasons for world hunger</u>, Penguin, London.

Gerhart, J., 1975: "The diffusion of hybrid maize in Western Kenya", CIMMYT, Mexico.

Gibbon, D., 1981a: "Rainfed farming systems in the mediterranean region", in <u>Plant and Soil</u>, vol. 58 (1-3).

Gilbert, E.H., Norman, D.W. and Winch, F.E., 1980: "Farming systems research: a critical appraisal", <u>MSU Rural Development Paper</u>, no. 6, Department of Agricultural Economics, Michigan State University.

General Office of Statistics, 1978: <u>Thirty years economic and cultural development of the Democratic Republic of Vietnam</u>, in Vietnamese, Su That, Hanoi.

Gomory, R.E., 1983: "Technology development", in <u>Science</u>, vol. 220, May.

Gordon, A., 1974: "The Green Revolution in North Vietnam", in <u>Journal of contemporary Asia</u>, vol.4, no.1.

__________, 1978: "The role of class struggle in North Vietnam", in <u>Monthly Review</u>, vol.29, no.8, January.

Gough, K., 1978: "The Green Revolution in South India and North Vietnam", in <u>Bulletin of concerned Asian scholars</u>, vol.X, no.1.

Grabowski, R., 1979: "Implications of an induced innovation model", in <u>Economic Development and Cultural Change</u>, vol. 27.

Griffin, K.B., 1974: <u>The political economy of agrarian change</u>, Macmillan, London.

Griliches, Z., 1959: "Research costs and social returns: hybrid corn and related innovations", in <u>Journal of Political Economy</u>, vol. 66.

__________, 1964: "Research expenditures, education and the aggregate agricultural production function", in <u>American Economic Review</u>, vol. 54.

Guttman, J.M., 1978: "Interest groups and the demand for agricultural research", <u>Journal of Political Economy</u>, vol. 86 (3).

Hardin, L.S., 1981: "Emerging roles of agricultural economists working in international research institutions such as IRRI and CIMMYT", in G. Johnson and A. Maunder (eds.): <u>Rural changes – the challenge for agricultural economists</u>, Gower, Aldershot.

Hargrove, T.R., 1977: <u>Genetic and sociological aspects of rice breeding in India</u>, International Rice Research Institute, Manila, no. 10, September.

__________, 1978: <u>Diffusion and adoption of genetic materials among rice breeding programmes in Asia</u>, International Rice Research Institute, Manila, no. 18, June.

Hargrove, T.R., Coffman, W.R. and Cabanilla, V.L., 1979: <u>Genetic interrelations of improved rice varieties in Asia</u>, International Rice Research Institute, Manila, no. 23, January.

Hariss, B., 1972: "Innovation adoption in Indian agriculture - the high- yielding variety programme", in <u>Modern Asian studies</u>, vol. 6, no. 1, November.

__________, 1982: "Stated and implicit priorities in agricultural research", paper prepared for the FAO/UNICEF/NPU Workshop on Nutrition in Agriculture, Hissar, Haryana, India, April.

Hart, H.C., 1961: <u>Campus India: an appraisal of American college programmes in India</u>, Michigan State University Press, East Lansing.

Harwood, R.R., 1979: <u>Small farm development, understanding and improving farming systems in the humid tropics</u>, Westview Press, Boulder, Colorado.

Hayami, Y., 1978: <u>Anatomy of a peasant economy</u>, International Rice Research Institute, Los Banos.

__________, 1982: "Growth and equity: is there a trade off?", paper prepared for the 18th International Conference of Agricultural Economists, Jakarta, Indonesia, August 24th to September 2nd.

Hayami, Y. and M. Akino, 1977: "Organisation and productivity of agricultural research systems in Japan", in T.M. Arndt, D.G. Dalrymple and V.W. Ruttan: <u>Resource allocation and productivity</u>, University of Minnesota Press, Minneapolis.

Hayami, Y. and V.W. Ruttan, 1971: <u>Agricultural development: an international perspective</u>, John Hopkins Press, Baltimore.

Hertford, R., J. Ardila, A. Rocha and C. Trujillo, 1977: "Productivity of agricultural research in Colombia", in T.M. Arndt, D.G. Dalrymple and V.W. Ruttan (eds.): <u>Resource allocation and productivity,</u>

University of Minnesota Press, Minneapolis.

Hewitt de Alcantara, C., 1976: <u>Modernising Mexican agriculture: socio-economic implications of technological change 1940-1970</u>, UNRISD, Geneva, report no.76.5.

Hicks, J.R., 1932: <u>The theory of wages</u>, Macmillan, London.

Hightower, J., 1976: "Hard tomatoes, hard times: the failure of the Land Grant College complex", in Merrill, R. (ed.) <u>Radical agriculture</u>, Harper and Row.

Hines, J., 1972: "The utilisation of research for development: Two case studies in rural modernisation and agriculture in Peru", Ph.D. dissertation, Princeton University, N.J.

Ho Chi Minh, 1977: <u>Ve Lien Minh Cong Nong</u>/On worker-peasant alliance, Su That, Hanoi.

Hoang Dinh Cao and Le Hung, 1979: <u>To Chuc Cung Ung Vat Tu Ky Thuat O Hop Tac Xa San Xuat Nong Nghiep</u>/The organisation of the supply of technical inputs in agricultural production co-operatives, Nong Nghiep, Hanoi.

Holdebrand, P.E., 1979: "Summary of the Sondeo methodology used by ICTA", paper given at the Rapid Rural Appraisal Conference, 4-7 December, Institute of Development Studies, University of Sussex, Brighton.

Holterman, S., 1979: <u>Intermediate technology in Ghana</u>, Intermediate Technology Publications Ltd., London.

Hopper, D.W., 1978: "Distortions of agricultural development resulting from government prohibitions", in T. Schultz (ed.): <u>Distortions of agricultural incentives</u>, Indiana University Press, Bloomington and London.

Horticultural Crops Development Authority, 1979: *Exports of fresh fruit, vegetables and cut flowers by air from Kenya to all countries*, Nairobi, mimeo.

Horticultural Crops Development Authority, 1974: See Ministry of Agriculture, HCDA, FAO.

Horton, D., 1986: "Assessing the impact of international agricultural research and development programs", in *World Development*, vol.14, no.4, April.

Hossain, M., 1980: "Foodgrain production in Bangladesh: performance, potential and constraints", in *The Bangladesh Development Studies*, vol. VIII, nos. 1 & 2 (special issue).

__________, 1981: *Land tenure and agricultural development in Bangladesh*, Institute of Developing Economies, Tokyo, V.R.F. Series no. 85.

Howell, J., 1982: "Managing agricultural extension: the T and V system in practice", *Discussion Paper 8*, Overseas Development Institute, London.

Huffman, W.E. and Miranowski, T.A., 1981: "The economic analysis of expenditures on agricultural experiment station research", in *American Journal of Agricultural Economics*, vol. 63 (1), February.

Hunter, G., 1969: *Modernising peasant societies*, Oxford University Press, London.

Hyami, Y. and Ruttan, V.W., 1971: *Agricultural development: an international perspective*, Johns Hopkins University Press, Baltimore.

ICAR, 1978: *Report of the Review Committee on Agricultural Universities*; Indian Council of Agricultural Research, New Delhi.

Idachaba, F.S., 1980: "Agricultural research policy in Nigeria", *IFPRI Research Paper no. 17*, Washington DC, August.

India, Government of, Ministry of Agriculture and Irrigation, 1976: <u>Report of the National Commission on Agriculture, Part XI</u>.

Indian Council of Agricultural Research, 1979: <u>50 years of agricultural research and education in India</u>, New Delhi.

International Rice Research Institute (IRRI), 1972: "IR 8 and Beyond" <u>IRRI Reporter</u>, no. 2, IRRI, Los Banos.

__________, 1975: <u>Changes in rice farming in selected areas of Asia</u>, IRRI, Los Banos.

__________, 1977: <u>Constraints to high yields on Asian rice farms: an interim report</u>, IRRI, Los Banos.

__________, 1978a: <u>Economic consequences of the new rice technology</u>, IRRI, Los Banos.

__________, 1978b: <u>Interpretive analysis of selected papers from changes in rice farming in selected areas of Asia</u>, IRRI, Los Banos.

__________, 1979: <u>Farm-level constraints to high yields in Asia: 1974-77</u>, IRRI, Los Banos.

International Service for National Agricultural Research and International Agricultural Development Service (ISNAR), 1982: "The role of international associations in strengthening national agricultural research", ISNAR, The Hague.

__________, 1982: <u>Annual Report</u>, ISNAR, The Hague.

Jamieson, B, 1978: <u>Resource allocation to agricultural research in Kenya from 1963 to 1978</u>, Institute of Development Studies, Working Paper no. 345, University of Nairobi.

de Janvry, A., 1977: "The Organisation and Productivity of National Research Systems" in Thomas M. Arndt, Dana G. Dalrymple and Vernon W. Ruttan (eds): <u>Resource Allocation and Productivity in National and International Agricultural Research</u>, University of Minnesota Press, Minneapolis.

_____________, 1977a: "Inducement of technical and institutional innovations: an interpretive framework", in T.M. Arndt et al. (eds.): <u>Resource allocation and productivity in national and international agricultural research</u>, University of Minnesota Press, Minneapolis.

_____________, 1982: "Introduction of technological and institutional innovations: an interpretative framework", in T.M. Arndt et al. (ed.): <u>Resource allocation and productivity in national and international agricultural research</u>, University of Minnesota Press, Minneapolis.

de Janvry, A. and Dethier, J.-J., 1985: <u>Technological innovation in agriculture: the political economy of its rate and bias</u>, Consultative Group in International Agricultural Research, study paper no.1, World Bank, Washington, D.C.

Jéquier, N. (ed.), 1976: <u>Appropriate technology problems and promises</u>, OECD, Paris.

Jéquier, N. and Blanc, G., 1983: <u>The world of appropriate technology: a quantitative analysis</u>, OECD, Paris.

Johnston, B. et al., 1972: <u>Criteria for the design of agricultural development strategies</u>, Food Research Institute Studies, Stanford University.

Johnson, B. and Blake, R.O., 1980: <u>The environment and bilateral development aid</u>, International Institute for Environment and Development, Washington, D.C.

Jones, R.W., 1970: "The role of technology in the theory of international trade", in R. Vernon <u>The technology factor in international trade</u>, Colombia University Press for the National Bureau of Economic Research, New York.

Joy, J.L., 1973: "Food and nutrition planning", in _Journal of Agricultural Economics_, vol. XXIV (1), January.

Judd, M.A., Boyce, J.K. and Evenson, R.E., 1986: "Investing in agricultural supply: the determinants of agricultural research and extension investment", in _Economic Development and Cultural Change_, vol. 35, no. 1, October.

Kahlon, A.S., 1977: _Cost-benefit analysis of agricultural research projects_, Punjab Agricultural University.

Kahlon, A.S., H.K. Bal, P.N. Saxena, and D. Jha, 1977: "Returns to investment in research in India", in Thomas Arndt et al (eds.): _Resource allocation and productivity in national and international agricultural research_, University of Minnesota Press, Minneapolis.

Karnataka State Council for Science and Technology, 1976: _Annual Report 1975-76_, Indian Institute of Science, Bangalore.

Kaufman, C. and Cook, J., 1982: _Portrait of a poison_.

Ker, A.D.P., 1979: _Food or famine: an account of the crop science programme supported by the International Development Research Centre_, IDRC-143C, International Development Research Centre, Ottawa.

Khan, A.R., 1977: "Poverty and inequality in rural Bangladesh", in _Poverty and landlessness in rural Asia_, ILO, Geneva.

___________, 1979: "The Comilla model and the IRDP of Bangladesh: an experiment in co-operative capitalism", in _World Development_, vol. 7.

___________ et al, 1981: _Employment, income and the mobilisation of local resources_, ILO-ARTEP, Bangkok.

__________, 1981: "Technical efficiency and fertiliser use in Bangladesh agriculture", in W. Mahmud (ed): <u>Development issues in an agrarian economy - Bangladesh</u>, Dhaka University.

Klepper, R., 1980a: "Agricultural research, development planning and the rural poor in Zambia", Institute of Development Studies, Brighton.

__________, 1980b: "The determinants of agricultural research priorities in Zambia", Institute of Development Studies, Brighton.

Knutson, M. and L.G. Tweeten, 1979: "Toward an optimal rate of growth in agricultural production research and extension", in <u>American Journal of Agricultural Economics</u>, vol. 61.

Koppel, B. and Oara, E., 1987: "Induced innovation theory and Asia's Green Revolution: A case study of an ideology of neutrality", in <u>Development and Change</u>, vol. 18.

Kuhn, T.S., 1970: <u>The structure of scientific revolutions</u>, University of Chicago Press, Chicago, 2nd edition.

Ladd, G.W., 1979: "Artistic research tools for scientific minds", in <u>American Journal of Agricultural Economics</u>, vol. 61 (1), Februaruy.

Lapidus, I. and Ostrovityanov, K., 1929: <u>An outline of political economy</u>, Martin Lawrence, London.

Latimer, R., 1964: "Some economic aspects of agricultural research and extension in the U.S.", Ph.D. dissertation, Purdue University, Lafayette.

Leys, C., 1975: <u>Underdevelopment in Kenya</u>, Heinemann, London.

Lijoodi, J.L. and Ruthenberg, H., 1978: "Income distribution in Kenya's agriculture", in <u>Q.J. Int. Agric.</u>, no.17, Göttingen.

Limbourg, M.L., 1956: L'économie actuelle du Vietnam Démocratique, Edition en Langues Etrangères, Hanoi.

Linder, R.K. and Jarrett, F.G., 1978: "Supply shifts and the size of research benefits", in American Journal of Agricultural Economics, vol. 60 (1), August.

Lipscomb, J.F., 1972: "The research services in East Africa", in L. Winston Cone and J.F. Lepscomb: The history of Kenyan Agriculture, University Press of Af, Nairobi.

Lipton, M., 1978: "Inter-farm, inter-regional and farm-non farm income distribution: the impact of the new cereal varieties", in World Development, vol. 6 (3).

Lipton, M., 1968: "The theory of optimising peasant", in Journal of development studies, vol. 4, no. 3.

Lipton, M. and Longhurst, R. 1985: Modern varieties, international agricultural reasearch, and the poor, World Bank, Washington DC.

Luong dinh Cua, 1980: "De Dat 5 Tan Thoc Mot Hec Ta Mot Nam Tren Dien Tich Rong", in Bo Nong Nghiep, Tuyen Tap Cac Cong Trinh Nghien Cuu Khoa Hoc Va Ky Thuat Trong Nong Ngiep/Selected research on agricultural sciences and technologies, Nong Nghiep, Hanoi.

Luu, N.N., undated: "The technological development of agriculture in the People's Republic of China", Research Report Series, no. 5, Institute of Social Studies, The Hague.

Luu, N.N., 1982: Institutional factors and technological innovation in North Vietnamese agriculture, ILO, Geneva, mimeo. World Employment Programme research working paper (restricted).

Mahmud, W., 1980: "Development strategy and the problem of food supply in Bangladesh", in The

<u>Bangladesh development studies</u>, vol. VIII, nos 1 & 2 (special issue).

__________________, 1983: <u>Poverty, landlessness and changing agrarian structure in rural Bangladesh</u>, Bureau of Economic Research, Dhaka University, mimeo.

Mahmud. W. and S.R. Osmani, 1980: <u>The macro-model for the Second Five Year Plan — some preliminary projections</u>, Bangladesh Planning Commission, Dhaka.

<u>Selected readings from the works of Mao tse Tung</u>, 1971, Foreign Languages Press, Peking.

Marsden, J.S., G.E. Martin, D.J. Parham, T.J. Ridsdill and B.G. Johnston, 1980: <u>Returns on Australian agricultural research: The joint industries assistance commission — CSIRO benefit-cost study of the CSIRO division of entomology</u>, Commonwealth Scientific and Industrial Research Organisation, Canberra.

Maxwell, S., 1980: "Differentiation in the colonies of Santa Cruz: causes and effects", Working Paper no. 13, Centro de Investigacion Agricola Tropical (CIAT), Santa Cruz, Bolivia.

__________________, 1982: "Harvest and post-harvest issues in farming systems research", in <u>Bulletin</u>, Institute of Development Studies, vol. 13 (3), Brighton.

Maxwell, S. and Pozo, M., 1981: "Farm systems in the colonisation crescent of Santa Cruz, Bolivia: results of a survey", volume 1, text. Working paper no. 22, Centro de Investigacion Agricola Tropical (CIAT), Santa Cruz.

McCalla, A.F., 1978: "Politics of the agricultural research establishment", in D.F. Hadwiger and W.P. Perowne, (eds.): <u>The new politics of food</u>, D.C. Heath and Co., Lexington.

McRobie, G., 1981: <u>Small is possible</u>, Jonathan Cape Ltd., London.

Meade, J., 1964: _Efficiency, equality and the ownership of property_, Allen & Unwin, London.

Melrose, D, 1982: _Bitter pills: medicine and the third world poor_, Oxfam, Oxford.

Metcalf, D., 1969: _The economics of agriculture_, Penguin Books Ltd., Harmondsworth.

Ministry of Agriculture, HCDA, FAO, 1974: _Horticultural development guidelines_, Nairobi.

Ministry of Economic Planning and Community Affairs, 1977: _Integrated rural survey 1974/75, basic report_, Republic of Kenya, Nairobi, Government printer.

__________, 1979: _Development Plan 1979–83_, Republic of Kenya, Nairobi, Government printer.

Ministry of Finance and Planning, 1974: _Development Plan 1974–78_, Republic of Kenya, Nairobi, Government printer.

Ministry of Lands and Settlement, 1978: _Human settlements in Kenya_, Republic of Kenya, Nairobi, Government printer.

Moise, E.E., 1977: _Land reform in China and North Vietnam: revolution at the village level_, Ph.D. dissertation, University of Michigan.

Mooney, P.R., 1975: _Seeds of the earth: a private or public resource?_, International Coalition for Development Action, London.

Morehouse, W, 1981: "Technology and equity in black holes – the refraction effect of technology on social change", draft editorial commentary for _Science and Public Policy_, 5 April.

Morehouse, W. and Sigurdson, J., 1977: "Science, technology and poverty", in _Bulletin of the Atomic Sciences_, December.

Moseman, A.H., 1970: <u>Building agricultural research systems in the developing nations</u>, Agricultural Development Council, New York.

Moseman, H., 1971: <u>National agricultural research systems in Asia</u>, report of the Regional Seminar New Delhi, March 8-13, Agricultural Development Council, New York.

Mosher, A.T., 1976: "Reflections on rural development in South and South East Asia", Chapter 1 of <u>Briefing about rural development</u>, Agricultural Development Council, New York.

___________, 1982: <u>Some critical requirements for productive agricultural research</u>, ISNAR, The Hague.

Muchiri, G., 1984: "Farm equipment innovations for smallholders in semi-arid Kenya", in I. Ahmed and B.H. Kinsey (eds.): <u>Farm equipment innovations in eastern and southern Africa</u>, Gower, Aldershot.

Mukhopadhyay, S.K., 1976: <u>Sources of variation in agricultural productivity</u>, Macmillan Co., New Delhi.

___________, 1980: "Constraints to technological progress in rice cultivation: a study of two regions in India", VRF series no. 74, Institute of Developing Economies, Tokyo.

___________, 1982: "Agricultural growth and potential in West Bengal", Planning Commission, Government of India, University of Kalyani, (mimeo).

Muller, J., 1980: "Liquidation or consolation of indigenous technology: a study of the changing conditions of production of village blacksmiths in Tanzania", <u>Development Research Series no. 1</u>, Aalborg University Press, Denmark.

Muqtada, M., 1975: "The seed fertiliser technology and surplus labour in Bangladesh agriculture", in <u>Bangladesh development studies</u>, vol. III, no. 4.

Murray, G.F., 1980: "Haitian peasant contour ridges: The evolution of indigenous erosion control technology", <u>Development discussion paper no. 86</u>, Harvard Institute for International Development, Cambridge, USA.

Myrdal, G., 1974: "The transfer of technology to underdeveloped countries", in <u>Scientific American</u>, vol. 231, no. 3, September.

Nagy, J.G. and W.H. Furtan, 1978: "Economic costs and returns from crop development research: The case of rapeseed breeding in Canada", in <u>Canadian Journal of Agricultural Economics</u>, vol. 26.

Nair, K., 1961: <u>Blossoms in the dust</u>, Duckworth.

National Academy of Science, 1972: <u>Genetic vulnerability of major crops</u>, Washington, D.C.

___________, 1975: <u>Plant studies in the People's Republic of China: trip report of the American Plant Studies Delegation</u>, Washington, D.C.

National Horticultural Research Station, 1979: <u>Grain legume project</u>, Ministry of Agriculture, Interim Report no.15, Thika, mimeo.

NATURE, 1980: "Pesticides in developing countries", editorial, <u>Nature</u>, Vol. 286, (5776), 28 August.

Neeley, D., McProud, W. and Yohe, J., 1979: "Diversity by breeding for genetic variability on the farmers field", Chapter 15 in J.A. Roumasset et al. (ed.): <u>Risk, uncertainty and agricultural development</u>, Agricultural Development Council, New York.

Nelson, R.R., 1974: "Less developed countries – technology transfer and adaptation: the role of the indigenous science community", in <u>Economic Development and Cultural Change</u>, vol. 23 (1), October.

Nettle, J.P., 1967: <u>Political mobilisation: a sociological analysis of methods and concepts</u>, Faber, London.

Nghien Cuu Dat Phan, 1979: (Hao Noi), Khoa Hoc Ky Thuat, vol.1-6.

Nguyen, N.L., "The technological development of agriculture in the People's Republic of China", Institute of Social Studies, The Hague: <u>Research Report series, no.5</u> (undated).

Nguyen Khac Vien, <u>et al</u>, 1967: "La politique de coopération agricole", en <u>Etudes Vietnamiennes</u>, no.13.

Nguyen Lai Vien, 1967: "La Bataille du Riz", in <u>Etudes Vietnamiennes</u>, no.13.

Nguyen Vy, 1980: <u>Nhung Di An Trong Dat Lua Nang Xuat Cao</u>/The secrets of lands which give high yields of rice, Nong Nghiep, Hanoi.

Nguyen Xuan Lai, 1967: "L'économie familiale des paysans coopérateurs", en <u>Etudes Vietnamiennes</u>, no.13.

Hordhaus, W.D., 1973: "Some sceptical thoughts on the theory of induced innovations", in <u>Quarterly Journal of Economics</u>, no.87.

Norman, D.W., 1980: "The farming systems approach: relevancy for the small farmer", <u>MSU Rural development paper no. 5</u>, Department of Agricultural Economics, Michigan State University, East Lansing.

Norris, R., 1982: <u>Pills, pesticides and profits: the international trade in toxic substance</u>, North River Press Inc., Croton-on-Hudson, New York.

Norton, G.W. and Davis, J.S., 1981: "Evaluating returns to agricultural research: a review", in <u>American Journal of Agricultural Economics</u>, vol. 63 (4), November.

Norton, G.W. et al. (eds), 1981: "Evaluation of agricultural research" <u>Miscellaneous Publication 8-1981</u>, Minnesota Agricultural Experiment Station, University of Minnesota.

Oram, P.A. and Bindlish, V., 1981: <u>Resource allocations to national agricultural research: trends in the 1970s</u>, A review of third world systems, ISNAR, The Hague.

Pain, A., 1981: Nutritional criteria in plant breeding: technical problems and constraints discussed in relationship to Sri Lanka's plant breeding programme, paper prepared for FAO/UNICEF/NPU Workshop on <u>Nutrition in Agriculture</u>, Hissar, Haryana, India, April.

Passmore, J., 1978: <u>Science and its criteria</u>, G. Duckworth and Co. Ltd., London.

___________, 1980: <u>Man's responsibility for nature</u>, 2nd edition, Duckworth, London.

Payne, P.R., 1976: "Nutrition planning and food policy", in <u>Food policy</u>, vol. 1 (2).

___________, 1976a: "Nutritional criteria for breeding a selection of crops: with special reference to protein quality", in <u>Plant foods for man</u>, vol. 2.

Patel, S.J. (ed.), 1983: "The pharmaceutical sector and health in the Third World", in <u>World Development,</u> vol. II, no. 3, March, special issue.

Pearse, A., 1980: <u>Seeds of plenty, seeds of want: social and economic implications of the green revolution</u>, Oxford University Press, Oxford.

Pee, T.Y., 1977: "Social returns from rubber research on peninsular Malaysia", Ph.D. dissertation, Michigan State University.

Perelman, M., 1976: "The Green Revolution: American agriculture in the Third World" Chapter 8 in Merrill, R. (ed.): <u>Radical agriculture</u>, Harper and Row.

Perrin, R.K., Winkelmann, E.R., Moscardi, E.R. and Anderson, J.R., 1979: "From agronomic data to farmer recommendations: an economic training manual", in Information bulletin, no. 27, CIMMYT.

Perrin, R.K. and Winkelmann, D., 1976: "Impediments to technical progress on small versus large farms", in American Journal of Agricultural Economics, vol. 58 (5), December.

Peterson, W.L., 1967: "Returns to poultry research in the United States", in Journal of Farm Economics, vol. 49.

Peterson, W.L. and J.C. Fitzharris, 1977: "The organisation and productivity of the Federal-State research system in the United States", in T.M. Arndt et al (eds.): Resource Allocation and Productivity, University of Minnesota Press, Minneapolis.

Pham Bai, 1979: Con Duong Vu Thang/The path of Vu Thang Co-operative, Su That, Hanoi.

Pham Cuong, 1970: "La formation des cadres techniques pour les coopératives agricoles", in Etudes Vietnamiennes, No. 27.

Pham Cuong and Nguyen Van Ba, 1976: Revolution in the village Nam Hong, 1945- 1975, FLPH, Hanoi.

Pineiro, M., Trigo, E. and Fiorentino, R., 1979: "Technical changes in Latin American agriculture - A conceptual framework for its interpretation", in Food Policy, vol. 4 (3), August.

Pinstrup-Anderson, P., 1982: Agricultural research and technology in economic development, Longman, London.

Pinstrup-Anderson, P. and Byrnes, F.C. (eds.), 1975: Methods for allocating resources in applied agricultural research in Latin America, papers at CIAT/ADC Workshop California, Colombia, November 26-29, Centro International de Agricultura Tropical (CIAT), Colombia.

Pinstrup-Andersen, P., Diaz, R.O., Infante, M. and de Londana, N.R., 1975: "A proposed model for improving the information base for research resources allocation", in Pinstrup-Anderson, P. and Byrnes, F.C. (eds.): <u>Methods for allocating resources in applied agricultural research in Latin America</u>, series CE-11, Centro International de Agricultura Tropical (CIAT), Colombia.

Pinstrup-Andersen, P., de Landona, N.R. and Hoover, E., 1976: "The impact of increasing food supply on human nutrition: implications for commodity priorities in agricultural research and policy", in <u>American Journal of Agricultural Economics</u>, vol. 58, May.

Poats, S.V. and Castillo, G.T., 1982: "Beyond the farmer: potato consumption in the Tropics", paper presented to CIP Decennial Anniversary, February 22-26, Lima.

Popper, K., 1959: <u>The logic of scientific discovery</u>, Harper Torchbooks, New York, reprinted 1965.

Porteous, M., 1981: "Schemes for peasants? Processes and problems in Third World agricultural research". Typed essay, History and Social Studies of Science, University of Sussex, Brighton.

Pray, C.E., 1978: "The economics of agricultural research in British Punjab and Pakistani Punjab, 1905-1975", Ph.D. dissertation, University of Pennsylvania.

__________, 1979: "The economics of agricultural research in Bangladesh", in <u>Bangladesh Journal of Agricultural Economics</u>, vol. II (2), December.

__________, 1983: "Private agricultural research in Asia", in <u>Food Policy</u>, vol. 8 (2), May.

Pray, C.E., 1980: "An assessment of the accuracy of the official agricultural statistics in Bangladesh", in <u>Bangladesh development studies</u>, vol. VIII, nos 1 & 2 (special issue).

Prescott-Allen, R. and Prescott-Allen, C., 1981: "Wild plants and crop improvement", <u>Occasional paper no. 1</u>, World Wildlife Fund, Godalming, Surrey.

Purchase, H.G., 1977: "The etiology and control of Marek's disease of chickens and the economic impact of a successful research program", in J.A. Romberger (ed.): <u>Virology in agriculture: Beltsville symposium in agricultural research-I</u>, Allanheid, USMUM, Montclair, N.J.

Rahman, A., 1979: <u>Agrarian structure and capital formation: a study of Bangladesh agriculture</u>, Ph.D thesis, Cambridge University.

___________, 1981: "Adoption of new technology in Bangladesh agriculture: testing some hypotheses", in W. Mahmud (ed): <u>Development issues in an agrarian economy – Bangladesh</u>, Dhaka University.

Raj, K.N., 1970: "Some questions concerning growth, transformation and planning of agriculture in the developing countries", in E.A.G. Robinson and M. Kidron (eds): <u>Economic development in South Asia</u>, proceedings of a conference held by the International Economic Association, Macmillan, London.

Ranada, C.G., 1980: "Impact of cropping pattern on agricultural production", in <u>Indian Journal of Agricultural Economics</u>, April–June.

Rao, C.R., 1960: <u>Linear statistical inference and its applications</u>, John Wiley and Sons Inc., New York.

Redclift, M, 1982: "Poor environment or environmental poverty? Towards a political economy of natural resource use", Paper delivered to the 1981 Annual Conference of the Development Studies Association, Trinity College, Dublin, 23–25 September.

Reddy, V.R., Kasryno, F. and Siregar, M., 1985: "Diffusion and commercialisation of technology prototypes: rice post-harvest in Indonesia", ILO mimeographed World Employment Programme research working paper, Geneva, (restricted).

Rhoades, R.E., 1982: "The art of the informal agricultural survey", <u>Social science department training document</u>, International Potato Centre (CIP), Lima, Peru.

Rhoades, R.E. and Booth, R.H.,: "Farmer-back-to-farmer: a model for generating acceptable agricultural technology", in <u>Agricultural Administration</u>, vol.11.

Richards, P., 1979: "Community environmental knowledge in African rural development", in <u>IDS Bulletin</u>, vol. 10 (2), January.

Rimlinger, G., 1982: "African rural social science study: consultant report to ISNAR", annex 4 in International Service for National Agricultural Research: <u>Strategies to meet demands for rural social scientists in Africa</u>, ISNAR, The Hague.

Rogers, E.M., 1976: "New product adoption and diffusion", in <u>Journal of Consumer Research</u>, vol. 2, March.

___________, 1976a: "The passing of the dominant paradigm: Reflections on diffusion research" in W. Schramm and D. Lerrier (eds.): <u>Communication and change: the last ten years and the next</u>, East West Centre Press, Honolulu.

___________, 1980: "The diffusion of technological innovations: application to renewable energy resources in developing nations", paper prepared for the panel on the introduction and diffusion of renewable energy technologies, National Academy of Sciences, Washington, D.C.

Rosenberg, N. (ed.), 1971: <u>The economics of technical change</u>, Penguin Books Ltd., Hammondsworth.

Roumasset, J.A., 1976: <u>Rice and risks: decision-making among low income farmers</u>, Amsterdam, North-Holland.

Ruttan, V.W., 1978: "Reviewing Agricultural Research Programmes", in <u>Agricultural Administration</u>, Vol. 5.

_______________, 1959: "Usher and schumpeter on invention, innovation and technical change", in <u>Quarterly Journal of Economics</u>, November. Reprinted in N. Rosenberg, N. (ed.): <u>The economics of technical change</u>, Penguin Books Ltd., Hammondsworth, 1971.

_______________, 1975: "Technical and institutional transfer in agricultural development", in <u>Research Policy</u>, vol. 4.

_______________, 1975a: "Technology transfer, institutional transfer and induced technical and institutional change in agricultural development", in L.G. Reynolds (ed.): <u>Agriculture in development theory</u>, Yale University Press.

_______________, 1979: "Factor productivity and growth: a historical interpretation", in H.P. Binswanger and V.W. Ruttan (eds.): <u>Induced innovation: technology, institutions and development</u>, Johns Hopkins University Press, Baltimore.

_______________, 1979a: "Induced institutional change", in H.P. Binswanger and V.W. Ruttan (eds.): <u>Induced innovation: technology, institutions and development</u>, Johns Hopkins University Press, Baltimore.

_______________, 1980: "Institutional factors affecting the generation and diffusion of agricultural technology: issues, concepts and analysis", ILO mimeo. World Employment Programme research working paper, Geneva (restricted).

_______________,: "Production programmes for small farmers: plan puebla as myth and reality", in <u>Economic Development and Cultural Change</u>, vol.31 (3), April.

_______________, 1982: "An induced innovation interpretation of technical change in agriculture in developed countries", Document prepared for a seminar organised by UNDP/IICA in Coronado, Costa Rica.

_______________, 1982a: "Changing role of public and private sectors in agricultural research", in _Science_, vol. 216.

_______________, 1982b: _Agricultural research policy_, University of Minnesota Press, Minneapolis.

_______________, 1983: "Agricultural research policy issues", B.Y. Morrison Memorial Lecture, in _Horticultural Science_, vol.18 (6), Alexandria, VA.

Ruttan, V.W. and Binswanger, H.P., 1978: "Induced innovation and the green revolution", in H.P. Binswanger and V.W. Ruttan, (eds.): _Induced innovation: technologies, institutions and development_, John Hopkins University Press, Baltimore.

Ruttan, V.W. and Hayami, Y., 1984: "Toward a theory of induced institutional innovation", in _The Journal of Development Studies_, vol.20, no.4, July.

Ryan, J.G., 1981: "Reviews of R.W. Frank and B.H. Chasin, Seeds of famine: ecological destruction and the development dilemma in the West African Sahel, Allanheld, Osman and Co., Montclair, N.J., in _American Journal of Agricultural Economics_, vol. 63 (3), August.

Sandbach, F., 1980: _Environment, ideology and policy_, Basil Blackwell, Oxford.

Saxena, P.N., 1979: "Returns on investment in research and education", in _50 years of agricultural research and education_, ICAR, New Delhi.

Schaffer, B., 1969: "Crisis in development administration" in C. Leys (ed.): _Politics and development_, Cambridge University Press.

_______________, 1974: "Policy decisions and institutional evaluation", in _Development and Change_, vol. V (3).

Schluter, M., 1971: <u>Differential rates of adoption of the new seed varieties in India: the problem of the small farm</u>, USAID Development Occasional Paper no. 47, Cornell University, Ithaca, N.Y.

Schmitz, A. and Seckler, D., 1970: "Mechanised agriculture and social welfare: the case of the tomato harvester", in <u>American Journal of Agricultural Economics</u>, vol. 52 (4).

Schon, D., 1971: <u>Beyond the stable state</u>, Temple Smith, London.

Schuh, G.E. and Tollini, H., 1979: "Costs and benefits of agricultural research: the state of the arts", <u>World Bank staff working paper no. 360</u>, World Bank, Washington, D.C.

Schulz, T.W., 1979: "The economics of research and agricultural productivity" <u>IADS Occasional Paper</u>, International Agricultural Development Service, New York.

Schumacher, E.F., 1973: <u>Small is beautiful: a study of economics as if people mattered</u>, Blond and Briggs.

Schumpeter, J.A., 1934: <u>The theory of economic development</u>, Harvard University Press, Cambridge.

__________, 1939: <u>Business cycles</u>, McGraw-Hill.

Scobie, G.M., 1979: "Investment in international agricultural research: some economic dimensions", <u>World Bank staff working paper no. 361</u>, World Bank, Washington, D.C.

Scobie, G.M. and R.T. Posada, 1978: "The impact of technical change on income distribution: the case of rice in Colombia", in <u>American Journal of Agricultural Economics</u>, vol. 60.

Sen, A.K., 1968: <u>Choice of techniques</u>, 3rd edition, Blackwell.

Shaner, W.W., Philipp, P.F. and Schmehl, W.R., 1982: _Farming systems research and development: guidelines for developing countries_, Westview Press, Boulder, Colorado.

Sheldrake, R., 1981: _A new science of life: the hypothesis of formative causation_, Blond and Briggs, London.

Sheridan, M., 1981: _Peasant innovation and diffusion of agricultural technology in China_, Rural Development Committee, Centre for International Studies, Cornell University.

Shumway, C.R., 1977: "Models and methods used to allocate resources", in T.M. Arndt et al.: _Resource allocation and productivity in national and international agricultural research_, University of Minnesota Press, Minneapolis.

__________, 1981: "Subjectivity in ex-ante research evaluation", in _American Journal of Agricultural Economics_, vol. 63 (1), February.

Sigurdon, J., 1982: _Vietnam's science and technology: a tentative description of structure and planning_, University of Lund, mimeo.

Sigurdson, V., 1980: _Technology and science in the People's Republic of China_, Pergamon Press, Oxford.

__________, 1981: "China's tortuous road to autonomy in technology and science", in _Endeavour_, vol. 5, no. 2.

Sim, R.J.R. and R. Gardner, 1978: _A review of research and extension evaluation in agriculture_, University of Idaho, Department of Agricultural Economics Research Series 214, Moscow.

Singh, D.P., 1982: "Agricultural universities and transfer of technology in India: the importance of management", in R.S. Anderson, et al. (ed.): _Science, politics and the agricultural revolution in Asia_,

Westview Press, Boulder, Colorado.

Sivaraman, B., 1978: Address to the Meeting of Vice Chancellors of Agricultural Universities, New Delhi, October 16th.

Smith, L.D., 1976: "An overview of agricultural development policy", in J. Heyer, J.K. Maitha and W.M. Senga: Agricultural development in Kenya, Oxford University Press.

Sprague, G.F., 1975: "Agriculture in China", in P. Abelson (ed.), Food: politics, economics, nutrition and research, American Association for the Advancement of Science, Washington, D.C.

Stavis, B., 1974: "Making Green Revolution – the politics of agricultural development in China", in Rural Development Monograph no. 1, Cornell University.

__________, 1978: "Agricultural research and extension services in China", in World Development, vol. 6.

__________, 1979: Agricultural extension for small farmers, MSU Rural Development Working Paper no. 3, Michigan State University.

Stewart, F., 1978: Technology and underdevelopment, Macmillan, London.

Stocking, M., 1981: "Conservation strategies for less developed countries", in R.P.C. Morgan (ed.): Soil conservation problems and prospects.

Swanson, B.E., 1975: Organising agricultural technology transfer: the effects of alternative arrangements, Indiana University, International Development Research Centre, PASITAM.

Swerdlow, I. (ed.), 1963: Development administration concepts and problems, Syracuse University Press.

Tang, A., 1963: "Research and education in Japanese agricultural development", in <u>Economic Studies Quarterly</u>, vol. 13.

Timmer, C.P., 1980: "Public policy for improving technology choice", revised version of a paper produced at a Harvard Institute for International Development Rural Development Workshop of "Appropriate Technology, Macro Concepts and Micro Applications" February 29th.

Todaro, M.P., 1977: <u>Economic Development in the Third World</u>, Longman Inc., New York.

Tong Cuc Thong Ke, 1970: <u>Nien Giam Thong Ke 15 Nam Xay Dung Nen Kinh Te Xa Hoi Chu Nghia</u>.

Tran Dang Khoa, 1964: "Les problèmes hydrauliques au North Vietnam", in <u>Etudes Vietnamiennes</u>, no.2.

Tran Phuong, 1960: <u>Mot So Y Kien Ve Chu Ngia Tu Ban O Nong Thon Mien Bac Ngay Sau Cai Cach Ruong Dat</u>/Some ideas on the capitalism in the rural areas of the North after land reform, Khoa Hoc Xa Hoi, Hanoi.

Tran Phuong (ed.), 1968: <u>Cach Mang Ruong Dat o Viet Nam</u>/Agrarian revolution in Vietnam, Khoa Hoc Xa Hoi, Hanoi.

Trigo, E.J. and Peneiro, M.E., 1981: "Dynamics of agricultural research organisation in Latin America, in <u>Food Policy</u>, vol. 6 (1), February.

le Trong, 1980: <u>To Chuc Lao Dong Trong Hop Tac Xa San Xuat Nong Nghiep</u>/The organisation of labour in agricultural production co-operatives, Nong Nghiep, Hanoi.

Truong Chinh, 1976: <u>Cach Mang Dan Toc Dan Chu Nhan Dan Viet Nam</u>/National Democratic Revolution in Viet Nam, Su That, Hanoi, vol. 1 and 2.

_______________, 1969: <u>Kien Quyet Sua Chua Khuyet Diem, Phat Huy Uu Diem, Dua Phong Trao Hop Tac Hoa Nong Nghiep Vung Buoc Tien Len</u>, Su That, Hanoi.

Ubevoi, J.S., Nandy, A., Bagner, A. (undated): "Ethno-agriculture" research proposal to Centre for the study of Developing Societies, Delhi.

U.N. Conference on Desertification, 1977: Desertification: its causes and consequences, Pergamon Press, Oxford.

USDA, 1979: Changes in Farm Production and Efficiency, Washington, D.C.

Usher, A.P., 1954: A history of mechanical inventions, Harvard University Press.

____________, 1955: "Technical change and capital formation", Capital formation and economic growth, National Bureau of Economic Research, reprinted in N. Rosenberg (ed.): The economics of technical change, Penguin.

Vaidyanathan, A., 1978: "Labour use in Indian agriculture: an analysis based on farm management survey data", in Labour absorption in Indian agriculture, International Labour Office (ARTEP), Bangkok.

Van Tao: "Nong Dan Viet Nam va Con Duong Tien Len Chu Nghia Xa Hoi Khong Qua Giai Doan Phat Trien Tu Ban Chu Bghia Duoi Su Lanh Dao cua Giai Cap Cong Nhan", in Vien Su Hoc: Nong Dan Viet Nam Tien Len.

Vien Kinh Te, 1960: Kinh Te Viet Nam, 1945-1975/The Vietnamese Economy, 1945- 1975, Su That, Hanoi.

Vien Su Hoc, 1979: Nong Dan Viet Nam Tien Len Chu Nghia Xa Hoi/The Vietnamese peasantry advance toward socialism, Khoa Hoc Xa Hoi, Hanoi.

Vo Nguyen Giap, 1978: Day Manh Cuoc Cach Mang Khoa Hoc Ky Thuat Trong Nong Nghiep Nuoc Ta/Pushing strongly the scientific and technological revolution in the agriculture of our country, Su That, Hanoi.

__________, 1979: <u>Les tâches essentielles de la révolution scientifique et technique dans l'agriculture</u>, Su That, Hanoi.

Vo Nhan Tri, 1967: <u>Croissance économique de la République Démocratique du Vietnam</u>, Edition en Langues Etrangères, Hanoi.

Vu Quoc Tuan and Nguyen Xuan Lai, 1976: "La Seconde Résistance", in <u>Etudes Vietnamiennes</u>, no.44.

Vyas, V.S. and Kulkarni, M.N.: "Integrating agricultural research and extension: adaptive research trials programme for rice farmers of Thanjavur", in Gaikwad et al.: <u>Development of intensive agriculture</u>, Indian Institute of Management, Ahmedabad, (undated).

Wade, N., 1975: "International agricultural research", in Philip Abelson (ed.): <u>Food politics, economics, nutrition and research</u>, American Association for the Advancement of Science, Washington, D.C.

Warui, C.M. and van Eijnatten, C.L.M., 1979: <u>Considerations on the link of the farmer with agricultural extension and research services in Coast Province – Kenya</u>, Coast Agricultural Research Station, Kikambala, mimeo.

Weber, A., 1981: <u>Energy use in Kenya's agricultural sector 1960-78: a statistical and economic analysis</u>, Forum reports on Current Research in Agricultural Economics and Agribusiness Management, no.2, Kieler Wissenschaftsverlag Vauk, Kiel.

Wennergen, E.B. and Whitaker, M.D., 1977: "Social return to United States technical assistance in Bolivian agriculture: the case of sheep and wheat", in <u>American Journal of Agricultural Economics</u>, vol. 59 (3).

Whitcombe, R. and Carr, M., 1982: <u>Appropriate technology institutions: a review</u>, Intermediate Technology, London.

Whyte, W.F., 1918: "Participatory approach to agricultural research and development: a state-of-the-art paper", draft 3, Rural Development Committee, Cornell University, CIS.

Williams, G., 1981: "The World Bank and the peasant problem" in J. Heyer, P. Roberts and G. Williams (eds.): Rural development in tropical Africa, Macmillan, London.

Wise, W.S., 1977: "Notes and comments: estimating the benefits from agricultural research", in R & D Management, vol. 7 (2), February.

Wisner, B., 1977: "Man-made famine in Eastern Kenya: the interrelationships of environment and development", in Phil O'Keefe and Ben Wisner (eds.): Land use and development, International African Institute, London.

Wood, A., 1980: "The neglect of basic food crop research in Mali: a study of agricultural research priorities and indigenous production processes", M.A. Rural Development dissertation (unpublished), School of Development Studies University of East Anglia, Norwich.

World Bank, 1981: "Agricultural Research", Sector Policy Paper, Washington D.C., June.

Yamada, S. and Y. Hayami, 1979: "Agricultural growth in Japan, 1880-1970", in Y. Hayami, V.W. Ruttan and H. Southworth: Agricultural growth in Japan, Taiwan, Korea and the Philippines, The University Press of Hawaii, Honolulu.

Yamada, S. and V.W. Ruttan, 1980: "International comparisons of productivity in agriculture", in John W. Kendrick and Beatrice N. Vaccara (eds.): New developments in productivity measurement and analysis. University of Chicago Press, Chicago.

Yotopoulos, P.A.: "Resource use in agriculture: applications of the profit function to selected countries", <u>Food Research Institute Studies</u>, vol. XVII, no.1.

Zandstra, H.G., 1980: "Notes on methodology in cropping and farming systems research" in H. Ruthenberg: <u>Farming systems in the tropics</u>, 3rd edition, Clarendon Press, Oxford.

Zelena, A.T., 1975: <u>Development Plan for Kenya's horticultural industry</u>, Horticultural Crops Development Project, Horticultural Crops Development Authority, Nairobi.

Index

Central Institute of Agronomy (Viet Nam) 373-4
Central Ministry of Food
  and Agriculture (India)
  139
centralisation 102-3
centre-periphery model
  26, 59
CGIAR 27
Chasin, B. 41-2
Chiem rice 328-32, 383,
  386
Chinese socialist models
  29
'circularity of causation'
  236
Clay, E.J. 19
Cleaver, H.M. 41
Coast Agricultural
  Research Station, Kenya
  282, 287, 297
Cobb-Douglas framework 173
Coffee Research Foundation
  292
collectivisation 320,
  341-60, 363, 403
College of Agronomy
  (Viet Nam) 374
College of Forestry
  (Viet Nam) 374
Comilla co-operatives 199,
  213, 231, 233-5, 243
commodity programmes,
  research allocation in
  32
Communist Party of Viet
  Nam 334, 336, 356
    see also Lao Dong
    Party
co-operatives 231, 233-6,
  243
  Comilla 199, 213, 231,
    233-5, 243
  in Viet Nam 343-60,

  363, 364-5, 401, 403
  advantages of 353, 355-6
  and consumption of
    electricity 367
  management of 356-60
  and mechanisation 367-8
  productivity 394
  specialisation in 396-7,
    402
crop investment work 26
cropping pattern 328-32,
  372, 386-7
crops
  cash 43
  export 8, 11, 20, 40,
    43, 261-2
  food 11, 19
  horticultural 16, 250-3
  non-food 20
  subsistence 11,
cultivars 91
cultural endowments 69,
  74, 90, 98, 127

Dalrymple, D.G. 29
de Janvry, A. 55, 104, 335
decentralisation 28-9,
  102-3, 404-5, 411
decentralised diffusion
  model 50-1, 59
Deer 99
deforestation 46
'demand-oriented' research
  programmes 118
desertification 46
development agencies 231
Dien Thanh 389
differential resource
  endowment 207-12
diffusion model 26
  decentralised 50-1, 59
diffusion studies 23
Directorate of
  Agricultural Extension

461

innovation 1, 2-5, 400-1
model 13, 16, 30
criticism of 2-3, 15,
    55, 400
definition 2
and institution change
    85-98
of technical change
    75-85
industrialisation 320,
    323-4, 346
and link with
    agricultural
    transformation 361
'input package' 237-8
Institute of Agricultural
    Research (Viet Nam) 374
Institute of Agronomic
    Research (Viet Nam) 373
institution, concept of
    248
institution building 23,
    26-7
institutional innovation
    74, 90, 249, 319
definition 85
in the Philippines 4
problems of, in Viet
    Nam 333-4
sources of demand for
    86-87
sources of supply of
    87-89
and technological
    change,
    interrelations
    between 133-4, 403
theory of 89-91
institutional transfer
    27-9
Integrated Rural Survey
    270
Interactive model 52
interactive research 37

Intermediate and
    Appropriate Technology
    (AT) 32-3
international agricultural
    research centres 26
International Agricultural
    Research Institutes 19,
    42, 70
International Centre for
    Corn and Wheat
    Improvement (CIMMYT) 70
International Rice
    Research Institute
    (IRRI) 35-7, 70, 123,
    385
International Services for
    National Agricultural
    Research (ISNAR) 27
international trade 409
IRDP 233, 235-6
irrigation 32, 46, 165,
    193, 195, 213, 230
in Philippines 4
schemes in Kenya 257-9
in Viet Nam 338-9, 378-
    83
Ishikawa 225

Jefferson 99
Jha, D, 170
Joint Research Services
    279
Journal of Elementary
    Sciences 396

Kabazi Canners Ltd. 261
Kabete Faculty of
    Agriculture 288
Kahlon, A.S. 170
Kenya Agricultural
    Research Institute
    (KARI) 284, 288-90,
    294, 297, 299, 302,
    311-12